T0211011

A GUIDE
TO PHYSICS
PROBLEMS
part 2

**Thermodynamics,
Statistical Physics, and
Quantum Mechanics**

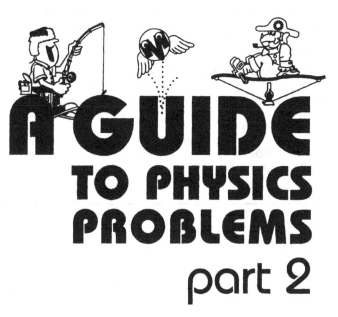

A GUIDE
TO PHYSICS
PROBLEMS
part 2

Thermodynamics, Statistical Physics, and Quantum Mechanics

Sidney B. Cahn

New York University
New York, New York

Gerald D. Mahan

University of Tennessee
Knoxville, Tennessee, and
Oak Ridge National Laboratory
Oak Ridge, Tennessee

and
Boris E. Nadgorny

Naval Research Laboratory
Washington, D.C.

PLENUM PRESS • NEW YORK AND LONDON

The Library of Congress cataloged the first volume of this title as follows:

Cahn, Sidney B.
   A guide to physics problems / Sidney B. Cahn and Boris E. Nadgorny.
   p.   cm.
   Includes bibliographical references and index.
   Contents: pt. 1. Mechanics, relativity, and electrodynamics.
   ISBN 0-306-44679-0
   1. Physics—Problems, exercises, etc. I. Nadgorny, Boris E. II. Title.
QC32.C25   1994   94-5210
530' .076—dc20   CIP

Front cover design by Vladimir Gitt
Back cover illustration by Yair Minsky

ISBN 0-306-45291-X

© 1997 Plenum Press, New York
A Division of Plenum Publishing Corporation
233 Spring Street, New York, N.Y. 10013-1578

http://www.plenum.com

10 9 8 7 6 5 4 3 2

Printed in the United States of America

Foreword

It is only rarely realized how important the design of suitable, interesting problems is in the educational process. This is true for the professor — who periodically makes up exams and problem sets which test the effectiveness of his teaching — and also for the student — who must match his skills and acquired knowledge against these same problems. There is a great need for challenging problems in all scientific fields, but especially so in physics. Reading a physics paper requires familiarity and control of techniques which can only be obtained by serious practice in solving problems. Confidence in performing research demands a mastery of detailed technology which requires training, concentration, and reflection — again, gained only by working exercises.

In spite of the obvious need, there is very little systematic effort made to provide balanced, doable problems that do more than gratify the ego of the professor. Problems often are routine applications of procedures mentioned in lectures or in books. They do little to force students to reflect seriously about new situations. Furthermore, the problems are often excruciatingly dull and test persistence and intellectual stamina more than insight, technical skill, and originality. Another rather serious shortcoming is that most exams and problems carry the unmistakable imprint of the teacher. (In some excellent eastern U.S. universities, problems are catalogued by instructor, so that a good deal is known about an exam even before it is written.)

In contrast, *A Guide to Physics Problems, Part 2* not only serves an important function, but is a pleasure to read. By selecting problems from different universities and even different scientific cultures, the authors have effectively avoided a one-sided approach to physics. All the problems are good, some are very interesting, some positively intriguing, a few are crazy; but all of them stimulate the reader to think about physics, not merely to train you to pass an exam. I personally received considerable pleasure in working the problems, and I would guess that anyone who wants to be a professional physicist would experience similar enjoyment. I must confess with some embarrassment that some of the problems gave me more trouble than I had expected. But, of course, this is progress. The coming generation can do with ease what causes the elder one trouble. This book will be a great help to students and professors, as well as a source of pleasure and enjoyment.

Max Dresden
Stanford

Preface

Part 2 of *A Guide to Physics Problems* contains problems from written graduate qualifying examinations at many universities in the United States and, for comparison, problems from the Moscow Institute of Physics and Technology, a leading Russian Physics Department. While Part 1 presented problems and solutions in Mechanics, Relativity, and Electrodynamics, Part 2 offers problems and solutions in Thermodynamics, Statistical Physics, and Quantum Mechanics.

The main purpose of the book is to help graduate students prepare for this important and often very stressful exam (see Figure P.1). The difficulty and scope of the qualifying exam varies from school to school, but not too dramatically. Our goal was to present a more or less universal set of problems that would allow students to feel confident at these exams, regardless of the graduate school they attended. We also thought that physics majors who are considering going on to graduate school may be able to test their knowledge of physics by trying to solve some of the problems, most of which are not above the undergraduate level. As in Part 1 we have tried to provide as many details in our solutions as possible, without turning to a trade expression of an exhausted author who, after struggling with the derivation for a couple of hours writes, "As it can be easily shown...."

Most of the comments to Part 1 that we have received so far have come not from the students but from the professors who have to give the exams. The most typical comment was, "Gee, great, now I can use one of your problems for our next comprehensive exam." However, we still hope that this does not make the book counterproductive and eventually it will help the students to transform from the state shown in Figure P.1 into a much more comfortable stationary state as in Figure P.2. This picture can be easily attributed to the present state of mind of the authors as well, who sincerely hope that Part 3 will not be forthcoming any time soon.

Some of the schools do not have written qualifying exams as part of their requirements: Brown, Cal-Tech, Cornell, Harvard, UT Austin, University of Toronto, and Yale. Most of the schools that give such an exam were

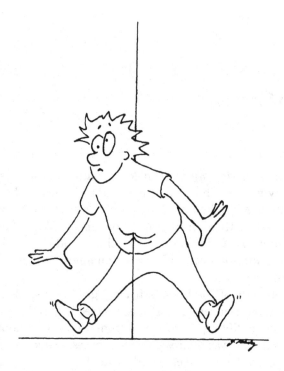

Figure **P.1**

Hapless physicist impaled on his own delta function
(demonstrating the perils of insufficient theoretical rigor)

happy to trust us with their problems. We wish to thank the Physics Depart-
ments of Boston University (Boston), University of Colorado at Boulder (Colorado), Columbia University (Columbia), University of Maryland (Maryland), Massachusetts Institute of Technology (MIT), University of Michigan (Michigan), Michigan State University (Michigan State), Michigan Tech-
nological University (Michigan Tech), Princeton University   (Princeton),
Rutgers University (Rutgers), Stanford University (Stanford), State University of New York at Stony Brook (Stony Brook), University of Tennessee at
Knoxville (Tennessee), and University of Wisconsin (Wisconsin-Madison).
The Moscow Institute of Physics and Technology (Moscow Phys-Tech) does
not give this type of qualifying exam in graduate school. Some of their problems came from the final written exam for the physics seniors, some of the
others, mostly introductory problems, are from their oral entrance exams or

Figure **P.2**

In an inertial frame for several days,
a physicist begins to long for absolute rest

magazines such as *Kvant*. A few of the problems were compiled by the authors
and have never been published before.

We were happy to hear many encouraging comments about Part 1 from
our colleagues, and we are grateful to everybody who took their time to review the book. We wish to thank many people who contributed some of the
problems to Part 2, or discussed solutions with us, in particular Dmitri Averin
(Stony Brook), Michael Bershadsky (Harvard), Alexander Korotkov (Stony
Brook), Henry Silsbee (Stony Brook), and Alexei Stuchebrukhov (UC Davis).
We thank Kirk McDonald (Princeton) and Liang Chen (British Columbia)
for their helpful comments to some problems in Part 1; we hope to include
them in the second edition of Part 1, coming out next year. We are indebted
to Max Dresden for writing the Foreword, to Tilo Wettig (Münich) who read
most of the manuscript, and to Vladimir Gitt and Yair Minsky who drew the
humorous pictures.

Sidney Cahn
New York

Gerald Mahan
Oak Ridge

Boris Nadgorny
Washington, D.C.

Textbooks Used in the Preparation of this Volume

Chapter 4 — Thermodynamics and Statistical Physics

1) Landau, L. D., and Lifshitz, E. M., *Statistical Physics*, Volume 5, part 1 of *Course of Theoretical Physics*, 3rd ed., Elmsford, New York: Pergamon Press, 1980

2) Kittel, C., *Elementary Statistical Physics*, New York: John Wiley and Sons, Inc., 1958

3) Kittel, C., and Kroemer, H., *Thermal Physics*, 2nd ed., New York: Freeman and Co., 1980

4) Reif, R., *Fundamentals of Statistical and Thermal Physics*, New York: McGraw-Hill, 1965

5) Huang, K., *Statistical Mechanics*, 2nd ed., New York: John Wiley and Sons, Inc., 1987

6) Pathria, R. K., *Statistical Mechanics*, Oxford: Pergamon Press, 1972

Chapter 5 — Quantum Mechanics

1) Liboff, R. L., *Introductory Quantum Mechanics*, 2nd ed., Reading, MA: Pergamon Press, 1977

2) Landau, L. D., and Lifshitz, E. M., *Quantum Mechanics, Nonrelativistic Theory*, Volume 3 of *Course of Theoretical Physics*, 3rd ed., Elmsford, New York: Pergamon Press, 1977

3) Sakurai, J. J., *Modern Quantum Mechanics*, Menlo Park: Benjamin/ Cummings, 1985

4) Sakurai, J. J., *Advanced Quantum Mechanics*, Menlo Park: Benjamin/Cummings, 1967

5) Schiff, L. I., *Quantum Mechanics*, 3rd ed., New York: McGraw-Hill, 1968

6) Shankar, R., *Principles of Quantum Mechanics*, New York: Plenum Press, 1980

Contents

PART I: PROBLEMS

PART II: SOLUTIONS

5. Quantum Mechanics **243**

PART III: APPENDIXES

PROBLEMS

Thermodynamics and Statistical Physics

Introductory Thermodynamics

4.1 Why Bother? (Moscow Phys-Tech)

A physicist and an engineer find themselves in a mountain lodge where the only heat is provided by a large woodstove. The physicist argues that

Figure **P.4.1**

they cannot increase the total energy of the molecules in the cabin, and therefore it makes no sense to continue putting logs into the stove. The engineer strongly disagrees (see Figure P.4.1), referring to the laws of thermodynamics and common sense. Who is right? Why do we heat the room?

4.2   Space Station Pressure (MIT)

A space station consists of a large cylinder of radius R_0 filled with air. The cylinder spins about its symmetry axis at an angular speed Ω providing an acceleration at the rim equal to g. If the temperature τ is constant inside the station, what is the ratio of air pressure P_c at the center of the station to the pressure P_0 at the rim?

4.3   Baron von Münchausen and Intergalactic Travel (Moscow Phys-Tech)

Recently found archives of the late Baron von Münchausen brought to light some unpublished scientific papers. In one of them, his calculations indicated that the Sun's energy would some day be exhausted, with the subsequent freezing of the Earth and its inhabitants. In order to avert this inevitable outcome, he proposed the construction of a large, rigid balloon, empty of all gases, 1 km in radius, and attached to the Earth by a long, light

Figure **P.4.3**

rope of extreme tensile strength. The Earth would be propelled through space to the nearest star via the Archimedes' force on the balloon, transmitted through the rope to the large staple embedded in suitable bedrock (see Figure P.4.3). Estimate the force on the rope (assuming a massless balloon). Discuss the feasibility of the Baron's idea (without using any general statements).

4.4 Railway Tanker (Moscow Phys-Tech)

A long, cylindrical tank is placed on a carriage that can slide without friction on rails (see Figure P.4.4). The mass of the empty tanker is $M = 180$ kg. Initially, the tank is filled with an ideal gas of mass $m = 120$ kg at a pressure $P_0 = 150$ atm at an ambient temperature $T_0 = 300$ K. Then one end of the tank is heated to 335 K while the other end is kept fixed at 300 K. Find the pressure in the tank and the new position of the center of mass of the tanker when the system reaches equilibrium.

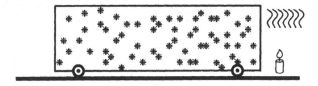

Figure **P.4.4**

4.5 Magic Carpet (Moscow Phys-Tech)

Once sitting in heavy traffic, Baron von Münchausen thought of a new kind of "magic carpet" type aircraft (see Figure P.4.5). The upper surface of the large flat panel is held at a constant temperature T_1 and the lower surface at a temperature $T_2 > T_1$. He reasoned that, during collision with the hot surface, air molecules acquire additional momentum and therefore will transfer an equal momentum to the panel. The back of the handkerchief estimates he was able to make quickly for 1 m^2 of such a panel showed that if $T_1 = 273$ K and $T_2 = 373$ K (air temperature 293 K) this panel would be able to levitate itself and a payload (the Baron) of about 10^3 kg. How did he arrive at this? Is it really possible?

Figure **P.4.5**

4.6 Teacup Engine (Princeton, Moscow Phys-Tech)

The astronaut from Problem 1.13 in Part I was peacefully drinking tea at five o'clock galactic time, as was his wont, when he had an emergency outside the shuttle, and he had to do an EVA to deal with it. Upon leaving the ship, his jetpack failed, and nothing remained to connect him to the shuttle. Fortunately, he had absentmindedly brought his teacup with him. Since this was the only cup he had, he did not want to throw it away in order to propel him back to the shuttle (besides, it was his favorite cup). Instead, he used the sublimation of the frozen tea to propel him back to the spaceship (see Figure P.4.6). Was it really possible? Estimate the time it might take him to return if he is a distance $L = 40$ m from the ship. Assume that the sublimation occurs at a constant temperature $T = 273$ K.

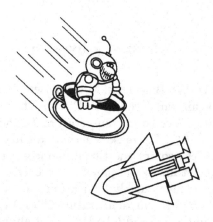

Figure **P.4.6**

The vapor pressure at this temperature is $P_0 = 600$ Pa, and the total mass of the astronaut $M = 110$ kg.

4.7 Grand Lunar Canals (Moscow Phys-Tech)

In one of his novels, H. G. Wells describes an encounter of amateur earthling astronauts with a lunar civilization living in very deep caverns beneath the surface of the Moon. The caverns are connected to the surface by long channels filled with air. The channel is dug between points A and B on the surface of the Moon so that the angle $AOB = 90°$ (see Figure P.4.7). Assume that the air pressure in the middle of a channel is $P_0 = 1$ atm. Estimate the air pressure in the channel near the surface of the Moon. The radius of the Moon $a \approx 1750$ km. The acceleration due to gravity on the surface of the Moon $g_m \approx g/6$, where g is the acceleration due to gravity on the surface of the Earth.

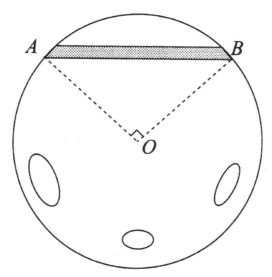

Figure **P.4.7**

4.8 Frozen Solid (Moscow Phys-Tech)

Estimate how long it will take for a small pond of average depth $D = 0.5$ m to freeze completely in a very cold winter, when the temperature is always below the freezing point of water (see Figure P.4.8). Take the thermal conductivity of ice to be $\kappa = 2.2$ W/m K, the latent heat of fusion

$q = 3.4 \cdot 10^5$ J/kg, and the density $\rho = 0.9 \cdot 10^3$ kg/m^3. Take the outside temperature to be a constant $T_0 \approx 263$ K.

Figure **P.4.8**

4.9   Tea in Thermos (Moscow Phys-Tech)

One liter of tea at 90°C is poured into a vacuum-insulated container (thermos). The surface area of the thermos walls $A = 600$ cm^2. The volume between the walls is pumped down to $P_0 \sim 5 \cdot 10^{-6}$ atm pressure (at room temperature). The emissivity of the walls $\epsilon = 0.1$, and the thermal capacity of water $C = 4.2 \cdot 10^3$ J/kg K. Disregarding the heat leakage through the stopper, estimate the

a) Net power transfer
b) Time for the tea to cool from 90°C to 70°C.

4.10   Heat Loss (Moscow Phys-Tech)

An immersion heater of power $J = 500$ W is used to heat water in a bowl. After 2 minutes, the temperature increases from $T_1 = 85°$C to $T_2 = 90°$C. The heater is then switched off for an additional minute, and the temperature drops by $\Delta T = 1°$C. Estimate the mass m of the water in the bowl. The thermal capacity of water $c = 4.2 \cdot 10^3$ J/kg K.

4.11 Liquid–Solid–Liquid (Moscow Phys-Tech)

A small amount of water of mass $m = 50$ g in a container at temperature $T = 273$ K is placed inside a vacuum chamber which is evacuated rapidly. As a result, part of the water freezes and becomes ice and the rest becomes vapor.

a) What amount of water initially transforms into ice? The latent heat of fusion (ice/water) $q_i = 80$ cal/g, and the latent heat of vaporization (water/vapor) $q_v = 600$ cal/g.

b) A piece of heated metal alloy of mass $M = 325$ g and original volume $V = 48$ cm^3 is placed inside the calorimeter together with the ice obtained as a result of the experiment in (a). The density of metal at $T = 273$ K is $\rho_0 = 6.8$ g/cm^3. The thermal capacity is $C = 0.12$ cal/g K, and the coefficient of linear expansion $\alpha = 1.1 \cdot 10^{-5}$ K^{-1}. How much ice will have melted when equilibrium is reached?

4.12 Hydrogen Rocket (Moscow Phys-Tech)

The reaction chamber of a rocket engine is supplied with a mass flow rate m of hydrogen and sufficient oxygen to allow complete burning of the fuel. The cross section of the chamber is A, and the pressure at the cross section is P with temperature T. Calculate the force that this chamber is able to provide.

4.13 Maxwell–Boltzmann Averages (MIT)

a) Write the properly normalized Maxwell–Boltzmann distribution $f(v)$ for finding particles of mass m with magnitude of velocity in the interval $[v, v + dv]$ at a temperature τ.

b) What is the most likely speed at temperature τ?

c) What is the average speed?

d) What is the average square speed?

4.14 Slowly Leaking Box (Moscow Phys-Tech, Stony Brook (a,b))

An ideal gas of atoms of number density n at an absolute temperature τ is confined to a thermally isolated container that has a small hole of area A in one of the walls (see Figure P.4.14). Assume a Maxwell velocity distribution for the atoms. The size of the hole is much smaller than the size of the container and much smaller than the mean free path of the atoms.

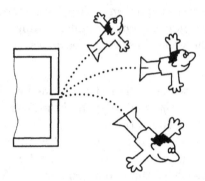

Figure **P.4.14**

a) Calculate the number of atoms striking the wall of the container per unit area per unit time. (Express your answer in terms of the mean velocity of the atoms.)

b) What is the ratio of the average kinetic energy of atoms leaving the container to the average kinetic energy of atoms initially occupying the container? Assume that there is no flow back to the container. Give a qualitative argument and compute this ratio.

c) How much heat must you transfer to/from the container to keep the temperature of the gas constant?

4.15   Surface Contamination (Wisconsin-Madison)

A surface scientist wishes to keep an exposed surface "clean" (≤ 0.05 adsorbed monolayer) for an experiment lasting for times $t \geq 5$ h at a temperature $T = 300$ K. Estimate the needed data and calculate a value for the required background pressure in the apparatus if each incident molecule sticks to the surface.

4.16   Bell Jar (Moscow Phys-Tech)

A vessel with a small hole of diameter d in it is placed inside a high-vacuum chamber (see Figure P.4.16). The pressure is so low that the mean free path $\lambda \gg d$. The temperature of the gas in the chamber is T_0, and the pressure is P_0. The temperature in the vessel is kept at a constant $T_1 = 4T_0$. What is the pressure inside the vessel when steady state is reached?

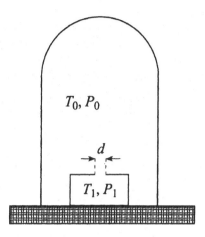

Figure **P.4.16**

4.17 Hole in Wall (Princeton)

A container is divided into two parts, I and II, by a partition with a small hole of diameter d. Helium gas in the two parts is held at temperatures $T_1 = 150$ K and $T_2 = 300$ K, respectively, through heating of the walls (see Figure P.4.17).

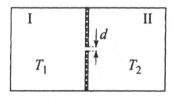

Figure **P.4.17**

a) How does the diameter d determine the physical process by which the gases come to steady state?

b) What is the ratio of the mean free paths λ_1/λ_2 between the two parts when $d \ll \lambda_1$, $d \ll \lambda_2$?

c) What is the ratio λ_1/λ_2 when $d \gg \lambda_1$, $d \gg \lambda_2$?

4.18 Ballast Volume Pressure (Moscow Phys-Tech)

Two containers, I and II, filled with an ideal gas are connected by two small openings of the same area, A, through a ballast volume B (see Figure P.4.18). The temperatures and pressures in the two containers are kept constant and equal to P, τ, and $P, 2\tau$, respectively. The volume B is thermally isolated. Find the equilibrium pressure and temperature in the ballast volume, assuming the gas is in the Knudsen regime.

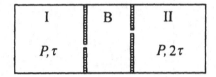

Figure **P.4.18**

4.19   Rocket in Drag (Princeton)

A rocket has an effective frontal area A and blasts off with a constant acceleration a straight up from the surface of the Earth (see Figure P.4.19).

Figure **P.4.19**

a) Use either dimensional analysis or an elementary derivation to find out how the atmospheric drag on the rocket should vary as some power(s) of the area A, the rocket velocity v, and the atmospheric density ρ (assuming that we are in the region of high Reynolds numbers).

b) Assume that the atmosphere is isothermal with temperature T. Derive the variation of the atmospheric density ρ with height z. Assume that the gravitational acceleration g is a constant and that the density at sea level is ρ_0.

c) Find the height h_0 at which the drag on the rocket is at a maximum.

4.20 Adiabatic Atmosphere (Boston, Maryland)

The lower 10–15 km of the atmosphere, the troposphere, is often in a convective steady state with constant entropy, not constant temperature (PV^γ is independent of the altitude, where $\gamma = C_P/C_V$.)

a) Find the change of temperature in this model with altitude dT/dz.
b) Estimate dT/dz in K/km. Consider the average diatomic molecule of air with molar mass $\mu = 29$ g/mole.

4.21 Atmospheric Energy (Rutgers)

The density of the Earth's atmosphere, $\rho(z)$, varies with height z above the Earth's surface. Assume that the "thickness" of the atmosphere is sufficiently small so that it is in a uniform gravitational field of strength g.

a) Write an equation to determine the atmospheric pressure $P(z)$, given the function $\rho(z)$.
b) In a static atmosphere, each parcel of air has an internal energy ΔE_i and a gravitational potential energy ΔE_g. To a very good approximation, the air in the atmosphere is an ideal gas with constant specific heat. Using this assumption, the result of part (a), and *classical* thermodynamics, show that the total energy in a vertical column of atmosphere of cross-sectional area A is given by

$$E \equiv E_i + E_g = \frac{AC_P}{\mu g} \int_0^{P_0} T(P)\, dP$$

and that the ratio of energies is

$$\frac{E_g}{E_i} = \gamma - 1$$

where T is the temperature, P_0 is the pressure at the Earth's surface, μ is the molar mass, C_P is the molar specific heat at constant pressure, and $\gamma \equiv C_P/C_V$ is the ratio of specific heats.

Hint: The above results do not depend on the specific way in which $P(z)$, $T(z)$ and $\rho(z)$ vary as a function of z (e.g., isothermal, adiabatic, or something intermediate). They depend only on the fact that $P(z)$ is monotonically decreasing. At some step of the derivation, you might find it useful to do an integration by parts.

4.22 Puncture (Moscow Phys-Tech)

A compressed ideal gas flows out of a small hole in a tire which has a pressure P_0 inside.

  a) Find the velocity of gas outside the tire in the vicinity of the hole if the flow is laminar and stationary and the pressure outside is P_1.
  b) Estimate this velocity for a flow of molecular hydrogen into a vacuum at a temperature $T = 1000$ K. Express this velocity in terms of the velocity of sound inside the tire, s_0.

Heat and Work

4.23 Cylinder with Massive Piston (Rutgers, Moscow Phys-Tech)

Consider n moles of an ideal monatomic gas placed in a vertical cylinder. The top of the cylinder is closed by a piston of mass M and cross section A (see Figure P.4.23). Initially the piston is fixed, and the gas has volume V_0 and temperature T_0. Next, the piston is released, and after several oscillations comes to a stop. Disregarding friction and the heat capacity of the piston and cylinder, find the temperature and volume of the gas at equilibrium. The system is thermally isolated, and the pressure outside the cylinder is P_a.

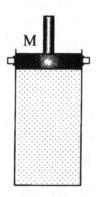

Figure **P.4.23**

4.24 Spring Cylinder (Moscow Phys-Tech)

One part of a cylinder is filled with one mole of a monatomic ideal gas at a pressure of 1 atm and temperature of 300 K. A massless piston separates the gas from the other section of the cylinder which is evacuated but has a spring at equilibrium extension attached to it and to the opposite wall of the cylinder. The cylinder is thermally insulated from the rest of the world, and the piston is fixed to the cylinder initially and then released (see Figure P.4.24). After reaching equilibrium, the volume occupied by the gas is double the original. Neglecting the thermal capacities of the cylinder, piston, and spring, find the temperature and pressure of the gas.

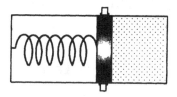

Figure **P.4.24**

4.25 Isothermal Compression and Adiabatic Expansion of Ideal Gas (Michigan)

An ideal gas is compressed at constant temperature τ from volume V_1 to volume V_2 (see Figure P.4.25).

a) Find the work done on the gas and the heat absorbed by the gas.

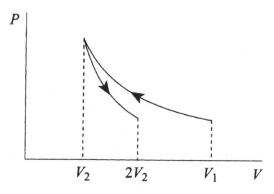

Figure **P.4.25**

b) The gas now expands adiabatically to volume $2V_2$. What is the final temperature T_f (derive this result from first principles)?

c) Estimate T_f for $T_i = 300$ K for air.

4.26 Isochoric Cooling and Isobaric Expansion (Moscow Phys-Tech)

An ideal gas of total mass m and molecular weight μ is isochorically (at constant volume) cooled to a pressure n times smaller than the initial pressure P_1. The gas is then expanded at constant pressure so that in the final state the temperature T_2 coincides with the initial temperature T_1. Calculate the work done by the gas.

4.27 Venting (Moscow Phys-Tech)

A thermally insulated chamber is pumped down to a very low pressure. At some point, the chamber is vented so that it is filled with air up to atmospheric pressure, whereupon the valve is closed. The temperature of the air surrounding the chamber is $T_0 = 300$ K. What is the temperature T of the gas in the chamber immediately after venting?

4.28 Cylinder and Heat Bath (Stony Brook)

Consider a cylinder 1 m long with a thin, massless piston clamped in such a way that it divides the cylinder into two equal parts. The cylinder is in a large heat bath at $T = 300$ K. The left side of the cylinder contains 1 mole of helium gas at 4 atm. The right contains helium gas at a pressure of 1 atm. Let the piston be released.

a) What is its final equilibrium position?

b) How much heat will be transmitted to the bath in the process of equilibration? (Note that $R = 8.3$ J/mole K.)

4.29 Heat Extraction (MIT, Wisconsin-Madison)

a) A body of mass M has a temperature-independent specific heat C. If the body is heated reversibly from a temperature T_i to a temperature T_f, what is the change in its entropy?

b) Two such bodies are initially at temperatures of 100 K and 400 K. A reversible engine is used to extract heat with the hotter body as a source and the cooler body as a sink. What is the maximum amount of heat that can be extracted in units of MC?

c) The specific heat of water is $C = 4.2$ J/g K, and its density is 1 g/cm^3. Calculate the maximum useful work that can be extracted, using as a source 10^3 m^3 of water at 100°C and a lake of temperature 10°C as a sink.

4.30 Heat Capacity Ratio (Moscow Phys-Tech)

To find C_P/C_V of a gas, one sometimes uses the following method. A certain amount of gas with initial temperature T_0, pressure P_0, and volume V_0 is heated by a current flowing through a platinum wire for a time t. The experiment is done twice: first at a constant volume V_0 with the pressure changing from P_0 to P_1, and then at a constant pressure P_0 with the volume changing from V_0 to V_1. The time t is the same in both experiments. Find the ratio C_P/C_V (the gas may be considered ideal).

4.31 Otto Cycle (Stony Brook)

The cycle of a highly idealized gasoline engine can be approximated by the Otto cycle (see Figure P.4.31). $1 \rightarrow 2$ and $3 \rightarrow 4$ are adiabatic compression and expansion, respectively; $2 \rightarrow 3$ and $4 \rightarrow 1$ are constant-volume processes. Treat the working medium as an ideal gas with constant $\gamma = c_p/c_v$.

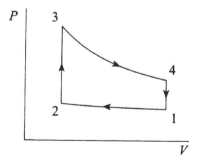

Figure **P.4.31**

a) Compute the efficiency of this cycle for $\gamma = 1.4$ and compression ratio $r = V_i/V_f = 10$.

b) Calculate the work done on the gas in the compression process $1 \rightarrow 2$, assuming initial volume $V_i = 2$ L and $P_i = 1$ atm.

4.32   Joule Cycle (Stony Brook)

Find the efficiency of the Joule cycle, consisting of two adiabats and two isobars (see Figure P.4.32). Assume that the heat capacities of the gas C_P and C_V are constant.

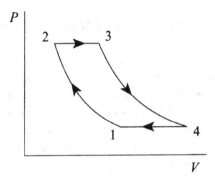

Figure **P.4.32**

4.33   Diesel Cycle (Stony Brook)

Calculate the efficiency of the Diesel cycle, consisting of two adiabats, $1 \rightarrow 2$ and $3 \rightarrow 4$; one isobar $2 \rightarrow 3$; and one constant-volume process $4 \rightarrow 1$ (see Figure P.4.33). Assume C_V and C_P are constant.

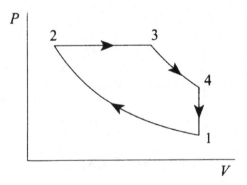

Figure **P.4.33**

4.34   Modified Joule–Thomson (Boston)

Figure P.4.34 shows container A of variable volume V controlled by a frictionless piston, immersed in a bath at temperature τ. This container is connected by a pipe with a porous plug to another container, B, of fixed volume V'. Container A is initially occupied by an ideal gas at pressure P while container B is initially evacuated. The gas is allowed to flow through the plug, and the pressure on the piston is maintained at the constant value P. When the pressure of the gas in B reaches P, the experiment is terminated. Neglecting any heat conduction through the plug, show that the final temperature of the gas in B is $\tau_1 = (C_P/C_V)\,\tau$, where C_P and C_V are the molar heats at constant pressure and volume of the gas.

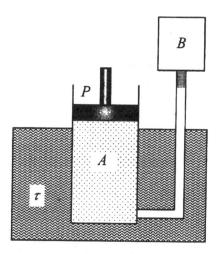

Figure **P.4.34**

Ideal Gas and Classical Statistics

4.35   Poisson Distribution in Ideal Gas (Colorado)

Consider a monatomic *ideal* gas of total $\tilde{N}$ molecules in a volume $\tilde{V}$. Show that the probability, P_N, for the number N of molecules contained in a small element of V is given by the Poisson distribution

$$P_N = \frac{e^{-\langle N \rangle}\,\langle N \rangle^N}{N!}$$

where $\langle N \rangle = \tilde{N}V/\tilde{V}$ is the average number of molecules found in the volume V.

4.36   Polarization of Ideal Gas (Moscow Phys-Tech)

Calculate the electric polarization $\mathbf{P}$ of an ideal gas, consisting of molecules having a constant electric dipole moment $\mathbf{p}$ in a homogeneous external electric field $\mathbf{E}$ at temperature τ. What is the dielectric constant of this gas at small fields?

4.37   Two-Dipole Interaction (Princeton)

Two classical dipoles with dipole moments μ_1 and μ_2 are separated by a distance R so that only the orientation of the magnetic moments is free. They are in thermal equilibrium at a temperature τ. Compute the mean force $\langle \mathbf{f} \rangle$ between the dipoles for the high-temperature limit $\mu_1 \mu_2/\tau R^3 \ll 1$. **Hint:** The potential energy of interaction of two dipoles is

$$U = \frac{\boldsymbol{\mu}_1 \cdot \boldsymbol{\mu}_2}{r^3} - 3\frac{(\boldsymbol{\mu}_1 \cdot \mathbf{r})(\boldsymbol{\mu}_2 \cdot \mathbf{r})}{r^5}$$

4.38   Entropy of Ideal Gas (Princeton)

A vessel of volume V_1 contains N molecules of an ideal gas held at temperature τ and pressure P_1. The energy of a molecule may be written in the form

$$E_k(p_x, p_y, p_z) = \frac{p_x^2}{2m} + \frac{p_y^2}{2m} + \frac{p_z^2}{2m} + \varepsilon_k$$

where ε_k denotes the energy levels corresponding to the internal states of the molecules of the gas.

a) Evaluate the free energy F. Explicitly display the dependence on the volume V_1.

Now consider another vessel, also at temperature τ, containing the same number of molecules of the identical gas held at pressure P_2.

b) Give an expression for the total entropy of the two gases in terms of P_1, P_2, τ, N.

c) The vessels are then connected to permit the gases to mix without doing work. Evaluate explicitly the change in entropy of the system. Check whether your answer makes sense by considering the special case $V_1 = V_2$ ($P_1 = P_2$).

4.39 Chemical Potential of Ideal Gas (Stony Brook)

Derive the expression for the Gibbs free energy and chemical potential of N molecules of an ideal gas at temperature τ, pressure P, and volume V. Assume that all the molecules are in the electronic ground state with degeneracy g. At what temperature is this approximation valid?

4.40 Gas in Harmonic Well (Boston)

A classical system of N distinguishable noninteracting particles of mass m is placed in a three-dimensional harmonic well:

$$U(r) = \frac{x^2 + y^2 + z^2}{2V^{2/3}}$$

a) Find the partition function and the Helmholtz free energy.
b) Regarding V as an external parameter, find the thermodynamic force $\tilde{P}$ conjugate to this parameter, exerted by the system; find the equation of state and compare it to that of a gas in a container with rigid walls.
c) Find the entropy, internal energy, and total heat capacity at constant volume.

4.41 Ideal Gas in One-Dimensional Potential (Rutgers)

a) An ideal gas of particles, each of mass m at temperature τ, is subjected to an external force whose potential energy has the form

$$U(x) = Ax^n$$

with $0 \le x \le \infty$, $A > 0$, and $n > 0$. Find the average potential energy per particle.
b) What is the average potential energy per particle in a gas in a uniform gravitational field?

4.42 Equipartition Theorem (Columbia, Boston)

a) For a classical system with Hamiltonian

$$H(q_1 \cdots p_N) = \sum_{i=1}^{N} \frac{p_i^2}{2m_i} + \sum_{i=1}^{N} \frac{k_i}{2} q_i^2$$

at a temperature τ, show that

$$\frac{\langle p_i^2 \rangle}{2m_i} = \frac{\tau}{2} \qquad \frac{k_i}{2} \langle q_i^2 \rangle = \frac{\tau}{2}$$

b) Using the above, derive the law of Dulong and Petit for the heat capacity of a harmonic crystal.

c) For a more general Hamiltonian,

$$H(q_1 \cdots p_N) = \sum_{i=1}^{N} \frac{p_i^2}{2m_i} + U(q_1 \cdots q_N)$$

prove the generalized equipartition theorem:

$$\left\langle x_i \frac{\partial H}{\partial x_j} \right\rangle = \tau \delta_{ij}$$

where $x_1 = q_1, \ldots, x_N = q_N, x_{N+1} = p_1, \ldots x_{2N} = p_N$. You will need to use the fact that U is infinite at $q_i = \pm\infty$.

d) Consider a system of a large number of classical particles and assume a general dependence of the energy of each particle on the generalized coordinate or momentum component q given by $\varepsilon(q)$, where

$$\lim_{q \to \pm\infty} \varepsilon(q) = +\infty$$

Show that, in thermal equilibrium, the generalized equipartition theorem holds:

$$\left\langle q \frac{\partial \varepsilon(q)}{\partial q} \right\rangle = \tau$$

What conditions should be satisfied for $\varepsilon(q)$ to conform to the equipartition theorem?

4.43 Diatomic Molecules in Two Dimensions (Columbia)

You have been transported to a two-dimensional world by an evil wizard who refuses to let you return to your beloved Columbia unless you can determine the thermodynamic properties for a rotating heteronuclear diatomic molecule constrained to move only in a plane (two dimensions). You may assume in what follows that the diatomic molecule does not undergo translational motion. Indeed, it only has rotational kinetic energy about its center of mass. The quantized energy levels of a diatomic in two dimensions are

$$\varepsilon_J = hcBJ^2 \qquad J = 0, 1, 2, 3, \ldots$$

with degeneracies $g_J = 2$, for J not equal to zero, and $g_J = 1$ when $J = 0$. As usual, $B = h/8\pi^2 Ic$, where I is the moment of inertia.

Hint: For getting out of the wizard's evil clutches, treat all levels as having the same degeneracy and then.... Oh, no! He's got me, too!

a) Assuming $\tau \gg hcB$, derive the partition function Z_{rot} for an individual diatomic molecule in two dimensions.

b) Determine the thermodynamic energy E and heat capacity C_V in the limit, where $\tau \gg hcB$, for a set of indistinguishable, independent, heteronuclear diatomic molecules constrained to rotate in a plane. Compare these results to those for an ordinary diatomic rotor in three dimensions. Comment on the differences and discuss briefly in terms of the number of degrees of freedom required to describe the motion of a diatomic rotor confined to a plane.

4.44 Diatomic Molecules in Three Dimensions (Stony Brook, Michigan State)

Consider the free rotation of a diatomic molecule consisting of two atoms of mass m_1 and m_2, respectively, separated by a distance a. Assume that the molecule is rigid with center of mass fixed.

a) Starting from the kinetic energy ε_k, where

$$\varepsilon_k = \frac{1}{2} \sum_{i=1}^{2} m_i \left(\dot{x}_i^2 + \dot{y}_i^2 + \dot{z}_i^2 \right)$$

derive the kinetic energy of this system in spherical coordinates and show that

$$\varepsilon_k = \frac{1}{2} I \left(\dot{\theta}^2 + \dot{\varphi}^2 \sin^2\theta \right)$$

where I is the moment of inertia. Express I in terms of m_1, m_2, and a.

b) Derive the canonical conjugate momenta p_θ and p_φ. Express the Hamiltonian of this system in terms of p_θ, p_φ, θ, φ, and I.

c) The classical partition function is defined as

$$Z_{\text{cl}} = \frac{1}{\hbar^2} \int_{-\infty}^{\infty} \int_{-\infty}^{\infty} \int_{0}^{2\pi} \int_{0}^{\pi} e^{-H/\tau} \, d\theta \, d\varphi \, dp_\theta \, dp_\varphi$$

Calculate Z_{cl}.   Calculate the heat capacity for a system of N molecules.

d) Assume now that the rotational motion of the molecule is described by quantum mechanics. Write the partition function in this case, taking into account the degeneracy of each state. Calculate the heat capacity of a system of N molecules in the limit of low and high temperatures and compare them to the classical result.

4.45 Two-Level System (Princeton)

Consider a system composed of a very large number N of distinguishable atoms at rest and mutually noninteracting, each of which has only two (nondegenerate) energy levels: 0, $\varepsilon > 0$. Let E/N be the mean energy per atom in the limit $N \to \infty$.

a) What is the maximum possible value of E/N if the system is not necessarily in thermodynamic equilibrium? What is the maximum attainable value of E/N if the system is in equilibrium (at positive temperature)?

b) For thermodynamic equilibrium compute the entropy per atom S/N as a function of E/N.

4.46 Zipper (Boston)

A zipper has N links; each link has a state in which it is closed with energy 0 and a state in which it is open with energy ε. We require that the zipper only unzip from one side (say from the left) and that the link can only open if all links to the left of it $(1, 2, \ldots, n-1)$ are already open. (This model is sometimes used for DNA molecules.)

a) Find the partition function.

b) Find the average number of open links $\langle n \rangle$ and show that for low temperatures $\tau \ll \varepsilon$, $\langle n \rangle$ is independent of N.

4.47 Hanging Chain (Boston)

The upper end of a hanging chain is fixed while the lower end is attached to a mass M. The (massless) links of the chain are ellipses with major axes $l + a$ and minor axes $l - a$, and can place themselves only with either the major axis or the minor axis vertical. Figure P.4.47 shows a four-link chain in which the major axes of the first and fourth links and the minor axes of the second and third links are vertical. Assume that the chain has N links and is in thermal equilibrium at temperature τ.

a) Find the partition function.
b) Find the average length of the chain.

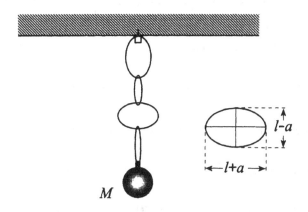

Figure **P.4.47**

4.48   Molecular Chain (MIT, Princeton, Colorado)

Consider a one-dimensional chain consisting of N molecules which exist in two configurations, α, β, with corresponding energies ε_α, ε_β, and lengths a and b. The chain is subject to a tensile force f.

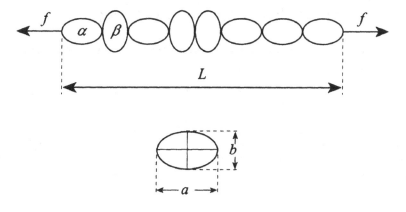

Figure **P.4.48**

a) Write the partition function Z_N for the system.

b) Calculate the average length $\langle L \rangle$ as a function of f and the temperature τ.

c) Assume that $\varepsilon_\alpha > \varepsilon_\beta$ and $a > b$. Estimate the average length $\langle L \rangle$ in the absence of the tensile force $f = 0$ as a function of temperature. What are the high- and low-temperature limits, and what is the characteristic temperature at which the changeover between the two limits occurs?

d) Calculate the linear response function

$$ \chi = \left(\frac{\partial \langle L \rangle}{\partial f} \right)_{f=0} $$

Produce a general argument to show that $\chi > 0$.

Nonideal Gas

4.49   Heat Capacities (Princeton)

Consider a gas with arbitrary equation of state $P = f(\tau, V)$, at a temperature τ, $\tau < \tau_{\mathrm{cr}}$, where τ_{cr} is a critical temperature of this gas.

a) Calculate $C_P - C_V$ for this gas in terms of f. Does $C_P - C_V$ always have the same sign?

b) Using the result of (a), calculate $C_P - C_V$ for one mole of a van der Waals gas.

4.50   Return of Heat Capacities (Michigan)

In a certain range of temperature τ and pressure p, the specific volume v of a substance is described by the equation

$$ v(\tau, P) = v_1 \exp \left(\frac{\tau}{\tau_1} - \frac{P}{P_1} \right) $$

where v_1, τ_1, P_1 are positive constants. From this information, determine (insofar as possible) as a function of temperature and pressure the following quantities:

a) $c_p - c_v$

b) c_p

c) c_v

4.51   Nonideal Gas Expansion (Michigan State)

A gas obeys the equation of state

$$P = \frac{\tau}{V} + \frac{B(\tau)}{V^2}$$

where $B(\tau)$ is a function of the temperature τ only. The gas is initially at temperature τ and volume V_0 and is expanded isothermally and reversibly to volume $V_1 = 2V_0$.

a) Find the work done in the expansion.
b) Find the heat absorbed in the expansion.

Some Maxwell relations:

$$\left(\frac{\partial \tau}{\partial V}\right)_S = -\left(\frac{\partial P}{\partial S}\right)_V \qquad\qquad \left(\frac{\partial P}{\partial \tau}\right)_V = \left(\frac{\partial S}{\partial V}\right)_\tau$$

$$\left(\frac{\partial V}{\partial \tau}\right)_P = -\left(\frac{\partial S}{\partial P}\right)_\tau \qquad\qquad \left(\frac{\partial \tau}{\partial P}\right)_S = \left(\frac{\partial V}{\partial S}\right)_P$$

4.52   van der Waals (MIT)

A monatomic gas obeys the van der Waals equation

$$P = \frac{N\tau}{V - Nb} - \frac{N^2 a}{V^2}$$

and has a heat capacity $C_V = 3N/2$ in the limit $V \to \infty$.

a) Prove, using thermodynamic identities and the equation of state, that

$$\left.\frac{\partial C_V}{\partial V}\right|_\tau = 0$$

b) Use the preceding result to determine the entropy of the van der Waals gas, $S(\tau, V)$, to within an additive constant.
c) Calculate the internal energy $\varepsilon(\tau, V)$ to within an additive constant.
d) What is the final temperature when the gas is adiabatically compressed from (V_1, τ_1) to final volume V_2?
e) How much work is done in this compression?

4.53   Critical Parameters (Stony Brook)

Consider a system described by the Dietrici equation of state

$$P(V - nB) = nN_A\tau e^{-nA/(N_A\tau V)}$$

where A, B, R are constants and P, V, τ, and n are the pressure, volume, temperature, and number of moles. Calculate the critical parameters, i.e., the values of P, V, and τ at the critical point.

Mixtures and Phase Separation

4.54   Entropy of Mixing (Michigan, MIT)

a) A 2-L container is divided in half: One half contains oxygen at 1 atm, the other nitrogen at the same pressure, and both gases may be considered ideal. The system is in an adiabatic enclosure at a temperature $T = 293$ K. The gases are allowed to mix. Does the temperature of the system change in this process? If so, by how much? Does the entropy change? If so, by how much?

b) How would the result differ if both sides contained oxygen?

c) Now consider one half of the enclosure filled with diatomic molecules of oxygen isotope ^{16}O and the other half with ^{18}O. Will the answer be different from parts (a) and (b)?

4.55   Leaky Balloon (Moscow Phys-Tech)

Sometimes helium gas in a low-temperature physics lab is kept temporarily in a large rubber bag at essentially atmospheric pressure. A physicist left a 40-L bag filled with He floating near the ceiling before leaving on vacation. When she returned, all the helium was gone (diffused through the walls of the bag). Find the entropy change of the gas. Assume that the atmospheric helium concentration is approximately $5 \cdot 10^{-4}\%$. What is the minimum work needed to collect the helium back into the bag?

4.56   Osmotic Pressure (MIT)

Consider an ideal mixture of N_0 monatomic molecules of type A and N_2 monatomic molecules of type B in a volume V.

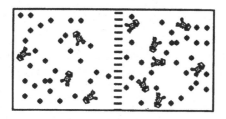

● - type A molecules (solvent)

🐟 - type B molecules (solute)

Figure P.4.56

a) Calculate the free energy $F(\tau, V, N_0, N_1)$. Calculate the Gibbs potential $G(\tau, P, N_0, N_1)$. G is the Legendre transform of F with respect to V.

b) If $N_0 \gg N_1$, the molecules of type A are called the solvent, and those of type B the solute. Consider two solutions with the same solvent (type A) and different concentrations of solute (type B molecules) separated by a partition through which solvent molecules can pass but solute molecules cannot (see Figure P.4.56). There are N_0 particles in volume V (or $2N_0$ in volume $2V$), and N_1 and N_2 particles in volume V on the left and right of the membrane, respectively. Calculate the pressure difference across the membrane at a given temperature and volume. Assume that the concentrations of the solutions are small; i.e.,

$$C_1 = \frac{N_1}{N_0} \ll 1$$

and

$$C_2 = \frac{N_2}{N_0} \ll 1$$

4.57 Clausius–Clapeyron (Stony Brook)

a) Derive the Clausius–Clapeyron equation for the equilibrium of two phases of a substance. Consider a liquid or solid phase in equilibrium with its vapor.

b) Using part (a) and the ideal gas law for the vapor phase, show that the vapor pressure follows the equation $\ln P_v = A - B/\tau$. Make reasonable assumptions as required. What is B?

4.58   Phase Transition (MIT)

The curve separating the liquid and gas phases ends in the critical point
V_c, τ_c, where $(\partial P/\partial V)_{\tau=\tau_c} = 0$. Using arguments based on thermodynamic
stability, determine

$$\left(\frac{\partial^2 P}{\partial V^2}\right)_{\tau=\tau_c}$$

at the critical point.

4.59   Hydrogen Sublimation in Intergalactic Space (Princeton)

A lump of condensed molecular hydrogen in intergalactic space would tend
to sublimate (evaporate) because the ambient pressure of hydrogen is well
below the equilibrium vapor pressure. Find an order-of-magnitude estimate
of the rate of sublimation per unit area at $T = 3$ K. The latent heat of
sublimation is $L \sim 450$ J/g, and the vapor pressure at the triple point
$T_t \sim 15$ K is $P_t \sim 50$ mm of Hg. (1 atm $\sim$ 760 mm Hg $\approx 10^6$ dyn/cm^2.)

4.60   Gas Mixture Condensation (Moscow Phys-Tech)

A mixture of $m_N = 100$ g of nitrogen and some oxygen is isothermally
compressed at $T = 77.4$ K. The result of this experiment is plotted as the
pressure dependence of the mixture versus volume in arbitrary units (see
Figure P.4.60). Find the mass of oxygen and the oxygen saturation vapor
pressure at this temperature.

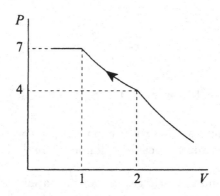

Figure P.4.60

Hint: $T = 77.4$ K is the boiling temperature of liquid nitrogen at atmospheric pressure. Oxygen boils at a higher temperature.

4.61 Air Bubble Coalescence (Moscow Phys-Tech)

A tightly closed jar is completely filled with water. On the bottom of the jar are two small air bubbles (see Figure P.4.61a) which sidle up to each other and become one bubble (see Figure P.4.61b). The pressure at the top of the jar is P_0, the radius of each original bubble is R_0, and the coefficient of surface tension is σ. Consider the process to be isothermal. Evaluate the change of pressure inside the jar upon merging of the two bubbles.

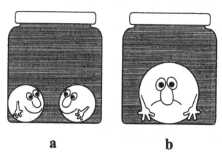

a   b

Figure **P.4.61**

4.62 Soap Bubble Coalescence (Moscow Phys-Tech)

Two soap bubbles B_1 and B_2 of radii R_1 and R_2 become one bubble B_0 of radius R_0. Find the surface tension coefficient for the soap solution. The ambient pressure is P_a.

4.63 Soap Bubbles in Equilibrium (Moscow Phys-Tech)

Two soap bubbles of radius R_0 are connected by a thin "straw" of negligible volume compared to the volume of the bubbles (see Figure P.4.63). The ambient pressure is P_a, the temperature is τ, and the surface tension coefficient is σ.

a) Is this system in stable equilibrium? What is the final state?
b) Calculate the entropy change between the final-state configuration and the configuration in Figure P.4.63. Assume $2\sigma/R_0 \ll P_a$.

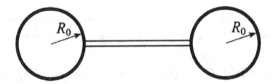

Figure **P.4.63**

Quantum Statistics

4.64 Fermi Energy of a 1D Electron Gas (Wisconsin-Madison)

Calculate the Fermi energy for a one-dimensional metal with one free electron per atom and an atomic spacing of 2.5 Å at $T = 0$.

4.65 Two-Dimensional Fermi Gas (MIT, Wisconson-Madison)

Consider a noninteracting nonrelativistic gas of N spin-1/2 fermions at $T = 0$ in a box of area A.

a) Find the Fermi energy.
b) Show that the total energy is given by

$$E = \frac{1}{2} n \varepsilon_F$$

c) Qualitatively discuss the behavior of the heat capacity of this system at low temperatures.

4.66 Nonrelativistic Electron Gas (Stony Brook, Wisconsin-Madison, Michigan State)

a) Derive the relation between pressure and volume of a free nonrelativistic electron gas at zero temperature.
b) The formula obtained in (a) is approximately correct for sufficiently low temperatures (the so-called strongly degenerate gas). Discuss the applicability of this formula to common metals.

4.67 Ultrarelativistic Electron Gas (Stony Brook)

Derive the relation between pressure and volume of a free ultrarelativistic electron gas at zero temperature.

4.68 Quantum Corrections to Equation of State (MIT, Princeton, Stony Brook)

Consider a noninteracting, one-component quantum gas at temperature τ, with a chemical potential μ in a cubic volume V. Treat the separate cases of bosons and fermions.

a) For a dilute system derive the equation of state in terms of temperature τ, pressure P, particle density n, and particle mass m. Do this derivation approximately by keeping the leading and next-leading powers of n. Interpret your results as an effective classical system.

b) At a given temperature, for which densities are your results valid?

4.69 Speed of Sound in Quantum Gases (MIT)

The sound velocity u in a spin-1/2 Fermi gas is given at $\tau = 0$ by

$$u^2 = \left(\frac{\partial P}{\partial \rho} \right)_{\tau=0}$$

where $\rho = mn$, m is the mass of the gas particles, and n is the number density.

a) Show that

$$\left(\frac{\partial P}{\partial \rho} \right)_\tau = \frac{n}{m} \frac{\partial \mu}{\partial n}$$

where μ is the chemical potential.

b) Calculate the sound velocity in the limit of zero temperature. Express your answer in terms of n, m.

c) Show that

$$\left(\frac{\partial P}{\partial \rho} \right)_{\tau=0} = 0$$

in a Bose gas below the Bose–Einstein temperature.

4.70   Bose Condensation Critical Parameters (MIT)

Consider an ideal Bose gas of N particles of mass m and spin zero in a volume V and temperature τ above the condensation point.

   a) What is the critical volume V_c below which Bose–Einstein condensation occurs? An answer up to a numerical constant will be sufficient.
   b) What is the answer to (a) in two dimensions?

4.71   Bose Condensation (Princeton, Stony Brook)

Consider Bose condensation for an arbitrary dispersion law in D dimensions (see Figure P.4.71). Assume a relation between energy and momentum of the form $\varepsilon \propto |p^\sigma|$. Find a relation between D and σ for Bose condensation to occur.

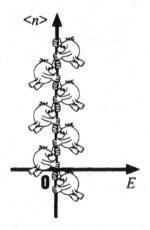

Figure **P.4.71**

4.72   How Hot the Sun? (Stony Brook)

The total radiant energy flux at the Earth from the Sun, integrated over all wavelengths, is observed to be approximately 0.14×10^7 erg cm^{-2}s^{-1}. The distance from the Earth to the Sun, d, is 1.5×10^{13} cm and the solar radius, R_S, is 7×10^{10} cm. Treating the Sun as a "blackbody," make a crude estimate of the surface temperature of the Sun (see Figure P.4.72). To make the numerical estimate, you are encouraged to ignore all factors of 2's and π's, to express any integrals that you might have in dimensionless form, and to take all dimensionless quantities to be unity.

Figure **P.4.72**

4.73 Radiation Force (Princeton, Moscow Phys-Tech, MIT)

Consider an idealized Sun and Earth, both blackbodies, in otherwise empty flat space. The Sun is at a temperature $T_S = 6000$ K, and heat transfer by oceans and atmosphere on the Earth is so effective as to keep the Earth's surface temperature uniform. The radius of the Earth is $R_E = 6.4 \cdot 10^6$ m, the radius of the Sun is $R_S = 7 \cdot 10^8$ m, and the Earth–Sun distance is $d = 1.5 \cdot 10^{11}$ m. The mass of Sun $M_S = 2 \cdot 10^{30}$ kg.

- a) Find the temperature of the Earth.
- b) Find the radiation force on the Earth.
- c) Compare these results with those for an interplanetary "chondrule" in the form of a spherical, perfectly conducting blackbody with a radius $R = 0.1$ cm, moving in a circular orbit around the Sun at a radius equal to the Earth–Sun distance d.
- d) At what distance from the Sun would a metallic particle melt (melting temperature $T_m = 1550$ K)?
- e) For what size particle would the radiation force calculated in (c) be equal to the gravitational force from the Sun at a distance d?

4.74 Hot Box and Particle Creation (Boston, MIT)

The electromagnetic radiation in a box of volume V can be treated as a noninteracting ideal Bose gas of photons. If the cavity also contains atoms capable of absorbing and emitting photons, the number of photons in the cavity is not definite. The box is composed of a special material that can withstand extremely high temperatures of order $\tau \gg \tau_e = m_e c^2$.

a) Derive the average number of photons in the box.
Hint:

$$\int_0^\infty \frac{x^2\,dx}{e^x - 1} \approx 2.4$$

b) What is the total energy of the radiation in the box for $\tau/\tau_e \ll 1$?

c) What is the entropy of the radiation for $\tau/\tau_e \ll 1$?

d) Assume that photons can create neutral particles of mass m and zero spin and that these neutral particles χ can create photons by annihilation or some other mechanism. The cavity now contains photons and particles in thermal equilibrium at a temperature τ. Find the particle density $n = N/V$. Consider only the process where a single photon is emitted or absorbed by making a single particle.
Hint: Minimize the free energy.

Now, instead of neutral particles, consider the creation of electron–positron pairs.

e) What is the total concentration of electrons and positrons inside the box when $\tau/\tau_e \ll 1$?

f) What is the total concentration of electrons and positrons when $\tau/\tau_e \gg 1$?
Hint:

$$\int_0^\infty \frac{x^2\,dx}{e^x + 1} \approx 1.8$$

4.75   *D*-Dimensional Blackbody Cavity (MIT)

Consider a D-dimensional hypercube blackbody cavity. What is the energy density as a function of temperature? It is not necessary to derive the multiplicative constant. Assume that the radiation is in quanta of energy $E(\omega) = \hbar\omega$.

4.76   Fermi and Bose Gas Pressure (Boston)

For a photon gas the entropy is

$$S = \frac{1}{\tau}\sum_i \frac{\hbar\omega_i}{e^{\hbar\omega_i/\tau} - 1} - \sum_i \ln\left(1 - e^{-\hbar\omega_i/\tau}\right) \qquad \text{(P.4.76.1)}$$

where ω_i is the angular frequency of the ith mode. Using (P.4.76.1):

a) Show that the isothermal work done by the gas is

$$dW = -\sum_i n_i \hbar \frac{d\omega_i}{dV} dV$$

where n_i is the average number of photons in the ith mode.

b) Show that the radiation pressure is equal to one third of the energy density:

$$P = \frac{1}{3}\frac{E}{V}$$

c) Show that for a nonrelativistic Fermi gas the pressure is

$$P = \frac{2}{3}\frac{E}{V}$$

4.77 Blackbody Radiation and Early Universe (Stony Brook)

The entropy of the blackbody radiation in the early universe does not change if the expansion is so slow that the occupation of each photon mode remains constant (or the other way around). To illustrate this consider the following problem. A one-dimensional harmonic oscillator has an infinite series of equally spaced energy states, with $\varepsilon_n = n\hbar\omega$, where n is a positive integer or zero and ω is the classical frequency of the oscillator.

a) Show that for a harmonic oscillator the free energy is

$$F = \tau \ln\left[1 - e^{-\hbar\omega/\tau}\right]$$

b) Find the entropy S. Establish the connection between entropy and occupancy of the modes by showing that for one mode of frequency ω the entropy is a function of photon occupancy $\langle n \rangle$ only:

$$S = \langle n + 1 \rangle \ln \langle n + 1 \rangle - \langle n \rangle \ln \langle n \rangle$$

4.78 Photon Gas (Stony Brook)

Consider a photon gas at temperature T inside a container of volume V. Derive the equation of state and compare it to that of the classical ideal gas (which has the equation $PV/\tau = $ const). Also compute the energy of the photon gas in terms of PV. You need not get all the numerical factors in this derivation.

4.79   Dark Matter (Rutgers)

From virial theorem arguments, the velocity dispersions of bright stars in
dwarf elliptical galaxies imply that most of the mass in these systems is in
the form of "dark" matter - possibly massive neutrinos (see Figure P.4.79).
The central parts of the Draco dwarf galaxy may be modeled as an isothermal gas sphere, with a phase-space distribution of mass of the form

$$f(\mathbf{r}, \mathbf{p}) = \frac{\rho(r)}{(2\pi m^2 \sigma^2)^{3/2}} \exp\left(-\frac{p^2}{m^2 \sigma^2}\right)$$

Here, $\rho(r)$ is the local mass density in the galaxy, σ is the velocity dispersion, and m is the mass of a typical "particle" in the galaxy. Measurements
on Draco yield $\sigma \approx 10$ km/s and $r_0 \approx 150$ pc (1 pc ≈ 3.3 light years). r_0 is
the "core" radius, where the density has decreased by close to a factor of 2
from its value at $r = 0$.

a) Using the virial theorem, write a very rough (order of magnitude)
   relation between r_0, σ, and $\rho(0)$.
b) Assume that most of the mass in Draco resides in one species of
   massive neutrino. Show how, if the Pauli exclusion principle is not to
   be violated, the distribution function above sets a lower limit on the
   mass of this neutrino.
c) Using the observations and the result of part (a), estimate this lower
   limit (in units of eV/c^2) and comment on whether current measurements of neutrino masses allow Draco to be held together in the manner suggested.

Figure **P.4.79**

4.80 Einstein Coefficients (Stony Brook)

You have two-state atoms in a thermal radiation field at temperature T. The following three processes take place:

1) Atoms can be promoted from state 1 to state 2 by absorption of a photon according to

$$\left(\frac{dN_1}{dt}\right)_{abs} = -B_{12}N_1\rho(\nu) \qquad (\text{P.4.80.1})$$

2) Atoms can decay from state 2 to state 1 by spontaneous emission according to

$$\left(\frac{dN_2}{dt}\right)_{spon} = -A_{21}N_2 \qquad (\text{P.4.80.2})$$

3) Atoms can decay from state 2 to state 1 by stimulated emission according to

$$\left(\frac{dN_2}{dt}\right)_{abs} = -B_{21}N_2\rho(\nu) \qquad (\text{P.4.80.3})$$

The populations N_1 and N_2 are in thermal equilibrium, and the radiation density is

$$\rho(\nu) = \frac{8\pi h\nu^3}{c^3}\frac{1}{e^{h\nu/\tau}-1} \qquad (\text{P.4.80.4})$$

a) What is the ratio N_2/N_1?
b) Calculate the ratios of coefficients A_{21}/B_{21} and B_{21}/B_{12}.
c) From the ratio of stimulated to spontaneous emission, how does the pump power scale with wavelength when you try to make short-wavelength lasers?

4.81 Atomic Paramagnetism (Rutgers, Boston)

Consider a collection of N identical noninteracting atoms, each of which has total angular momentum J. The system is in thermal equilibrium at temperature τ and is in the presence of an applied magnetic field $\mathbf{H} = H\hat{\mathbf{z}}$. The magnetic dipole moment associated with each atom is given by $\boldsymbol{\mu} = -g\mu_B\mathbf{J}$, where g is the gyromagnetic ratio and μ_B is the Bohr magneton. Assume the system is sufficiently dilute so that the local magnetic field at each atom may be taken as the applied magnetic field.

a) For a typical atom in this system, list the possible values of μ, the magnetic moment along the magnetic field, and the corresponding magnetic energy associated with each state.

b) Determine the thermodynamic mean value of the magnetic moment μ and the magnetization of the system M, and calculate it for $J = 1/2$ and $J = 1$.
c) Find the magnetization of the system in the limits $H \to \infty$ and $H \to 0$, and discuss the physical meaning of the results.

4.82   Paramagnetism at High Temperature (Boston)

a) Show that for a system with a discrete, finite energy spectrum ε_n, the specific heat per particle at high temperatures ($\tau \gg \varepsilon_n$ for all n) is

$$c = \frac{\sigma^2}{\tau^2}$$

where σ is the spectrum variance

$$\sigma^2 = \langle \varepsilon^2 \rangle - \langle \varepsilon \rangle^2$$

b) Use the result of (a) to derive the high-temperature specific heat for a paramagnetic solid treated both classically and quantum mechanically.
c) Compare your quantum mechanical result for $J = 1/2$ with the exact formula for c.

4.83   One-Dimensional Ising Model (Tennessee)

Consider N spins in a chain which can be modeled using the one-dimensional Ising model

$$H = -J \sum_{n=1}^{N-1} s_n s_{n+1} \qquad \text{(P.4.83.1)}$$

where the spin has the values $s_n = \pm 1$.

a) Find the partition function.
b) Find the heat capacity per spin.

4.84   Three Ising Spins (Tennessee)

Assume three spins are arranged in an equilateral triangle with each spin interacting with its two neighbors (see Figure P.4.84). The energy expression for the Ising model in a magnetic field ($F = \mu \cdot H$) is

$$H = -J (s_1 s_2 + s_2 s_3 + s_3 s_1) - F (s_1 + s_2 + s_3) \qquad \text{(P.4.84.1)}$$

Derive expressions for the

a) Partition function
b) Average spin $\langle s \rangle$
c) Internal energy ε

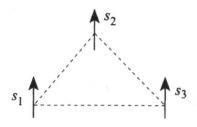

Figure **P.4.84**

4.85 N Independent Spins (Tennessee)

Consider a system of N independent spin-1/2 particles. In a magnetic field H, in the $\hat{z}$ direction, they can point either up or down with energy $\pm\mu H$, where μ is the magnetic moment. Derive expressions for the

a) Partition function
b) Internal energy
c) Entropy

4.86 N Independent Spins, Revisited (Tennessee)

Consider a system of N independent spin-1/2 particles. In a magnetic field H, in the $\hat{z}$ direction, they can point either up or down with energy $s_i\mu H$, where μ is the magnetic moment and $s_i = \pm 1$. Derive expressions for the entropy $S(f)$ in the case of a microcanonical ensemble, where the number of particles N and the magnetization $M = \mu f N$ ($f = \sum s_i/N$) are fixed.

4.87 Ferromagnetism (Maryland, MIT)

The spins of a regular Ising lattice interact by the energy

$$H = -J\sum_{i\neq j}' \sigma_i\sigma_j + \mu B \sum_i \sigma_i \qquad (\text{P.4.87.1})$$

where B is an external field, μ is the magnetic moment, and the prime indicates that the summation is only over the nearest neighbors. Each spin σ_i has z nearest neighbors. The spins are restricted to equal ± 1. The coupling constant J is positive. Following Weiss, represent the effect on σ_i of the spin–spin interaction in (P.4.87.1) by the mean field set up by the neighboring spins σ_j. Calculate the linear spin susceptibility $\chi(\tau)$ using this mean field approximation. Your expression should diverge at some temperature $\tau = \tau_c$. What is the physical significance of this divergence? What is happening to the spin lattice at $\tau = \tau_c$?

4.88 Spin Waves in Ferromagnets (Princeton, Colorado)

Consider the quantum mechanical spin-1/2 system with Hamiltonian

$$H = -J \sum_{\langle ij \rangle} \mathbf{s}_i \cdot \mathbf{s}_j \qquad (\text{P.4.88.1})$$

where the summation is over nearest-neighbor pairs in three dimensions.

a) Derive the equation of motion for the spin $\mathbf{s}_i$ at site i of the lattice.
b) Convert the model to a classical microscopic model by inserting the classical spin field $\mathbf{s}(\mathbf{r}, t)$ into the equation of motion. Express $\dot{\mathbf{s}}(\mathbf{r}, t)$ to lowest order in its gradients, considering a simple cubic lattice with lattice constant a.
c) Consider the ferromagnetic case with uniform magnetization $\mathbf{M} = M\hat{\mathbf{z}}$. Derive the frequency-versus-wave vector relation of a small spin-wave fluctuation $\mathbf{s}(\mathbf{r}, t) = \mathbf{M} + \mathbf{m}_0(t) \sin(\mathbf{k} \cdot \mathbf{r})$.
d) Quantize the spin waves in terms of magnons which are bosons. Derive the temperature dependence of the heat capacity.

Fluctuations

4.89 Magnetization Fluctuation (Stony Brook)

Consider N moments μ with two allowed orientations $\pm\mu$ in an external field H at temperature τ. Calculate the fluctuation of magnetization M, i.e.,

$$\frac{\sqrt{\langle M^2 \rangle - \langle M \rangle^2}}{\langle M \rangle}$$

4.90 Gas Fluctuations (Moscow Phys-Tech)

A high-vacuum chamber is evacuated to a pressure of 10^{-11} atm. Inside the chamber there is a thin-walled ballast volume filled with helium gas at a pressure $P = 10^{-6}$ atm and a temperature $T = 293$ K. On one wall of this ballast volume, there is a small hole of area $A = 10^{-3}$ cm^2. A detector counts the number of particles leaving the ballast volume during time intervals $\Delta t = 1$ ms.

a) Find the average number of molecules counted by the detector.
b) Find the mean square fluctuation of this number.
c) What is the probability of not counting any particles in one of the measurements?

4.91 Quivering Mirror (MIT, Rutgers, Stony Brook)

a) A very small mirror is suspended from a quartz strand whose elastic constant is D. (Hooke's law for the torsional twist of the strand is $\tau = -D\theta$, where θ is the angle of the twist.) In a real-life experiment the mirror reflects a beam of light in such a way that the angular fluctuations caused by the impact of surrounding molecules (Brownian motion) can be read on a suitable scale. The position of the equilibrium is $\langle \theta \rangle = 0$. One observes the average value $\langle \theta^2 \rangle$, and the goal is to find Avogadro's number (or, what is the same thing, determine the Boltzmann constant). The following are the data: At $T = 287$ K, for a strand with $D = 9.43 \cdot 10^{-9}$ dyn·cm, it was found that $\langle \theta^2 \rangle = 4.20 \cdot 10^{-6}$. You may also use the universal gas constant $R = 8.31 \cdot 10^7$ erg/K mole. Calculate Avogadro's number.
b) Can the amplitude of these fluctuations be reduced by reducing gas density? Explain your answer.

4.92 Isothermal Compressibility and Mean Square Fluctuation (Stony Brook)

a) Derive the relation

$$\kappa_\tau = \frac{V}{N^2} \left(\frac{\partial N}{\partial \mu} \right)_{V,\tau}$$

where κ_τ is the isothermal compressibility:

$$\kappa_\tau = -\frac{1}{V} \left(\frac{\partial V}{\partial p} \right)_{\tau,N}$$

b) From (a), find the relation between κ_T and the mean square fluctuation of N in the grand canonical ensemble. How does this fluctuation depend on the number of particles?

4.93 Energy Fluctuation in Canonical Ensemble (Colorado, Stony Brook)

Show that for a canonical ensemble the fluctuation of energy in a system of constant volume is related to the specific heat and, hence, deduce that the specific heat at constant volume is nonnegative.

4.94 Number Fluctuations (Colorado (a,b), Moscow Phys-Tech (c))

Show that for a grand canonical ensemble the number of particles N and occupational number n_j in an ideal gas satisfy the conditions:

a) $\langle n_k^2 \rangle - \langle n_k \rangle^2 = \langle n_k \rangle (1 \pm \langle n_k \rangle)$ quantum statistics
b) $\langle N^2 \rangle - \langle N \rangle^2 = \langle N \rangle$ classical statistics

For an electron spin $s = 1/2$ Fermi gas at temperature $\tau \ll \varepsilon_F$,

c) Find $\langle (\Delta N)^2 \rangle$.

4.95 Wiggling Wire (Princeton)

A wire of length L and mass per unit length ρ is fixed at both ends and tightened to a tension f. What is the rms fluctuation, in classical statistics, of the midpoint of the wire when it is in equilibrium with a heat bath at temperature τ? A useful series is

$$\sum_{m=0}^{\infty} (2m+1)^{-2} = \frac{\pi^2}{8}$$

4.96 *LC* Voltage Noise (MIT, Chicago)

The circuit in Figure P.4.96 consists of a coil of inductance L and a capacitor of capacitance C. What is the rms noise voltage across AB at temperature τ in the limit where

a) τ is very large?
b) τ is very small?

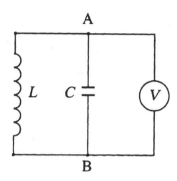

Figure **P.4.96**

Applications to Solid State

4.97 Thermal Expansion and Heat Capacity (Princeton)

a) Find the temperature dependence of the thermal expansion coefficient if the interaction between atoms is described by a potential

$$V_0(x) = \frac{1}{2}K_0x^2 - \frac{1}{3}m\lambda x^3$$

where λ is a small parameter.

b) Derive the anharmonic corrections to the Dulong–Petit law for a potential

$$V(x) = V_0(x) + \frac{1}{4}\eta x^4$$

where η is a small parameter.

4.98 Schottky Defects (Michigan State, MIT)

N atoms from a perfect crystal of total number of atoms $\tilde{N}$ are displaced to the surface of the crystal. Let ε_0 be the energy needed to displace one atom from the bulk of the crystal to the surface. Find the equilibrium number of defects N at low temperatures $\tau \ll \varepsilon_0$, assuming $N \ll \tilde{N}$.

4.99 Frenkel Defects (Colorado, MIT)

N atoms are arranged regularly to form a perfect crystal. If one replaces n atoms among them from lattice sites to interstices of the lattice, this becomes an imperfect crystal with n defects (of the Frenkel type). The number $\tilde{N}$ of interstitial sites into which an atom can enter is of the same order as N. Let ε_0 be the energy necessary to remove an atom from a lattice site to an interstitial site. Show that, in the equilibrium state at temperature τ such that $\tau \ll \varepsilon_0$, the following relation is valid:

$$n = \sqrt{N\tilde{N}}\ e^{-\varepsilon_0/2\tau}$$

4.100   Two-Dimensional Debye Solid (Columbia, Boston)

An atom confined to a surface may be thought of as an object "living" in a two-dimensional world. There are a variety of ways to look at such an atom. Suppose that the atoms adsorbed on the surface are not independent but undergo collective oscillations as do the atoms in a Debye crystal. Unlike the atoms in a Debye crystal, however, there are only two dimensions in which these collective vibrations can occur.

a) Derive an expression for the number of normal modes between ω and $\omega + d\omega$ and, by thinking carefully about the total number of vibrational frequencies for N atoms confined to a surface, rewrite it in terms of N and ω_D, the maximum vibration frequency allowed due to the discreteness of the atoms.

b) Obtain an integral expression for the energy E for the two-dimensional Debye crystal. Use this to determine the limiting form of the temperature dependence of the heat capacity (analogous to the Debye τ^3 law) as $\tau \to 0$ for the two-dimensional Debye crystal up to dimensionless integrals.

4.101   Einstein Specific Heat (Maryland, Boston)

a) Derive an expression for the average energy at a temperature τ of a quantum harmonic oscillator having natural frequency ω.

b) Assuming unrealistically (as Einstein did) that the normal-mode vibrations of a solid all have the same natural frequency (call it ω_E), find an expression for the heat capacity of an insulating solid.

c) Find the high-temperature limit for the heat capacity as calculated in (b) and use it to obtain a numerical estimate for the heat capacity of a $V = 5$ cm^3 piece of an insulating solid having a number density of $n = 6 \cdot 10^{28}$ atoms/m^3. Would you expect this to be a poor or a good estimate for the high-temperature heat capacity of the material? Please give reasons.

d) Find the low-temperature limit of the heat capacity and explain why it is reasonable in terms of the model.

4.102 Gas Adsorption (Princeton, MIT, Stanford)

Consider a vapor (dilute monatomic gas) in equilibrium with a submono-layer (i.e., less than one atomic layer) of atoms adsorbed on a surface. Model the binding of atoms to the surface by a potential energy $V = -\varepsilon_0$. Assume there are N_0 possible sites for adsorption, and find the vapor pressure as a function of surface concentration $\theta = N/N_0$ (N is the number of adsorbed particles).

4.103 Thermionic Emission (Boston)

a) Assume that the evaporation of electrons from a hot wire (Richardson's effect) is thermodynamically equivalent to the sublimation of a solid. Find the pressure of the electron gas, provided that the electrons outside the metal constitute an ideal classical monatomic gas and that the chemical potential μ_s of the electrons in the metal (the solid phase) is a constant.

b) Derive the same result by using the Clausius–Clapeyron equation

$$\frac{dp}{d\tau} = \frac{L}{\tau\,\Delta V}$$

where L is the latent heat of electron evaporation. Neglect the volume occupied by the electrons in the metal.

4.104 Electrons and Holes (Boston, Moscow Phys-Tech)

a) Derive a formula for the concentration of electrons in the conduction band of a semiconductor with a fixed chemical potential (Fermi level) μ, assuming that in the conduction band $\varepsilon - \mu \gg \tau$ (nondegenerate electrons).

b) What is the relationship between hole and electron concentrations in a semiconductor with arbitrary impurity concentration and band gap E_g?

c) Find the concentration of electrons and holes for an intrinsic semi-conductor (no impurities), and calculate the chemical potential if the electron mass is equal to the mass of the hole: $m_e = m_h$.

4.105 Adiabatic Demagnetization (Maryland)

A paramagnetic sample is subjected to magnetic cooling.

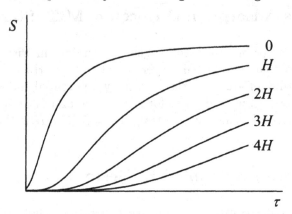

Figure **P.4.105**

a) Show that

$$\left(\frac{\partial S}{\partial H}\right)_\tau = \left(\frac{\partial M}{\partial \tau}\right)_H \qquad \text{(P.4.105.1)}$$

Assume χ is independent of H. Show that

$$\left(\frac{\partial C_H}{\partial H}\right)_\tau = V\tau H\left(\frac{\partial^2 \chi}{\partial \tau^2}\right)_H \qquad \text{(P.4.105.2)}$$

where $M = V\chi H$ is the magnetization, χ is the isothermal magnetic susceptibility per unit volume, H is the magnetic field, and C_H is the heat capacity at constant H.

b) For an adiabatic process, show that

$$\left(\frac{\partial \tau}{\partial H}\right)_S = -\frac{V\tau H}{C_H}\left(\frac{\partial \chi}{\partial \tau}\right)_H \qquad \text{(P.4.105.3)}$$

c) Assume that χ can be approximated by Curie's law $\chi = a/\tau$ and that the heat capacity at zero magnetic field is given by

$$C_H(\tau, 0) = \frac{Vb}{\tau^2} \qquad \text{(P.4.105.4)}$$

where a and b are constants. Show that

$$C_H(\tau, H) = \frac{V}{\tau^2}\left(b + aH^2\right) \qquad \text{(P.4.105.5)}$$

For an adiabatic process, show that the ratio of final and initial temperatures is given by

$$\frac{T_f}{T_i} = \sqrt{\frac{b + aH_f^2}{b + aH_i^2}} \qquad \text{(P.4.105.6)}$$

d) Explain and indicate in the S–τ diagram given in Figure P.4.105 a possible route for the adiabatic demagnetization cooling process to approach zero temperature.

4.106 Critical Field in Superconductor (Stony Brook, Chicago)

Consider a massive cylinder of volume V made of a type I superconducting material in a magnetic field parallel to its axis.

a) Using the fact that the superconducting state displays perfect diamagnetism, whereas the normal state has negligible magnetic susceptibility, show that the entropy discontinuity across the phase boundary is at zero field H:

$$S_n - S_s = -\frac{V}{4\pi} H_c \frac{\partial H_c}{\partial \tau}$$

where $H_c(\tau)$ is the critical H field for suppressing superconductivity at a temperature τ.

b) What is the latent heat when the transition occurs in a field?
c) What is the specific heat discontinuity in zero field?

Quantum Mechanics

One-Dimensional Potentials

5.1 Shallow Square Well I (Columbia)

A particle of mass m moving in one dimension has a potential $V(x)$ which is a shallow square well near the origin:

$$V(x) = \begin{cases} -V_0 & |x| < a \\ 0 & |x| > a \end{cases} \qquad (P.5.1.1)$$

where V_0 is a positive constant. Derive the eigenvalue equation for the state of lowest energy, which is a bound state (see Figure P.5.1).

Figure **P.5.1**

5.2 Shallow Square Well II (Stony Brook)

A particle of mass m is confined to move in one dimension by a potential $V(x)$ (see Figure P.5.2):

$$V(x) = \begin{cases} \infty & x < 0 \\ -V_0 & 0 < x < a \\ 0 & a < x \end{cases} \qquad \text{(P.5.2.1)}$$

Figure **P.5.2**

a) Derive the equation for the bound state.
b) From the results of part (a), derive an expression for the minimum value of V_0 which will have a bound state.
c) Give the expression for the eigenfunction of a state with positive energy $E > 0$.
d) Show that the results of (c) define a phase shift for the potential, and derive an expression for the phase shift.

5.3 Attractive Delta Function Potential I (Stony Brook)

A particle of mass m moves in one dimension under the influence of an attractive delta function potential at the origin. The Schrödinger equation

is

$$\left\{ -\frac{\hbar^2}{2m}\frac{\partial^2}{\partial x^2} - C\delta(x) - i\hbar\frac{\partial}{\partial t} \right\} \Psi(x,t) = 0 \qquad \text{(P.5.3.1)}$$

Figure **P.5.3**

a) Find the eigenvalue and eigenfunction of the bound state.
b) If the system is in the bound state and the strength of the potential is changed suddenly $C \to C'$, what is the probability that the particle remains bound?

5.4 Attractive Delta Function Potential II (Stony Brook)

A particle of mass m is confined to the right half-space, in one dimension, by an infinite potential at the origin. There is also an attractive delta function potential $V(x) = -V_0\, a\delta(x - a)$, where $a > 0$ (see Figure P.5.4).

a) Find the expression for the energy of the bound state.
b) What is the minimum value of V_0 required for a bound state?

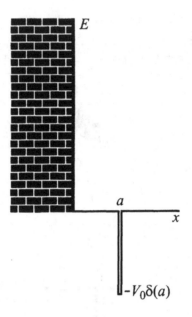

Figure **P.5.4**

5.5   Two Delta Function Potentials (Rutgers)

A particle of mass m moves in a one-dimensional potential of the form

$$V(x) = -\frac{\hbar^2 P}{m} \delta \left(x^2 - a^2 \right) \qquad \text{(P.5.5.1)}$$

where P is a positive dimensionless constant and a has units of length. Discuss the bound states of this potential as a function of P.

5.6   Transmission Through a Delta Function Potential (Michigan State, MIT, Princeton)

A particle of mass m moves in one dimension where the only potential $V(x) = C\delta(x)$ is at the origin with $C > 0$. A free particle of wave vector k approaches the origin from the left. Derive an expression for the amplitude T of the transmitted wave as a function of k, C, m, and $\hbar$.

5.7   Delta Function in a Box (MIT)

A particle of mass m is confined to a box, in one dimension, between $-a < x < a$, and the box has walls of infinite potential. An attractive delta function $V(x) = -aC\delta(x)$ is at the center of the box.

a) What are the eigenvalues of odd-parity states?
b) Find the value of C for which the lowest eigenvalue is zero.
c) Find the ground state wave function for the case that the lowest eigenvalue is less than zero energy.

5.8 Particle in Expanding Box (Michigan State, MIT, Stony Brook)

A particle of mass m is contained in a one-dimensional impenetrable box extending from $x = -L/2$ to $x = +L/2$. The particle is in its ground state.

a) Find the eigenfunctions of the ground state and the first excited state.
b) The walls of the box are moved outward instantaneously to form a box extending from $-L \le x \le L$. Calculate the probability that the particle will stay in the ground state during this sudden expansion.
c) Calculate the probability that the particle jumps from the initial ground state to the first excited final state.

5.9 One-Dimensional Coulomb Potential (Princeton)

An electron moves in one dimension and is confined to the right half-space $(x > 0)$ where it has a potential energy

$$V(x) = -\frac{e^2}{4x} \qquad (\text{P.5.9.1})$$

where e is the charge on an electron. This is the image potential of an electron outside a perfect conductor.

a) Find the ground state energy.
b) Find the expectation value $\langle x \rangle$ in the ground state.

5.10 Two Electrons in a Box (MIT)

Two electrons are confined in one dimension to a box of length a. A clever experimentalist has arranged that both electrons have the same spin state. Ignore the Coulomb interaction between electrons.

a) Write the ground state wave function $\psi(x_1, x_2)$ for the two-electron system.
b) What is the probability that both electrons are found in the same half of the box?

5.11   Square Well (MIT)

A particle of mass m is confined to a space $0 < x < a$ in one dimension by infinitely high walls at $x = 0, a$. At $t = 0$ the particle is initially in the left half of the well with constant probability

$$\Psi(x,0) = \begin{cases} \sqrt{2/a} & 0 < x < a/2 \\ 0 & a/2 < x < a \end{cases} \qquad \text{(P.5.11.1)}$$

a) Find the time-dependent wave function $\Psi(x,t)$.
b) What is the probability that the particle is in the nth eigenstate?
c) Write an expression for the average value of the particle energy.

5.12   Given the Eigenfunction (Boston, MIT)

A particle of mass m moves in one dimension. It is remarked that the exact eigenfunction for the ground state is

$$\psi(x) = \frac{A}{\cosh \lambda x} \qquad \text{(P.5.12.1)}$$

where λ is a constant and A is the normalization constant. Assuming that the potential $V(x)$ vanishes at infinity, derive the ground state eigenvalue and $V(x)$.

5.13   Combined Potential (Tennessee)

A particle of mass m is confined to $x > 0$ in one dimension by the potential

$$V(x) = V_0 \left[\frac{b^2}{x^2} - \frac{b}{x} \right] \qquad \text{(P.5.13.1)}$$

where V_0 and b are constants. Assuming there is a bound state, derive the exact ground state energy.

Harmonic Oscillator

5.14   Given a Gaussian (MIT)

A particle of mass m is coupled to a simple harmonic oscillator in one dimension. The oscillator has frequency ω and distance constant $x_0^2 = \hbar/m\omega$.

At time $t = 0$ the particle's wave function $\Psi(x, t)$ is given by

$$\Psi(x, 0) = \frac{e^{-x^2/2\sigma^2}}{(\pi\sigma^2)^{1/4}} \qquad \text{(P.5.14.1)}$$

The constant σ is unrelated to any other parameters. What is the probability that a measurement of energy at $t = 0$ finds the value of $E_0 = \hbar\omega/2$?

5.15 Harmonic Oscillator ABCs (Stony Brook)

Consider the harmonic oscillator given by

$$H = \frac{p^2}{2} + \frac{x^2}{2}$$

Define

$$a = \frac{1}{\sqrt{2}}(p - ix)$$

$$a^\dagger = \frac{1}{\sqrt{2}}(p + ix)$$

a) Show that $[a, a^\dagger] = 1$.
b) Show that $H = a^\dagger a + 1/2$.
c) Show that $[a^\dagger a, H] = 0$.
d) Show that if $|n\rangle$ is an eigenstate of $N = a^\dagger a$ with eigenvalue n:

$$a^\dagger a |n\rangle = n |n\rangle$$

then $a^\dagger |n\rangle$ and $a |n\rangle$ are also eigenstates of N with eigenvalues $n + 1$ and $n - 1$, respectively.
e) Define $|0\rangle$ such that $a |0\rangle = 0$. What is the energy eigenvalue of $|0\rangle$?
f) How can one construct other eigenstates of H starting from $|0\rangle$?
g) What is the energy spectrum of H? Are negative eigenvalues possible?

5.16 Number States (Stony Brook)

Consider the quantum mechanical Hamiltonian for a harmonic oscillator with frequency ω:

$$H = \frac{p^2}{2m} + \frac{m\omega^2 q^2}{2}$$

and define the operators

$$a = \frac{1}{\sqrt{2m\hbar\omega}}(p - im\omega q)$$

$$a^\dagger = \frac{1}{\sqrt{2m\hbar\omega}}(p + im\omega q)$$

a) Suppose we define a state $|0\rangle$ to obey

$$a\,|0\rangle = 0$$

Show that the states

$$|n\rangle = \frac{1}{\sqrt{n!}}\left(a^\dagger\right)^n |0\rangle$$

are eigenstates of the number operator, $N = a^\dagger a$, with eigenvalue n:

$$a^\dagger a\,|n\rangle = n\,|n\rangle$$

b) Show that $|n\rangle$ is also an eigenstate of the Hamiltonian and compute its energy.
   Hint: You may assume $\langle n\,|n\rangle = 1$

c) Using the above operators, evaluate the expectation value $\langle n\,|q^2|\,n\rangle$ in terms of $E(n)$, m, and ω.

5.17   Coupled Oscillators (MIT)

Two identical harmonic oscillators in one dimension each have mass m and frequency ω. Let the two oscillators be coupled by an interaction term Cx_1x_2, where C is a constant and x_1 and x_2 are the coordinates of the two oscillators. Find the exact spectrum of eigenvalues for this coupled system.

5.18   Time-Dependent Harmonic Oscillator I (Wisconsin-Madison)

Consider a simple harmonic oscillator in one dimension:

$$H = \frac{p^2}{2m} + \frac{m\omega^2 x^2}{2} \tag{P.5.18.1}$$

At $t = 0$ the wave function is

$$\Psi(x,0) = \sqrt{\frac{1}{3}}\,\psi_0(x) + \sqrt{\frac{2}{3}}\,\psi_2(x) \tag{P.5.18.2}$$

where $\psi_n(x)$ is the exact eigenstate of the harmonic oscillator with eigenvalue $\hbar\omega(n + 1/2)$.

a) Give $\Psi(x, t)$ for $t \geq 0$.
b) What is the parity of this state? Does it change with time?
c) What is the average value of the energy for this state? Does it change with time?

5.19   Time-Dependent Harmonic Oscillator II (Michigan State)

Consider a simple harmonic oscillator in one dimension. Introduce the raising and lowering operators, $a^\dagger$ and a, respectively. The Hamiltonian H and wave function Ψ at $t = 0$ are

$$H = \hbar\omega\left(a^\dagger a + \frac{1}{2}\right) \qquad \text{(P.5.19.1)}$$

$$\Psi(0) = \frac{1}{\sqrt{5}}|1\rangle + \frac{2}{\sqrt{5}}|2\rangle \qquad \text{(P.5.19.2)}$$

where $|n\rangle$ denotes the eigenfunction of energy $E_n = \hbar\omega(n + 1/2)$.

a) What is wave function $\Psi(t)$ at positive times?
b) What is the expectation value for the energy?
c) The position x can be represented in operators by $x = X_0(a + a^\dagger)$ where $X_0 = \sqrt{\hbar/2m\omega}$ is a constant. Derive an expression for the expectation of the time-dependent position

$$\langle x(t)\rangle = \langle \Psi(t)|x|\Psi(t)\rangle \qquad \text{(P.5.19.3)}$$

You may need operator expressions such as $a|n\rangle = \sqrt{n}\,|n-1\rangle$ and $a^\dagger|n\rangle = \sqrt{n+1}\,|n+1\rangle$.

5.20   Switched-on Field (MIT)

Consider a simple harmonic oscillator in one dimension with the usual Hamiltonian

$$H_0 = \frac{p^2}{2m} + \frac{m\omega^2}{2}x^2 \qquad \text{(P.5.20.1)}$$

a) The eigenfunction of the ground state can be written as

$$\psi_0(x) = Ne^{-\alpha^2 x^2/2} \qquad \text{(P.5.20.2)}$$

Determine the constants N and α.

b) What is the eigenvalue of the ground state?

c) At time $t = 0$, an electric field $|\mathbf{E}|$ is switched on, adding a perturbation of the form $H' = e|\mathbf{E}|x$. What is the new ground state energy?

d) Assuming that the field is switched on in a time much faster than $1/\omega$, what is the probability that the particle stays in the ground state?

5.21   Cut the Spring! (MIT)

A particle is allowed to move in one dimension. It is initially coupled to two identical harmonic springs, each with spring constant K. The springs are symmetrically fixed to the points $\pm a$ so that when the particle is at $x = 0$ the classical force on it is zero.

a) What are the eigenvalues of the particle while it is connected to both springs?

b) What is the wave function in the ground state?

c) One spring is suddenly cut, leaving the particle bound to only the other one. If the particle is in the ground state before the spring is cut, what is the probability it is still in the ground state after the spring is cut?

Angular Momentum and Spin

5.22   Given Another Eigenfunction (Stony Brook)

A nonrelativistic particle of mass m moves in a three-dimensional central potential $V(r)$ which vanishes at $r \to \infty$. We are given that an exact eigenstate is

$$\psi(\mathbf{r}) = Cr^{\sqrt{3}}e^{-\alpha r}\cos\theta \qquad\qquad (\text{P.5.22.1})$$

where C and α are constants.

a) What is the angular momentum of this state?

b) What is the energy?

c) What is $V(r)$?

5.23   Algebra of Angular Momentum (Stony Brook)

Given the commutator algebra

$$[J_1, J_2] = iJ_3$$

$$[J_2, J_3] = i J_1$$

$$[J_3, J_1] = i J_2$$

a) Show that $J^2 = J_1^2 + J_2^2 + J_3^2$ commutes with J_3.
b) Derive the spectrum of $\{J^2, J_3\}$ from the commutation relations.

5.24   Triplet Square Well (Stony Brook)

Consider a two-electron system in one dimension, where both electrons have spins aligned in the same direction (say, up). They interact only through the attractive square well in relative coordinates

$$V(|x_1 - x_2|) = \begin{cases} -V_0 & |x_1 - x_2| < a \\ 0 & |x_1 - x_2| > a \end{cases} \qquad (P.5.24.1)$$

What is the lowest energy of the two-electron state? Assume the total momentum is zero.

5.25   Dipolar Interactions (Stony Brook)

Two spin-1/2 particles are separated by a distance $\mathbf{a} = a\hat{z}$ and interact only through the magnetic dipole energy

$$H = \frac{\boldsymbol{\mu}_1 \cdot \boldsymbol{\mu}_2}{a^3} - 3 \frac{(\boldsymbol{\mu}_1 \cdot \mathbf{a})(\boldsymbol{\mu}_2 \cdot \mathbf{a})}{a^5} \qquad (P.5.25.1)$$

where $\boldsymbol{\mu}_j$ is the magnetic moment of spin j. The system of two spins consists of eigenstates of the total spin (S^2) and total S_z.

a) Write the Hamiltonian in terms of spin operators.
b) Write the Hamiltonian in terms of S^2 and S_z.
c) Give the eigenvalues for all states.

5.26   Spin-Dependent $1/r$ Potential (MIT)

Consider two identical particles of mass m and spin 1/2. They interact only through the potential

$$V = \frac{g}{r} \boldsymbol{\sigma}_1 \cdot \boldsymbol{\sigma}_2 \qquad (P.5.26.1)$$

where $g > 0$ and $\boldsymbol{\sigma}_j$ are Pauli spin matrices which operate on the spin of particle j.

a) Construct the spin eigenfunctions for the two particle states. What is the expectation value of V for each of these states?
b) Give the eigenvalues of all of the bound states.

5.27   Three Spins (Stony Brook)

Consider three particles of spin $1/2$ which have no motion. The raising
$(s^+ = s_x + is_y)$ and lowering $(s^- = s_x - is_y)$ operators of the individual
spins have the property

$$s^- |\uparrow\rangle = |\downarrow\rangle \qquad\qquad\qquad (P.5.27.1)$$

$$s^+ |\downarrow\rangle = |\uparrow\rangle \qquad\qquad\qquad (P.5.27.2)$$

where the arrows indicate the spin orientation with regard to the z-
direction.

    a) Write explicit wave functions for the four $J = 3/2$ states: $(M = 3/2, 1/2, -1/2, -3/2)$.

    b) Using the definition that $J^{\pm} = \sum_i s_i^{\pm}$, construct the 4×4 matrices which represent the J^+ and J^- operators.

    c) Construct the 4×4 matrices which represent J_x and J_y.

    d) Construct from J_x, J_y, J_z the value of the matrix J^2.

5.28   Constant Matrix Perturbation (Stony Brook)

Consider a system described by a Hamiltonian

$$H = H_0 + V \qquad\qquad\qquad (P.5.28.1)$$

$$H_0 = \varepsilon \begin{pmatrix} 1 & 0 & 0 \\ 0 & 1 & 0 \\ 0 & 0 & 1 \end{pmatrix} \qquad\qquad\qquad (P.5.28.2)$$

$$V = -G \begin{pmatrix} 1 & 1 & 1 \\ 1 & 1 & 1 \\ 1 & 1 & 1 \end{pmatrix} \qquad\qquad\qquad (P.5.28.3)$$

where ε and G are positive.

    a) Find the eigenvalues and eigenvectors of this Hamiltonian.

    b) Consider the two states

$$|i\rangle = \begin{pmatrix} 1 \\ 0 \\ 0 \end{pmatrix} \qquad\qquad\qquad (P.5.28.4)$$

$$|f\rangle = \begin{pmatrix} 0 \\ 0 \\ 1 \end{pmatrix} \qquad\qquad\qquad (P.5.28.5)$$

At $t = 0$, the system is in state $|i\rangle$. Derive the probability that at any later time it is in state $|f\rangle$.

5.29 Rotating Spin (Maryland, MIT)

A spin-1/2 particle interacts with a magnetic field $\mathbf{B} = B_0\hat{\mathbf{z}}$ through the Pauli interaction $H = \mu\boldsymbol{\sigma} \cdot \mathbf{B}$, where μ is the magnetic moment and $\boldsymbol{\sigma} = (\sigma_x, \sigma_y, \sigma_z)$ are the Pauli spin matrices. At $t = 0$ a measurement determines that the spin is pointing along the positive x-axis. What is the probability that it will be pointing along the negative y-axis at a later time t?

5.30 Nuclear Magnetic Resonance (Princeton, Stony Brook)

A spin-1/2 nucleus is placed in a large magnetic field B_0 in the z-direction. An oscillating field $B_1 \ll B_0$ of radio frequency ω is applied in the xy-plane, so the total magnetic field is

$$\mathbf{B} = (B_1 \cos \omega t, B_1 \sin \omega t, B_0) \qquad (P.5.30.1)$$

The Hamiltonian is $H = \mu\mathbf{B} \cdot \boldsymbol{\sigma}$, where μ is the magnetic moment. Use the notation $\hbar\Omega_\| = \mu B_0$, $\hbar\Omega_\perp = \mu B_1$.

a) If the nucleus is initially pointing in the $+z$-direction at $t = 0$, what is the probability that it points in the $-z$-direction at later times?
b) Discuss why most NMR experiments adjust B_0 so that $\Omega_\| \sim \omega/2$.

Variational Calculations

5.31 Anharmonic Oscillator (Tennessee)

Use variational methods in one dimension to estimate the ground state energy of a particle of mass m in a potential $V(x) = Ax^4$.

5.32 Linear Potential I (Tennessee)

A particle of mass m is bound in one dimension by the potential $V(x) = |Fx|$, where F is a constant. Use variational methods to estimate the energy of the ground state.

5.33   Linear Potential II (MIT, Tennessee)

A particle of mass m moves in one dimension in the right half-space. It has a potential energy $V(x)$ given by

$$V(x) = \begin{cases} \infty & x \le 0 \\ Fx & x > 0 \end{cases} \qquad\qquad \text{(P.5.33.1)}$$

where F is a positive real constant. Use variational methods to obtain an estimate for the ground state energy. How does the wave function behave in the limits $x \to 0$ or $x \to \infty$?

5.34   Return of Combined Potential (Tennessee)

A particle of mass m moves in one dimension according to the potential

$$V(x) = V_0 \left[\frac{b^2}{x^2} - \frac{b}{x} \right] \qquad\qquad \text{(P.5.34.1)}$$

where V_0 and b are both constants.

a) Show that the wave function must vanish at $x = 0$ so that a particle on the right of the origin never gets to the left.
b) Use variational methods to estimate the energy of the ground state.

5.35   Quartic in Three Dimensions (Tennessee)

A particle of mass m is bound in three dimensions by the quartic potential $V(r) = Ar^4$. Use variational methods to estimate the energy of the ground state.

5.36   Halved Harmonic Oscillator (Stony Brook, Chicago (b), Princeton (b))

Consider a particle of mass m moving in one dimension (see Figure P.5.36) in a potential

$$V(x) = \begin{cases} \infty & x < 0 \\ \dfrac{1}{2} m\omega^2 x^2 & x > 0 \end{cases}$$

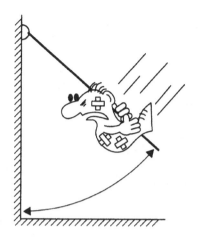

Figure **P.5.36**

a) Using the normalized trial function

$$\psi(x) = \begin{cases} 0 & x < 0 \\ 2\lambda^{3/2}xe^{-\lambda x} & x > 0 \end{cases}$$

find the value of λ which minimizes the ground state energy and the resulting estimate of the ground state energy. How is this value related to the true ground state energy?

b) What is the *exact* ground state wave function and energy for this system (neglect the normalization of the wave function)? Do not solve the Schrödinger equation directly. Rather, state the answer and justify it.

Hint: You may need the integral

$$\int_0^\infty x^n e^{-x}\,dx = n!$$

5.37 Helium Atom (Tennessee)

Do a variational calculation to estimate the ground state energy of the electrons in the helium atom. The Hamiltonian for two electrons, assuming the nucleus is fixed, is

$$H = -\frac{\hbar^2}{2m}\left[\nabla_1^2 + \nabla_2^2\right] - 2e^2\left(\frac{1}{r_1} + \frac{1}{r_2}\right) + \frac{e^2}{|\mathbf{r}_1 - \mathbf{r}_2|} \qquad \text{(P.5.37.1)}$$

Assume a wave function of the form

$$\psi(\mathbf{r}_1, \mathbf{r}_2) = \frac{\alpha^3}{\pi a_0^3} e^{-\alpha(r_1 + r_2)/a_0} \chi_s \qquad \text{(P.5.37.2)}$$

where a_0 is the Bohr radius, α is the variational parameter, and χ_s is the spin state of the two electrons.

Perturbation Theory

5.38 Momentum Perturbation (Princeton)

A particle of mass m moves in one dimension according to the Hamiltonian

$$H_0 = \frac{p^2}{2m} + V(x) \qquad \text{(P.5.38.1)}$$

$$H_0 \psi_n(x) = E_n \psi_n(x) \qquad \text{(P.5.38.2)}$$

All eigenfunctions $\psi_n(x)$ and eigenvalues E_n are known. Suppose we add a term to the Hamiltonian, where λ and m are constants and p is the momentum operator:

$$H = H_0 + \frac{\lambda}{m} p \qquad \text{(P.5.38.3)}$$

Derive an expression for the eigenvalues and eigenstates of the new Hamiltonian H.

5.39 Ramp in Square Well (Colorado)

A particle of mass m is bound in a square well where $-a/2 < x < a/2$.

a) What are the energy and eigenfunction of the ground state?
b) A small perturbation is added, $V(x) = 2\varepsilon|x|/a$. Use perturbation theory to calculate the change in the ground state energy to $O(\varepsilon)$.

5.40 Circle with Field (Colorado, Michigan State)

A particle with charge e and mass m is confined to move on the circumference of a circle of radius r. The only term in the Hamiltonian is the kinetic energy, so the eigenfunctions and eigenvalues are

$$\psi_n(\phi) = \frac{1}{\sqrt{2\pi}} e^{in\phi} \qquad \text{(P.5.40.1)}$$

$$E_n = \frac{\hbar^2 n^2}{2mr^2} \qquad \text{(P.5.40.2)}$$

where ϕ is the angle around the circle. An electric field $\mathbf{E}$ is imposed in the plane of the circle. Find the perturbed energy levels up to $O(|\mathbf{E}|^2)$.

5.41 Rotator in Field (Stony Brook)

Consider a rigid body with moment of inertia I, which is constrained to rotate in the xy-plane, and whose motion is given by the Schrödinger equation

$$-\frac{\hbar^2}{2I}\frac{d^2\psi}{d\phi^2} = E\psi \qquad (P.5.41.1)$$

a) Find the eigenfunctions and eigenvalues.
b) Assume the rotator has a fixed dipole moment $\mathbf{p}$ in the plane. An electric field $\mathbf{E}$ is applied to the plane. Find the changes in the energy levels to first and second order in the field.

5.42 Finite Size of Nucleus (Maryland, Michigan State, Princeton, Stony Brook)

Regard the nucleus of charge Z as a sphere of radius R_0 with a uniform charge density. Assume that $R_0 \ll a_0$ where a_0 is the Bohr radius of the hydrogen atom.

a) Derive an expression for the electrostatic potential $V(r)$ between the nucleus and the electrons in the atom. If $V_0(r) = -Ze^2/r$ is the potential from a point charge, find the difference $\delta V = V(r) - V_0(r)$ due to the size of the nucleus.
b) Assume one electron is bound to the nucleus in the lowest bound state. What is its wave function when calculated using the potential $V_0(r)$ from a point nucleus?
c) Use first-order perturbation theory to derive an expression for the change in the ground state energy of the electron due to the finite size of the nucleus.

5.43 U and U^2 Perturbation (Princeton)

A particle is moving in the three-dimensional harmonic oscillator with potential energy $V(r)$. A weak perturbation δV is applied:

$$V(r) = \frac{m\omega^2}{2}(x^2 + y^2 + z^2) \qquad (P.5.43.1)$$

$$\delta V = Uxyz + \frac{U^2}{\hbar\omega}x^2y^2z^2 \qquad (P.5.43.2)$$

The same small constant U occurs in both terms of δV. Use perturbation theory to calculate the change in the ground state energy to order $O(U^2)$.

5.44 Relativistic Oscillator (MIT, Moscow Phys-Tech, Stony Brook (a))

Consider a spinless particle in a one-dimensional harmonic oscillator potential:

$$V(x) = \frac{1}{2}m\omega^2 x^2$$

a) Calculate leading relativistic corrections to the ground state to first order in perturbation theory.
b) Consider an anharmonic classical oscillator with

$$H = \frac{p^2}{2m} + \frac{m\omega^2 x^2}{2} + \alpha x^3$$

For what values of α will the leading corrections be the same as in (a)?

5.45 Spin Interaction (Princeton)

Consider a spin-1/2 particle which is bound in a three-dimensional harmonic oscillator with frequency ω. The ground state Hamiltonian H_0 and spin interaction are

$$H = H_0 + H' \tag{P.5.45.1}$$

$$H_0 = \frac{p^2}{2m} + \frac{1}{2}m\omega^2 r^2 \tag{P.5.45.2}$$

$$H' = \lambda \mathbf{r} \cdot \boldsymbol{\sigma} \tag{P.5.45.3}$$

where λ is a constant and $\boldsymbol{\sigma} = (\sigma_x, \sigma_y, \sigma_z)$ are the Pauli matrices. Neglect the spin–orbit interaction. Use perturbation theory to calculate the change in the ground state energy to order $O(\lambda^2)$.

5.46 Spin–Orbit Interaction (Princeton)

Consider in three dimensions an electron in a harmonic oscillator potential which is perturbed by the spin–orbit interaction

$$H = H_0 + V_{\text{so}} \tag{P.5.46.1}$$

$$H_0 = \frac{p^2}{2m} + V(r) \qquad\qquad \text{(P.5.46.2)}$$

$$V(r) = \frac{m\omega^2 r^2}{2} \qquad\qquad \text{(P.5.46.3)}$$

$$V_{so} = \frac{\hbar^2}{2m^2c^2} \frac{1}{r} \frac{\partial V(r)}{\partial r} \mathbf{L} \cdot \mathbf{S} \qquad\qquad \text{(P.5.46.4)}$$

a) What are the eigenvalues of the ground state and the lowest excited states of the three-dimensional harmonic oscillator?

b) Use perturbation theory to estimate how these eigenvalues are altered by the spin–orbit interaction.

5.47   Interacting Electrons (MIT)

Consider two electrons bound to a proton by Coulomb interaction. Neglect the Coulomb repulsion between the two electrons.

a) What are the ground state energy and wave function for this system?

b) Consider that a weak potential exists between the two electrons of the form

$$V(\mathbf{r}_1 - \mathbf{r}_2) = V_0 \delta^3(\mathbf{r}_1 - \mathbf{r}_2) \mathbf{s}_1 \cdot \mathbf{s}_2 \qquad\qquad \text{(P.5.47.1)}$$

where V_0 is a constant and $\mathbf{s}_j$ is the spin operator for electron j (neglect the spin–orbit interaction). Use first-order perturbation theory to estimate how this potential alters the ground state energy.

5.48   Stark Effect in Hydrogen (Tennessee)

Consider a single electron in the $n = 2$ state of the hydrogen atom. We ignore relativistic corrections, so the $2s$ and $2p$ states are initially degenerate. Then we impose a small static electric field $\mathbf{E} = |\mathbf{E}|\hat{z}$. Use perturbation theory to derive how the $n=2$ energy levels are changed to lowest order in powers of $|\mathbf{E}|$.

5.49 $n = 2$ Hydrogen with Electric and Magnetic Fields (MIT)

Consider an electron in the $n = 2$ state of the hydrogen atom. We ignore relativistic corrections, so the $2s$ and $2p$ states are initially degenerate.

Then we impose two simultaneous perturbations: an electric field $\mathbf{E}$ in the x-direction ($\mathbf{E} = |\mathbf{E}|\hat{\mathbf{x}}$) and a magnetic field B, which is given by the vector potential $\mathbf{A} = (B/2)(-y, x, 0)$. Ignore the magnetic moment of the electron. Calculate how the $n = 2$ states are altered by these simultaneous perturbations.

5.50 Hydrogen in Capacitor (Maryland, Michigan State)

A hydrogen atom in its ground state is placed between the parallel plates of a capacitor. For times $t < 0$, no voltage is applied. Starting at $t = 0$, an electric field $\mathbf{E}(t) = \mathbf{E}_0 e^{-t/\tau}$ is applied, where τ is a constant. Derive the formula for the probability that the electron ends up in state j due to this perturbation. Evaluate the result for j:

a) a $2s$ state
b) a $2p$ state

5.51 Harmonic Oscillator in Field (Maryland, Michigan State)

A particle of mass m and charge e moves in one dimension. It is attached to a spring of constant k and is initially in the ground state ($n = 0$) of the harmonic oscillator. An electric field $\mathbf{E}$ is switched on during the interval $0 < t < \tau$, where τ is a constant.

a) What is the probability of the particle ending up in the $n = 1$ state?
b) What is the probability of the particle ending up in the $n = 2$ state?

5.52 β-Decay of Tritium (Michigan State)

Tritium is an isotope of hydrogen with one proton and two neutrons. A hydrogen-like atom is formed with an electron bound to the tritium nucleus. The tritium nucleus undergoes β-decay, and the nucleus changes its charge state suddenly to $+2$ and becomes an isotope of helium. If the electron is initially in the ground state in the tritium atom, what is the probability that the electron remains in the ground state after the sudden β-decay?

WKB

5.53   Bouncing Ball (Moscow Phys-Tech, Chicago)

A ball of mass m acted on by uniform gravity (let g be the acceleration of gravity) bounces up and down elastically off a floor. Take the floor to be at the zero of potential energy. Working in the WKB approximation, compute the quantized energy levels of the bouncing ball system.

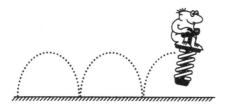

Figure P.5.53

5.54   Truncated Harmonic Oscillator (Tennessee)

A truncated harmonic oscillator in one dimension has the potential

$$V(x) = \begin{cases} \frac{1}{2}m\omega^2(x^2 - b^2) & |x| < b \\ 0 & |x| > b \end{cases} \qquad \text{(P.5.54.1)}$$

a) Use WKB to estimate the energies of the bound states.
b) Find the condition that there is only one bound state: it should depend on m, ω and b.

5.55   Stretched Harmonic Oscillator (Tennessee)

Use WKB in one dimension to calculate the eigenvalues of a particle of mass m in the following potential (see Figure P.5.55):

$$V(x) = \begin{cases} 0 & |x| < a \\ \frac{K}{2}(|x| - a)^2 & |x| > a \end{cases} \qquad \text{(P.5.55.1)}$$

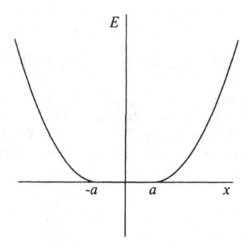

Figure **P.5.55**

5.56   Ramp Potential (Tennessee)

Use WKB in one dimension to find the eigenvalues of a particle of mass m in the potential $V(x) = F|x|$, where $F > 0$.

5.57   Charge and Plane (Stony Brook)

A particle moving in one dimension feels the potential

$$V(x) = \gamma|x| + C\delta(x) \qquad (\gamma > 0)$$

(This potential would be appropriate for an electron moving in the presence of a uniformly charged sheet where C is the transparency of the sheet.)

a) Using the WKB approximation, find the energy spectrum, E_n, for this one-dimensional problem for all n for $C = 0$.

b) Find the energy spectrum, E_n, for $C \to \infty$.

c) Derive an equation that describes the energies E_n for even wave functions for an arbitrary value of C. What can you say about the energies E_n for odd wave functions?

5.58 Ramp Phase Shift (Tennessee)

Use WKB to calculate the phase shift in one dimension of a particle of mass m confined by the ramp potential

$$V(x) = \begin{cases} 0 & x > 0 \\ |Fx| & x < 0 \end{cases} \qquad (P.5.58.1)$$

5.59 Parabolic Phase Shift (Tennessee)

Use WKB to calculate the phase shift in one dimension of a particle of mass m confined by the parabolic potential

$$V(x) = \begin{cases} 0 & x > 0 \\ \dfrac{1}{2}Kx^2 & x < 0 \end{cases} \qquad (P.5.59.1)$$

5.60 Phase Shift for Inverse Quadratic (Tennessee)

A particle of mass m moves in one dimension in the right half-space ($x > 0$) with the potential

$$V(x) = \lambda^2 \frac{\hbar^2}{2mx^2} \qquad (P.5.60.1)$$

where the dimensionless constant λ determines the strength of the potential. Use WKB to calculate the phase shift as a function of energy.

Scattering Theory

5.61 Step-Down Potential (Michigan State, MIT)

A particle of mass m obeys a Schrödinger equation with a potential

$$V(x) = \begin{cases} V_0 & x < 0 \\ 0 & x > 0 \end{cases} \qquad (P.5.61.1)$$

Since $V_0 > 0$, the potential is higher on the left of zero than on the right. Find the reflection coefficient $R(p)$ for a particle coming in from the left with momentum p (see Figure P.5.61).

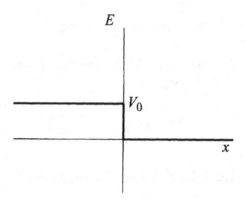

Figure P.5.61

5.62 Step-Up Potential (Wisconsin-Madison)

Consider a particle scattering in one dimension from a potential $V(x)$ which is a simple step at $x = 0$:

$$V(x) = \begin{cases} 0 & x < 0 \\ V_0 & x > 0 \end{cases} \qquad (\text{P.5.62.1})$$

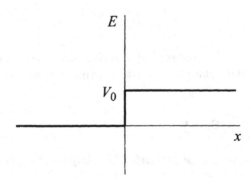

Figure P.5.62

where $V_0 > 0$. A particle with kinetic energy $E > V_0$ is incident from the left (see Figure P.5.62).

a) Find the intensity of the reflected (R) and transmitted (T) waves.
b) Find the currents (J_R, J_T), and the sum $J_R + J_T$, of the reflected and transmitted waves.

5.63   Repulsive Square Well (Colorado)

Consider in three dimensions a repulsive ($V_0 > 0$) square well at the origin
of width a. The potential is

$$V(r) = \begin{cases} V_0 & r < a \\ 0 & r > a \end{cases} \qquad (P.5.63.1)$$

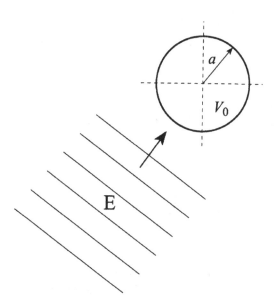

Figure P.5.63

A particle of energy $E = \hbar^2 k^2 / 2m < V_0$ is incident upon the square well
(see Figure P.5.63).

a) Derive the phase shift for s-waves.
b) How does the phase shift behave as $V_0 \to \infty$?
c) Derive the total cross section in the limit of zero energy.

5.64   3D Delta Function (Princeton)

Consider a particle of mass m which scatters in three dimension from a
potential which is a shell at radius a:

$$V(r) = -C\delta(|\mathbf{r}| - a) \qquad (P.5.64.1)$$

Derive the exact s-wave expression for the scattering cross section in the
limit of very low particle energy.

5.65   Two-Delta-Function Scattering (Princeton)

A free particle of mass m, traveling with momentum p parallel to the z-axis, scatters off the potential

$$V(\mathbf{r}) = V_0 \left[\delta \left(\mathbf{r} - \varepsilon \hat{\mathbf{z}} \right) - \delta \left(\mathbf{r} + \varepsilon \hat{\mathbf{z}} \right) \right]$$

Calculate the differential scattering cross section, $d\sigma/d\Omega$, in the Born approximation. Does this approximation provide a reasonable description for scattering from this potential? In other words, is it valid to use unperturbed wave functions in the scattering amplitude?

5.66   Scattering of Two Electrons (Princeton)

Two electrons scatter in a screened environment where the effective potential is

$$V(r) = \frac{e^2}{r} e^{-\lambda r} \qquad\qquad (P.5.66.1)$$

where λ is a constant. Consider both electrons in the center-of-mass frame, where both electrons have energy E. This energy is much larger than a Rydberg but much less than mc^2, so use nonrelativistic kinematics. Derive an approximate differential cross section for scattering through an angle θ when the two electrons are

a) in a total spin state of $S = 0$,
b) in a total spin state of $S = 1$.

5.67   Spin-Dependent Potentials (Princeton)

Consider the nonrelativistic scattering of an electron of mass m and momentum k through an angle θ. Calculate the differential cross section in the Born approximation for the spin-dependent potential

$$V = e^{-\mu r^2} [A + B\boldsymbol{\sigma} \cdot \mathbf{r}] \qquad\qquad (P.5.67.1)$$

where $\boldsymbol{\sigma} = (\sigma_x, \sigma_y, \sigma_z)$ are the Pauli spin matrices and (μ, A, B) are constants. Assume the initial spin is polarized along the incident direction, and sum over all final spins. (Note: Ignore that the potential violates parity.)

5.68 Rayleigh Scattering (Tennessee)

Rayleigh scattering is the elastic scattering of photons. Assume there is a matrix element $M(\mathbf{k}, \mathbf{k}')$ which describes the scattering from $\mathbf{k}$ to $\mathbf{k}'$. It has the dimensions of J m^3.

a) Derive an expression for the differential cross section $d\sigma/d\Omega$ for Rayleigh scattering. Ignore the photon polarization.

b) Assume the specific form for the matrix element

$$M(\mathbf{k}, \mathbf{k}') = 2\pi\hbar\omega\xi_{\mathbf{k}} \cdot \tilde{\alpha}(\omega) \cdot \xi_{\mathbf{k}'} \qquad (P.5.68.1)$$

where $\tilde{\alpha}(\omega)$ is the polarizability tensor and $\xi_{\mathbf{k}}$ are the polarization vectors of the photons. What is the result if the initial photons are unpolarized and the final photon polarizations are summed over? Assume the polarizability is isotropic: $\tilde{\alpha} = \alpha\tilde{I}$ where $\tilde{I}$ is the unit tensor.

5.69 Scattering from Neutral Charge Distribution (Princeton)

Consider the nonrelativistic scattering of a particle of mass m and charge e from a fixed distribution of charge $\rho(\mathbf{r})$. Assume that the charge distribution is neutral: $\int d^3r\, \rho(r) = 0$; it is spherically symmetric; and the second moment is defined as

$$A = \int d^3r\; r^2\rho(r) \qquad (P.5.69.1)$$

a) Use the Born approximation to derive the differential cross section $d\sigma/d\Omega$ for the scattering of a particle of wave vector k.

b) Derive the expression for forward scattering $(\theta \to 0)$.

c) Assume that $\rho(r)$ is for a neutral hydrogen atom in its ground state. Calculate A in this case. Neglect exchange effects and assume that the target does not recoil.

General

5.70 Spherical Box with Hole (Stony Brook)

A particle is confined to a spherical box of radius R. There is a barrier in the center of the box, which excludes the particle from a radius a. So the particle is confined to the region $a < r < R$. Assume that the wave function vanishes at both $r = a$ and $r = R$ and derive an expression for the eigenvalues and eigenfunctions of states with angular momentum $\ell = 0$.

5.71   Attractive Delta Function in 3D (Princeton)

A particle moves in three dimensions. The only potential is an attractive delta function at $r = a$ of the form

$$V(r) = -\frac{\hbar^2}{2mD}\delta(r-a) \qquad\qquad (P.5.71.1)$$

where D is a parameter which determines the strength of the potential.

   a) What are the matching conditions at $r = a$ for the wave function and its derivative?
   b) For what values of D do bound states exist for s-waves $(\ell = 0)$?

5.72   Ionizing Deuterium (Wisconsin-Madison)

The hydrogen atom has an ionization energy of $E_H = 13.5983$ eV when an electron is bound to a proton. Calculate the ionization energy of deuterium: an electron bound to a deuteron. Give your answer as the difference between the binding energy of deuterium and hydrogen $(\delta E = E_D - E_H)$. The deuteron has unit charge. The three masses are, in atomic mass units, $m_e = 5.4858 \cdot 10^{-4}$, $m_p = 1.00728$, $m_d = 2.01355$.

5.73   Collapsed Star (Stanford)

In a very simple model of a collapsed star a large number $A = N + Z$ of nucleons (N neutrons and Z protons) and Z electrons (to ensure electric neutrality) are placed in a *one-dimensional* box (i.e., an infinite square well) of length L. The neutron and proton have equal mass 1 GeV, and the electron has mass 0.5 MeV. Assume the nucleon number density is $\lambda = 0.5$ fm^{-1} (1 fm $= 10^{-13}$ cm), neglect all interactions between the particles in the well, and approximate $Z \gg 1$.

   a) Which particle species are relativistic?
   b) Calculate the ground state energy of the system as a function of Z for all possible configurations Z/A with fixed A.
   c) What value of Z/A (assumed small) minimizes the total energy of the system?

5.74 Electron in Magnetic Field (Stony Brook, Moscow Phys-Tech)

An electron is in free space except for a constant magnetic field B in the z-direction.

a) Show that the magnetic field can be represented by the vector potential $\mathbf{A} = B(0, x, 0)$.

b) Use this vector potential to derive the exact eigenfunctions and eigenvalues for the electron.

5.75 Electric and Magnetic Fields (Princeton)

Consider a particle of mass m and charge e which is in perpendicular electric and magnetic fields: $\mathbf{E} = |\mathbf{E}|\hat{z}, \mathbf{B} = |\mathbf{B}|\hat{y}$.

a) Write the Hamiltonian, using a convenient gauge for the vector potential.

b) Find the eigenfunctions and eigenvalues.

c) Find the average velocity in the x-direction for any eigenstate.

5.76 Josephson Junction (Boston)

Consider superconducting metals I and II separated by a very thin insulating layer, such that that electron wave functions can overlap between the metals (Josephson junction). A battery V is connected across the junction to ensure an average charge neutrality (see Figure P.5.76). This situation can be described by means of the coupled Schrödinger equations:

$$i\hbar\frac{\partial \Psi_1}{\partial t} = U_1\Psi_1 + K\Psi_2 + K\frac{\Psi_1\Psi_2^*}{\Psi_1^*} \qquad \text{(P.5.76.1)}$$

$$i\hbar\frac{\partial \Psi_2}{\partial t} = U_2\Psi_2 + K\Psi_1 + K\frac{\Psi_2\Psi_1^*}{\Psi_2^*}$$

Here Ψ_1 and Ψ_2 are the probability amplitudes for an electron in I and II, U_2 and U_2 are the electric potential energies in I and II, K is the coupling constant due to the insulating layer, and $K\Psi_1\Psi_2^*/\Psi_1^*$ and $K\Psi_2\Psi_1^*/\Psi_2^*$ describe the battery as a source of electrons.

a) Show that $\rho_1 = |\Psi_1|^2$ and $\rho_2 = |\Psi_2|^2$ are constant in time.

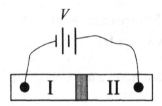

Figure P.5.76

b) Assuming $\rho_1 = \rho_2 = \rho_0$ (same metals) and expressing the probability amplitudes in the form

$$\Psi_1 = \sqrt{\rho_0}\,e^{i\theta_1} \tag{P.5.76.2}$$

$$\Psi_2 = \sqrt{\rho_0}\,e^{i\theta_2}$$

find the differential equations for θ_1 and θ_2.

c) Show that the battery current

$$I = \frac{K}{i\hbar}\,(\Psi_1\Psi_2^* - \Psi_1^*\Psi_2) \tag{P.5.76.3}$$

oscillates, and find the frequency of these oscillations.

SOLUTIONS

4

Thermodynamics and Statistical Physics

Introductory Thermodynamics

4.1 Why Bother? (Moscow Phys-Tech)

The physicist is right in saying that the total energy of the molecules in the room cannot be changed. Indeed, the total energy of an ideal gas is

$$E = Nc_v\tau \tag{S.4.1.1}$$

where N is the number of molecules, c_v is the heat capacity at constant volume per particle, and τ is the absolute temperature in energy units. In these units,

$$\tau \to k_{\mathrm{B}}T \qquad c_v \to \frac{c_v}{k_{\mathrm{B}}}$$

Since the pressure P in the room stays the same (as does the volume V) and equal to the outside air pressure, we have

$$PV = N\tau = \mathrm{const} \tag{S.4.1.2}$$

So, the total energy of the gas does not change. However, the average energy of each molecule does, of course, increase, and that is what defines the temperature (and part of the comfort level of the occupants). At the same time, the total number of molecules in the room decreases. In essence, we burn wood to force some of the molecules to shiver outside the room (this problem was first discussed in *Nature* **141**, 908 (1938)).

4.2   Space Station Pressure (MIT)

The rotation of the station around its axis is equivalent to the appearance of an energy $U = -m\Omega^2 R^2/2$, where m is the mass of an air particle and R is the distance from the center. Therefore, the particle number density satisfies the Boltzmann distribution (similar to the Boltzmann distribution in a gravitational field):

$$n(R) = n_c e^{-U(R)/\tau} = n_c e^{m\Omega^2 R^2/2\tau} \tag{S.4.2.1}$$

where n_c is the number density at the center and $\tau \equiv k_B T$ is the temperature in energy units. The pressure is related to the number density n simply by $P = n\tau$. So, at constant temperature,

$$P_0 = P(R_0) = P_c e^{m\Omega^2 R_0^2/2\tau} \tag{S.4.2.2}$$

Using the condition that the acceleration at the rim is $\Omega^2 R_0 = g$, we have

$$\frac{P_c}{P_0} = e^{-mgR_0/2\tau} \tag{S.4.2.3}$$

4.3   Baron von Münchausen and Intergalactic Travel (Moscow Phys-Tech)

The general statement that a closed system cannot accelerate as a whole in the absence of external forces is not usually persuasive to determined inventors. In this case, he would make the point that the force on the rope is real. To get an estimate of this force, assume that the balloon is just above the surface of the Earth and that the density of air is approximately constant to 2 km. Archimedes tells us that the force on the rope will equal the weight of the air, mass m, excluded by the empty balloon (given a massless balloon material). We then may use the ideal gas law

$$PV = \frac{m}{\mu}RT \tag{S.4.3.1}$$

to find the force F_a:

$$F_a = mg = \frac{PV\mu}{RT}g \qquad (S.4.3.2)$$

$$= \frac{10^5 \cdot 4\pi/3 \cdot 10^9 \cdot 29 \cdot 10}{8.31 \cdot 10^3 \cdot 300} \approx 5 \cdot 10^{10} \text{ N}$$

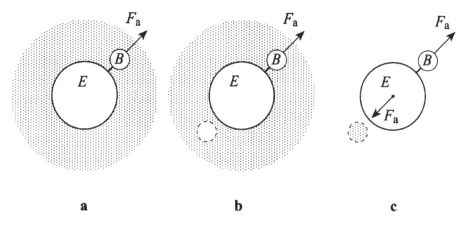

a   b   c

Figure S.4.3

where we approximate the acceleration due to gravity as constant up to 2 km; i.e., $g \sim 10 \text{ m/s}^2$. However, there will be no force acting on the Earth. The system (Earth + surrounding air) is no longer symmetric (see Figure S.4.3a). The symmetric system would be the one with no air on the opposite side of the Earth (see Figure S.4.3b). Therefore, there will be a force between this additional air, which can be treated as a "negative" mass, and the Earth (see Figure S.4.3c):

$$F_g = G\frac{mM_E}{R_E^2} = mg = F_a$$

where M_E and R_E are the mass and radius of the Earth, respectively, and G is the gravitational constant. So, the Archimedes force is completely canceled by the gravitational force from the air. Perhaps that is why the Baron shelved his idea.

4.4 Railway Tanker (Moscow Phys-Tech)

The new equilibrium pressure of the gas will be the same throughout the tanker, whereas the temperature across its length will vary: higher at the heated wall, and cooler at other end (see Problem 4.5). Expanding the temperature T along the length of the tanker in a Taylor series and keeping the first two terms (since the temperature difference between the walls is small compared to T_0), we have

$$T = T|_{x=0} + \left.\frac{dT(x)}{dx}\right|_{x=0} x = T_0 + \frac{\Delta T}{l} x \qquad (S.4.4.1)$$

We may write the ideal gas law as a function of position in the tanker:

$$p = n(x)k_B T(x) \qquad (S.4.4.2)$$

where $n(x)$ is the gas concentration. Rearranging, we have

$$n(x) = \frac{p}{k_B T(x)} \approx \frac{p}{k_B T_0} \left(1 - \frac{1}{T_0}\frac{\Delta T}{l}x\right) \qquad (S.4.4.3)$$

The total number N of molecules in the cylinder is given by

$$N = \int_0^l A n(x) \, dx \qquad (S.4.4.4)$$

$$= \frac{Ap}{k_B T_0}\left(l - \frac{1}{T_0}\int_0^l \frac{\Delta T}{l}x \, dx\right) = \frac{Alp}{k_B T_0}\left(1 - \frac{\Delta T}{2T_0}\right)$$

where A is the cross-sectional area of the tanker. Alternatively, we can integrate (S.4.4.3) exactly and expand the resulting logarithm, which yields the same result. The total number of molecules originally in the tank is

$$N_0 = Aln_0 = \frac{Alp_0}{k_B T_0} \qquad (S.4.4.5)$$

Since the total number of molecules in the gas before and after heating is the same, $N = N_0$ (no phase transitions), we may equate (S.4.4.4) and (S.4.4.5), yielding

$$p = \frac{p_0}{1 - \Delta T/2T_0} \approx p_0\left(1 + \frac{1}{2}\frac{\Delta T}{T_0}\right) \qquad (S.4.4.6)$$

$$= 150\left(1 + \frac{1}{2}\frac{35}{300}\right) = 159 \text{ atm}$$

The center of mass (inertia) X_0 of the gas found with the same accuracy is given by

$$X_0 = \frac{A}{N} \int_0^l dx\ x n(x) = \frac{A}{N} \int_0^l dx\ x\ \frac{p}{k_B T_0} \left(1 - \frac{\Delta T}{T_0} \frac{x}{l}\right)$$

$$= \frac{A p_0}{N k_B T_0} \left(1 + \frac{1}{2} \frac{\Delta T}{T_0}\right) \int_0^l dx\ x \left(1 - \frac{\Delta T}{T_0} \frac{x}{l}\right) \qquad (\text{S}.4.4.7)$$

$$= \frac{l}{2} - \frac{l}{12} \frac{\Delta T}{T_0}$$

As we have assumed that the tanker slides on frictionless rails, the center of mass of the system will not move but the center of the tanker will move by an amount ΔX such that

$$m\left(X_0 + \Delta X\right) + M \left(\frac{l}{2} + \Delta X\right) = (m + M) \frac{l}{2} \qquad (\text{S}.4.4.8)$$

Substituting (S.4.4.7) into (S.4.4.8) and rearranging give

$$\Delta X = \frac{1}{12} \frac{m}{m + M} \frac{\Delta T}{T_0} l = \frac{1}{12} \frac{120}{300} \frac{35}{300} \cdot 10 \approx 0.04\ \text{m} \qquad (\text{S}.4.4.9)$$

4.5 Magic Carpet (Moscow Phys-Tech)

First let us try to reproduce the line of reasoning the Baron was likely to follow. He must have argued that in the z direction the average velocity of a molecule of mass m is

$$v_z \approx \sqrt{\frac{\langle v^2 \rangle}{3}} = \sqrt{\frac{\tau}{m}} \qquad (\text{S}.4.5.1)$$

If we consider that during the collision the molecules thermalize, then the average velocities after reflection from the upper and lower surfaces become

$$v_{z1} = \sqrt{\frac{\tau_1}{m}} \qquad (\text{S}.4.5.2)$$

$$v_{z2} = \sqrt{\frac{\tau_2}{m}}$$

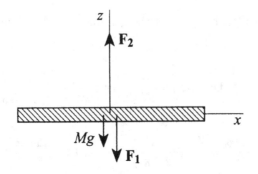

Figure **S.4.5**

The forces due to the striking of the molecules on the upper and lower surfaces are, respectively, $|\mathbf{F}_1|$ and $|\mathbf{F}_2|$ (see Figure S.4.5):

$$|\mathbf{F}_1| = mnv_z\,(v_{z1} + v_z) = mn\sqrt{\frac{\tau}{m}}\left(\sqrt{\frac{\tau_1}{m}} + \sqrt{\frac{\tau}{m}}\right) \qquad \text{(S.4.5.3)}$$

$$|\mathbf{F}_2| = mnv_z\,(v_{z2} + v_z) = mn\sqrt{\frac{\tau}{m}}\left(\sqrt{\frac{\tau_2}{m}} + \sqrt{\frac{\tau}{m}}\right)$$

where n is the concentration of the air molecules, and we have used the fact that the number of molecules colliding with 1 m^2 of the surface per second is approximately $n\langle v_z \rangle$ (the exact number is $n\langle v \rangle/4$; see Problem 4.14). The net resulting force ΔF is

$$\Delta F = |\mathbf{F}_2| - |\mathbf{F}_1|$$

$$= mn\sqrt{\frac{\tau}{m}}\left(\sqrt{\frac{\tau_2}{m}} - \sqrt{\frac{\tau_1}{m}}\right) = n\sqrt{\tau}\,(\sqrt{\tau_2} - \sqrt{\tau_1})$$

Substituting for $n = P/\tau$, we have

$$F = \frac{\sqrt{\tau_2} - \sqrt{\tau_1}}{\sqrt{\tau}}P = \frac{\sqrt{T_2} - \sqrt{T_1}}{\sqrt{T}}P \qquad \text{(S.4.5.4)}$$

$$= \frac{\sqrt{373} - \sqrt{273}}{\sqrt{293}}10^5 \approx 1.5 \cdot 10^4 \text{ N}$$

Unfortunately, this estimate is totally wrong since it assumes that the concentration of molecules is the same above and below the panel, whereas it would be higher near the cold surface and lower near the hot surface (see Problem 4.4) to ensure the same pressure above and below. That's why irons don't fly.

4.6 Teacup Engine (Princeton, Moscow Phys-Tech)

If the cup were vacuum tight, the number of molecules leaving the surface would be the same as the number of molecules returning to the surface. The mass flow rate of the molecules hitting the surface (see Problem 4.14) is

$$w = \frac{\rho_0 \langle v \rangle A}{4} \tag{S.4.6.1}$$

where ρ_0 is the vapor density corresponding to the saturation, $\langle v \rangle$ is the average velocity of the molecules, and A is the surface area of the ice. The mass flow rate of the molecules actually *returning* to the surface is

$$\tilde{w} = \frac{\rho_0 \langle v \rangle A \eta}{4} \tag{S.4.6.2}$$

where η is the sticking coefficient (the probability that the molecule hitting the surface will stick to it). Let us assume for now that $\eta = 1$ (we will see later that this is not true, but that actually gives us the lower limit of the distance). If the cup is open we can assume that the number of molecules leaving the surface is the same as in the closed cup, but there are few returning molecules. We then find that the time τ for complete evaporation of the ice is

$$\tau \approx \frac{m}{w} = \frac{4m}{\rho_0 \langle v \rangle A} \tag{S.4.6.3}$$

where we take $m \approx 200$ g as the mass of the ice, $A \approx 30$ cm^2, and

$$\langle v \rangle = 2\sqrt{\frac{2RT_0}{\pi \mu}} \tag{S.4.6.4}$$

from Problem 4.13. Substituting (S.4.6.4) into (S.4.6.3), we obtain

$$\tau \approx \frac{4m}{\rho_0 \langle v \rangle A} = \frac{m}{\rho_0 A}\sqrt{\frac{2\pi \mu}{RT_0}} \tag{S.4.6.5}$$

Once again using the ideal gas law, we may obtain ρ_0:

$$\rho_0 = \frac{P_0 \mu}{RT_0} \tag{S.4.6.6}$$

Substituting (S.4.6.6) into (S.4.6.5) yields

$$\tau = \frac{m}{P_0 A}\frac{RT_0}{\mu}\sqrt{\frac{2\pi \mu}{RT_0}} = \frac{m}{P_0 A}\sqrt{\frac{2\pi RT_0}{\mu}} \tag{S.4.6.7}$$

$$= \frac{0.2}{600 \cdot 30 \cdot 10^{-4}}\sqrt{\frac{2 \cdot 3.14 \cdot 8.3 \cdot 273}{18 \cdot 10^{-3}}} \approx 100 \text{ s}$$

During the sublimation of the ice, the acceleration of the astronaut is

$$a = \frac{P_0 A}{2M} \qquad (S.4.6.8)$$

where $P_0 A/2$ corresponds to the momentum transferred by the molecules leaving the surface. Using the time τ calculated above, he will cover a distance

$$L = \frac{a\tau^2}{2} = \frac{P_0 A \tau^2}{4M} = \frac{P_0 A}{4M}\frac{m^2}{P_0^2 A^2}\frac{2\pi RT_0}{\mu} = \frac{\pi RT_0 m^2}{2M P_0 A\mu} \qquad (S.4.6.9)$$

$$= \frac{\pi \cdot 8.3 \cdot 273 \cdot 0.2^2}{2 \cdot 110 \cdot 600 \cdot 30 \cdot 10^{-4} \cdot 18 \cdot 10^{-3}} \approx 40 \text{ m}$$

Note that this is the lower limit because we assumed $\eta = 1$ and that all the molecules that are leaving go to infinity. So, it seems that the astronaut can cover the distance to the ship by using his cup as an engine. Moreover, the sticking coefficient η, which is often assumed to be close to unity, could be much smaller (for water, $\eta \approx 0.03$ at 0°C). That explains why the water in a cup left in a room does not evaporate in a matter of minutes but rather in a few hours. For a detailed discussion see E. Mortensen and H. Eyring, J. Phys. Chem. **64**, 846 (1960). The physical reason for such a small sticking coefficient in water is based on the fact that in the liquid phase the rotational degrees of freedom are hindered, leading to a smaller rotational partition function. So, the molecules whose rotation cannot pass adiabatically into the liquid will be rejected and reflect back into the gaseous phase. These effects are especially strong in asymmetric polar molecules (such as water). The actual time the teacup engine will be working is significantly longer (about 30 times, if we assume that the sticking coefficient for ice is about the same as for water at 0°C).

4.7   Grand Lunar Canals (Moscow Phys-Tech)

Consider the atmosphere to be isothermal inside the channel. The pressure depends only on the distance from the center of the moon r (see Figure S.4.7), and as in Problem 4.19 we have

$$\rho g_{\rm m}(r)\, {\rm d}r = -{\rm d}P \qquad (S.4.7.1)$$

So

$$\frac{{\rm d}P}{{\rm d}r} = -\rho g_{\rm m}(r) = -\frac{\mu}{RT}P g_{\rm m}(r) \qquad (S.4.7.2)$$

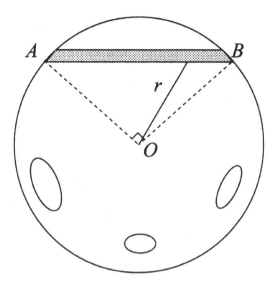

Figure **S.4.7**

The acceleration of gravity (see also Problem 1.10, Part I)

$$g_{\mathrm{m}}(r) = \frac{GM}{r^2} = \frac{G\rho_0 \frac{4}{3}\pi r^3}{r^2} = \frac{4}{3}\pi G\rho_0 r \qquad (S.4.7.3)$$

where M is the mass of the Moon and ρ_0 is the average density of the Moon (which we consider to be uniform). Therefore,

$$g_{\mathrm{m}}(r) = g_{\mathrm{m}}(a)\frac{r}{a} = \frac{gr}{6a} \qquad (S.4.7.4)$$

where we have set $g_{\mathrm{m}}(a) = g/6$. Now, from (S.4.7.2) and (S.4.7.4), we have

$$\frac{\mathrm{d}P}{P} = -\frac{\mu g}{6RTa} r \, \mathrm{d}r$$

$$\ln P|_{P_0}^{P_1} = -\frac{\mu g}{12RTa} r^2 \Big|_{a/\sqrt{2}}^{a}$$

where P_1 is the pressure on the surface of the Moon.

$$\ln \frac{P_1}{P_0} = -\frac{\mu g a}{24RT} = -\frac{29 \cdot 10 \cdot 1750 \cdot 10^3}{24 \cdot 8.3 \cdot 10^3 \cdot 273} \approx -9.33 \qquad (S.4.7.5)$$

$$P_1 = P_0 e^{-\mu g a/24RT} \approx e^{-9.33} \approx 10^{-4} \text{ atm} \qquad (S.4.7.6)$$

which implies that it is not impossible to have such cavities inside the Moon filled with gas (to say nothing of the presence of lunars).

4.8   Frozen Solid (Moscow Phys-Tech)

If the ice does not freeze too fast (which is usually the case with lakes), we can assume that the temperature is distributed linearly across the ice. Suppose that the thickness of the ice at a time t is x. Then the heat balance can be written in the form

$$\kappa \frac{T_m - T_0}{x} dt = q\rho \, dx \qquad (S.4.8.1)$$

where T_m is the melting temperature of ice. The left side represents the flow of heat through one square meter of ice surface due to the temperature gradient, and the right side the amount of heat needed to melt (freeze) an amount of ice $\rho \, dx$. Integrating (S.4.8.1), we obtain

$$\kappa \left(T_m - T_0\right) t_0 = \frac{1}{2} q\rho x^2 + \alpha \qquad (S.4.8.2)$$

where t_0 and α are integration constants. If we assume that there is no ice ($x = 0$) initially ($t_0 = 0$), then $\alpha = 0$, and we find that the time t to freeze solid is

$$t = \frac{q\rho D^2}{2\kappa \left(T_m - T_0\right)} = \frac{3.4 \cdot 10^5 \cdot 0.9 \cdot 10^3 \cdot 0.5^2}{2 \cdot 2.2 \cdot 10} \text{ s} \qquad (S.4.8.3)$$

$$\approx 1.7 \cdot 10^6 \text{ s} \approx 20 \text{ days}$$

4.9   Tea in Thermos (Moscow Phys-Tech)

There are two main sources of power dissipation: radiation from the walls of the thermos and thermal conductance of the air between the walls. Let us first estimate the radiative loss. The power radiated from the hotter inner wall minus the power absorbed from the outer wall is given by (see Problem 4.73)

$$J_r(0) = \epsilon \sigma A \left(T^4 - T_0^4\right) \qquad (S.4.9.1)$$

where T is the temperature of the tea, T_0 is room temperature, and the Stefan–Boltzmann constant $\sigma = 5.7 \cdot 10^{-8}$ W/m²K⁴. Initially, $T = 363$ K, $T_0 = 293$ K. So

$$P_r = 0.10 \cdot 5.7 \cdot 10^{-8} \cdot 600 \cdot 10^{-4} \left(363^4 - 293^4\right) = 3.4 \text{ W}$$

The power dissipation due to the thermal conductivity of the air can be estimated from the fact that, at that pressure, the mean free path of the air molecules is about $\lambda \approx 1$ cm. Therefore, there are very few collisions between the molecules that travel from one wall of the thermos to the other. We can assume that we are in the Knudsen regime of $\lambda \geq d$ (d is the distance between the walls). In this regime the thermal conductance is proportional to the pressure (if $\lambda \ll d$, it is independent of the pressure). Let us assume that after a molecule strikes the wall, it acquires the temperature of the wall. Initially after it hits the wall, a molecule will take away the energy

$$\varepsilon = c_v(T - T_0) \tag{S.4.9.2}$$

where we can take for air $c_v \approx 5/2 \, k_B$. The number of molecules striking the inner wall per time interval dt is

$$dN = \frac{n\langle v \rangle}{4} A \, dt \tag{S.4.9.3}$$

where n is the concentration of molecules and $\langle v \rangle$ is their average velocity (see Problem 4.14). The power due to the thermal conductance is

$$J_t = \varepsilon \left(\frac{dN}{dt} \right) = \frac{5}{2} k_B (T - T_0) \frac{n\langle v \rangle}{4} A \tag{S.4.9.4}$$

We can substitute $n = P/k_B T_0$ and

$$\langle v \rangle = \sqrt{\frac{8RT_0}{\pi\mu}} \approx \sqrt{\frac{8 \cdot 8.3 \cdot 293}{\pi \cdot 0.029}} \approx 460 \text{ m/s}$$

Then (S.4.9.4) becomes

$$J_t \approx \frac{5}{8} \left(\frac{T}{T_0} - 1 \right) PA \sqrt{\frac{8RT_0}{\mu}} \tag{S.4.9.5}$$

$$J_t(0) = \frac{5}{8}(0.24) \cdot 0.5 \cdot 0.06 \cdot 460 \approx 2.1 \text{ W}$$

So, we can see that radiation loss has about the same order of magnitude as thermal conductance at these parameters. Therefore the properties of the thermos can only be improved significantly by decreasing both the emissivity and the residual pressure between the walls. The energy dissipated is equal to the energy change of the mass m of the tea:

$$-Cm \, dT = (J_r + J_t) \, dt \tag{S.4.9.6}$$

So

$$-Cm\frac{\mathrm{d}T}{\mathrm{d}t} = \epsilon\sigma A\left(T^4 - T_0^4\right) + \frac{5}{8}\left(T - T_0\right)PA\sqrt{\frac{8R}{\pi\mu T_0}} \qquad \text{(S.4.9.7)}$$

$$\approx \epsilon\sigma A\left(T - T_0\right)4\tilde{T}^3 + \frac{5}{8}\left(T - T_0\right)PA\sqrt{\frac{8R}{\pi\mu T_0}}$$

where we used for an estimate the fact that T does not change significantly and $\tilde{T} = (T + T_0)/2 \approx 330$ K. Then the time t for the tea to cool from the initial temperature T_i to the final temperature T_f is given by

$$t \approx \frac{Cm\ln\left|\dfrac{T_i - T_0}{T_f - T_0}\right|}{A\left(4\epsilon\sigma\tilde{T}^3 + \dfrac{5}{8}P\sqrt{\dfrac{8R}{\pi\mu T_0}}\right)}$$

$$= \frac{4.2\cdot 10^3 \cdot 1 \cdot \ln\left|\dfrac{70}{50}\right|}{600\cdot 10^{-4}\left[4\cdot 0.10\cdot 5.7\cdot 10^{-8}(330)^3 + \dfrac{5}{8}\cdot 0.5\cdot 460\right]} \qquad \text{(S.4.9.8)}$$

$$\approx 5 \text{ h}$$

4.10 Heat Loss (Moscow Phys-Tech)

Let $\tau_1 = 2$ min be the time the heater is operating. The energy added to the water and bowl will heat the water as well as the environment. We will assume that the heat loss Q_1 to the surroundings is proportional to the elapsed time τ_1 and that the changing temperature difference $\tilde{T}$ between the water and room temperature $T_0 \approx 23°$C. During this phase, we may write

$$J\tau_1 = Cm\left(T_2 - T_1\right) + Q_1 \qquad \text{(S.4.10.1)}$$

The heat loss Q_1 is actually a time integral of some proportionality constant λ times the temperature difference as the water heats up. However, $\tilde{T}$ only varies by 5°C out of an average 65°C, so we will ignore the variation. The heat loss during the second phase is given by

$$Q_2 = Cm\,\Delta T \qquad \text{(S.4.10.2)}$$

Just as in the heating phase, the heat loss is proportional to the elapsed time $\tau_2 = \tau_1/2$. Since T again only changes a little, we have

$$Q_1 = 2Q_2 = 2Cm\,\Delta T \tag{S.4.10.3}$$

We may now eliminate Q_1 from (S.4.10.1), yielding

$$J\tau_1 = Cm\,(T_2 - T_1 + 2\,\Delta T)$$

$$m = \frac{J\tau_1}{C\,(T_2 - T_1 + 2\,\Delta T)} = \frac{500 \cdot 120}{4.2 \cdot 10^3 \cdot (5+2)} \approx 2 \text{ kg}$$

4.11 Liquid–Solid–Liquid (Moscow Phys-Tech)

a) Since the evaporation is very rapid, the heat to vaporize can only be obtained from the heat of fusion. Therefore, if m_i of water becomes solid and m_v vaporizes, we may write

$$m_i q_i = m_v q_v \tag{S.4.11.1}$$

Since the total mass $m = m_i + m_v$, we have

$$m_i = \frac{q_v}{q_i + q_v}\,m = \frac{1}{1 + q_i/q_v}\,m \approx 0.88m = 44 \text{ g} \tag{S.4.11.2}$$

If we continue pumping, the ice would, of course, gradually sublimate, but this process takes much longer, so we can neglect it.

b) The metal cools from its initial temperature by transferring heat q_m to melt some ice:

$$q_m = CM\,\Delta T \tag{S.4.11.3}$$

where ΔT is the temperature change. This may be determined from the sample's density ρ before it was placed in the calorimeter. Using the thermal coefficient of volume expansion β, where $\beta = 3\alpha$, we have

$$\rho = \frac{\rho_0}{1 + \beta\,\Delta T} \tag{S.4.11.4}$$

The temperature difference ΔT may be found from (S.4.11.4)

$$\Delta T = \frac{\rho_0 - \rho}{\beta\rho} = \frac{\rho_0 - M/V}{\beta M/V} = \frac{\rho_0 V - M}{\beta M} \tag{S.4.11.5}$$

Equating the amount of heat required to melt a mass $\tilde{m}_i$ of ice with the heat available in the metal, we have

$$\tilde{m}_i = \frac{MC\,\Delta T}{q_i} = \frac{MC(\rho_0 V - M)}{q_i \beta M} = \frac{C\,(\rho_0 V - M)}{3\alpha q_i} \qquad \text{(S.4.11.6)}$$

$$= \frac{0.12\,(6.8 \cdot 48 - 325)}{3 \cdot 1.1 \cdot 10^{-5} \cdot 80} \approx 64 \text{ g}$$

This mass exceeds the amount of ice from part (a), so all of it would melt.

4.12 Hydrogen Rocket (Moscow Phys-Tech)

Find the amount of water vapor produced in the reaction

$$2H_2 + O_2 = 2H_2O \qquad \text{(S.4.12.1)}$$

One mole of hydrogen yields one mole of water, or in mass

$$\frac{\mu_w}{\mu_h} = \frac{18}{2} = 9 \qquad \text{(S.4.12.2)}$$

Since m is the mass of fuel intake per second, $m' = 9m$ is the mass of water ejected from the engine per second. If the water vapor density is ρ, this rate may be expressed as

$$m' = \rho A v \qquad \text{(S.4.12.3)}$$

where v is the velocity of the gas ejected from the engine. Therefore,

$$v = \frac{m'}{\rho A} = \frac{9m}{\rho A} \qquad \text{(S.4.12.4)}$$

Express the density as

$$\rho = \frac{m'}{V} = \frac{P\mu_w}{RT} \qquad \text{(S.4.12.5)}$$

From (S.4.12.4) and (S.4.12.5), we then have

$$v = \frac{9mRT}{P\mu_w A} \qquad \text{(S.4.12.6)}$$

The mass ejected per second from the engine provides the momentum per second $m'v$, which will be equal to the force $F_r = m'v$ supplied by the engine. Apart from this reactive force, there is a static pressure from the engine providing a force $F_s = PA$, so the total force

$$F = F_r + F_s = 9m\frac{9mRT}{P\mu_w A} + PA = \frac{81m^2 RT}{P\mu_w A} + PA \qquad \text{(S.4.12.7)}$$

In real life the second term is usually small (P is not very high), so the force by an engine is determined by the reactive force.

4.13 Maxwell–Boltzmann Averages (MIT)

a) We may write the unnormalized Maxwell–Boltzmann distribution immediately as

$$f(v_x, v_y, v_z) = Ce^{-mv_x^2/2\tau}e^{-mv_y^2/2\tau}e^{-mv_z^2/2\tau} \qquad \text{(S.4.13.1)}$$

We would like to write (S.4.13.1) as $f(v)$, so we must integrate over all velocities in order to find the proper normalization:

$$1 = \int_{-\infty}^{\infty}\int_{-\infty}^{\infty}\int_{-\infty}^{\infty} Ce^{-m(v_x^2+v_y^2+v_z^2)/2\tau}\, dv_x\, dv_y\, dv_z \qquad \text{(S.4.13.2)}$$

Rewriting (S.4.13.2) in spherical coordinates v, θ, φ, we have

$$1 = C\int_0^{2\pi}\int_{-1}^{1}\int_0^{\infty} v^2 e^{-mv^2/2\tau}\, dv\, d\cos\theta\, d\varphi = 4\pi C\int_0^{\infty} v^2 e^{-mv^2/2\tau}\, dv \qquad \text{(S.4.13.3)}$$

A variety of problems contain the definite integral (S.4.13.3) and its variations. A particularly easy way to derive it is to start by writing the integral as

$$I = \int_{-\infty}^{\infty} e^{-ax^2}\, dx \qquad \text{(S.4.13.4)}$$

Now multiply I by itself, replacing x by y, yielding

$$I^2 = \int_{-\infty}^{\infty}\int_{-\infty}^{\infty} e^{-ax^2}e^{-ay^2}\, dx\, dy \qquad \text{(S.4.13.5)}$$

Rewriting (S.4.13.5) in polar coordinates gives

$$I^2 = \int_0^{2\pi}\int_0^{\infty} e^{-au^2}u\, du\, d\theta = 2\pi\int_0^{\infty} e^{-au^2}u\, du = \frac{\pi}{a}\int_0^{\infty} e^{-v}\, dv = \frac{\pi}{a} \qquad \text{(S.4.13.6)}$$

where we have substituted $v = au^2$ in (S.4.13.6). So we have $I = \sqrt{\pi/a}$ Integrating instead from 0 to ∞ then gives

$$\int_0^{\infty} e^{-au^2}\, du = \frac{\sqrt{\pi/a}}{2} \qquad \text{(S.4.13.7)}$$

The integral required here may be found by differentiating (S.4.13.7) once with respect to a:

$$\int_0^\infty u^2 e^{-au^2}\, du = -\frac{d}{da}\int_0^\infty e^{-au^2}\, du = -\frac{d}{da}\frac{\sqrt{\pi/a}}{2} = \frac{\sqrt{\pi/a^3}}{4} \qquad \text{(S.4.13.8)}$$

Using (S.4.13.8) in (S.4.13.3), where $a = m/2\tau$, gives

$$1 = 4\pi C\frac{\sqrt{\pi(2\tau/m)^3}}{4} = C\left(\frac{2\pi\tau}{m}\right)^{3/2} \qquad \text{(S.4.13.9)}$$

so

$$C = \left(\frac{m}{2\pi\tau}\right)^{3/2} \qquad \text{(S.4.13.10)}$$

$$f(v)\, dv = 4\pi\left(\frac{m}{2\pi\tau}\right)^{3/2} v^2 e^{-mv^2/2\tau}\, dv \qquad \text{(S.4.13.11)}$$

b) The most likely speed u occurs when (S.4.13.11) is a maximum. This may be found by setting its derivative or, simply the derivative of $\ln f(v)$, equal to 0:

$$\frac{d}{dv^2}\left(\ln v^2 e^{-mv^2/2\tau}\right) = \frac{1}{v^2} - \frac{m}{2\tau} = 0 \qquad \text{(S.4.13.12)}$$

$$u = \sqrt{v^2} = \sqrt{\frac{2\tau}{m}} \approx 1.4\sqrt{\frac{\tau}{m}}$$

c) The average speed is given by

$$\langle v\rangle = \int_0^\infty v f(v)\, dv = 4\pi C\int_0^\infty v^3 e^{-mv^2/2\tau}\, dv$$

$$= 4\pi C\frac{1}{2}\int_0^\infty v^2 e^{-mv^2/2\tau}\, dv^2 = 2\pi C\int_0^\infty x e^{-mx/2\tau}\, dx$$

$$= -2\pi C\frac{d}{da}\int_0^\infty e^{-ax}\, dx = -2\pi C\frac{d}{da}\frac{1}{a} \qquad \text{(S.4.13.13)}$$

$$= 2\pi C\frac{1}{a^2} = 2\pi\left(\frac{m}{2\pi\tau}\right)^{3/2}\left(\frac{2\tau}{m}\right)^2$$

$$= \sqrt{\frac{8\tau}{\pi m}} \approx 1.6\sqrt{\frac{\tau}{m}}$$

d) The mean square speed of the atoms may be found immediately by recalling the equipartition theorem (see Problem 4.42) and using the fact that there is $\tau/2$ energy per degree of freedom. So

$$\left\langle \frac{mv^2}{2} \right\rangle = \frac{3\tau}{2}$$

$$\langle v^2 \rangle = \frac{3\tau}{m} \qquad (S.4.13.14)$$

$$\sqrt{\langle v^2 \rangle} = \sqrt{\frac{3\tau}{m}} \approx 1.7\sqrt{\frac{\tau}{m}}$$

For completeness, though, the integral may be shown:

$$\langle v^2 \rangle = 4\pi C \int_0^\infty v^2 e^{-mv^2/2\tau} v^2 \, dv = 4\pi C \frac{d^2}{da^2} \int_0^\infty e^{-av^2} \, dv$$

$$= 4\pi C \frac{d^2}{da^2} \frac{\sqrt{\pi/a}}{2} \qquad (S.4.13.15)$$

$$= 4\pi \left(\frac{m}{2\pi\tau} \right)^{3/2} \frac{3}{8} \frac{\sqrt{\pi}}{(m/2\tau)^{5/2}} = \frac{3\tau}{m}$$

4.14 Slowly Leaking Box (Moscow Phys-Tech, Stony Brook (a,b))

a) The number of atoms per unit volume moving in the direction normal to the wall (in spherical coordinates) is

$$dn = n f(v) v^2 \, dv \sin \theta \, d\phi \, d\theta \qquad (S.4.14.1)$$

where ϕ is the azimuth angle, θ is the polar angle, n is the number density of atoms, and $f(v)$ is the speed distribution function (Maxwellian). To determine the number of atoms striking the area of the hole A on the wall per time dt, we have to multiply (S.4.14.1) by $Av \cos \theta \, dt$ (only the atoms within a distance $v \cos \theta \, dt$ reach the wall). To obtain the total atomic flow rate R through the hole, we have to integrate the following expression:

$$R = nA \int_0^{2\pi} d\phi \int_0^\infty f(v) v^3 \, dv \int_0^{\pi/2} \sin \theta \cos \theta \, d\theta = \pi A n \int_0^\infty v^3 f(v) \, dv \quad (S.4.14.2)$$

We integrate from 0 to $\pi/2$ since we only consider the atoms moving toward the wall. On the other hand, by definition, the average velocity $\langle v \rangle$ is given by

$$\langle v \rangle = \int_0^{2\pi} d\phi \int_0^\infty v f(v) v^2 \, dv \int_0^\pi \sin\theta \, d\theta = 4\pi \int_0^\infty v^3 f(v) \, dv \qquad (S.4.14.3)$$

Comparing (S.4.14.2) and (S.4.14.3), we see that

$$R = n \frac{\langle v \rangle}{4} A \qquad (S.4.14.4)$$

This result applies for any type of distribution function $f(v)$. We only consider a flow from the inside to the outside of the container. Since the hole is small, we can assume that the distribution function of the atoms inside the container does not change appreciably.

b) The average kinetic energy of the atoms leaving the container should be somewhat higher than the average kinetic energy in the container because faster atoms strike the wall more often than the ones moving more slowly. So, the faster atoms leave the container at a higher rate. Let us compute this energy $\langle \varepsilon_1 \rangle$. For a Maxwellian distribution we have $f(v) = Ce^{-mv^2/2\tau}$, where C is a normalizing constant:

$$\langle \varepsilon_1 \rangle = \frac{\int_0^{2\pi} \int_0^\infty \int_0^{\pi/2} nA \frac{mv^2}{2} v \cos\theta \, Ce^{-mv^2/2\tau} v^2 \, dv \sin\theta \, d\phi \, d\theta}{\int_0^{2\pi} \int_0^\infty \int_0^{\pi/2} nAv \cos\theta \, Ce^{-mv^2/2\tau} v^2 \, dv \sin\theta \, d\phi \, d\theta} \qquad (S.4.14.5)$$

The numerator is the total energy of the atoms leaving the container per second, and the denominator is the total number of atoms leaving the container per second. Define $1/\tau \equiv \xi$. From part (a), we can express this integral in terms of the average velocity $\langle v \rangle$. Then we have

$$\langle \varepsilon_1 \rangle = -\frac{d}{d\xi} \left(\ln \int_0^{2\pi} \int_0^\infty \int_0^{\pi/2} v^3 e^{-(mv^2/2)\xi} dv \cos\theta \sin\theta \, d\phi \, d\theta \right)$$

$$= -\frac{d}{d\xi} \ln \left(\frac{\langle v \rangle}{4C} \right) \qquad (S.4.14.6)$$

We know that $C \propto 1/\langle v \rangle^3$ (since it is a normalizing factor, see Problem 4.13), and

$$\langle v \rangle \propto \tau^{1/2} \propto \frac{1}{\xi^{1/2}}, \quad \text{so} \quad \langle \varepsilon_1 \rangle = -\frac{d}{d\xi} \ln \frac{1}{\xi^2} = \frac{2}{\xi} = 2\tau \qquad (S.4.14.7)$$

So $\langle \varepsilon_l \rangle$ is indeed higher than the average energy of the atoms:

$$\langle \varepsilon \rangle = \frac{3}{2}\tau \qquad (S.4.14.8)$$

c) From (b) we know that each atom leaving the container takes with it an additional energy $\Delta\varepsilon = \langle \varepsilon_l \rangle - \langle \varepsilon \rangle = \tau/2$. The flow rate of the atoms leaving the container (from (a)) is

$$R = n\frac{\langle v \rangle}{4}A \qquad (S.4.14.9)$$

The energy flow rate from the container becomes

$$\frac{dE}{dt} = R\,\Delta\varepsilon = \frac{1}{8}n\tau\langle v \rangle A \qquad (S.4.14.10)$$

To keep the temperature of the atoms inside the container constant, we have to transfer some heat to it at the same rate:

$$\frac{dE}{dt} = \frac{dQ}{dt} = \frac{1}{8}n\tau\langle v \rangle A \qquad (S.4.14.11)$$

Equating the flow rate to the decrease of the number of atoms inside gives

$$R = -\frac{d(nV)}{dt} = -V\frac{dn}{dt} = \frac{n\langle v \rangle}{4}A \qquad (S.4.14.12)$$

We then obtain

$$\frac{dn}{dt} = -n\frac{\langle v \rangle A}{4V} \qquad (S.4.14.13)$$

Solving this differential equation, we can find the change in number density:

$$n(t) = ne^{-t/t_0}, \quad \text{where} \quad t_0 \equiv \frac{4V}{A\langle v \rangle} \qquad (S.4.14.14)$$

is the time constant and n is the initial number density. Therefore, the heat flow rate is

$$\frac{dQ}{dt} = \frac{1}{8}n\tau A\langle v \rangle e^{-t/t_0} \qquad (S.4.14.15)$$

4.15 Surface Contamination (Wisconsin-Madison)

The number of molecules striking a unit area of the surface N during the time of the experiment t (see Problem 4.14) is given by

$$N = \frac{n\langle v \rangle}{4}t \qquad (S.4.15.1)$$

For an estimate we can assume that the adsorbed molecules are closely packed and that the number of adsorption sites on a surface of area A is just

$$N_0 = \frac{4A}{\pi d^2} \tag{S.4.15.2}$$

where d is the average diameter of the adsorbed atoms, and we take $d \approx 3$ Å. The total number of adsorption sites may actually be smaller (these data can be obtained from the time to create one monolayer at lower pressure). We may write

$$N \leq \gamma N_0$$

or, for 1 m^2 of surface,

$$\frac{n \langle v \rangle}{4} t \leq \gamma \frac{4}{\pi d^2}$$

Using the average velocity from Problem 4.13 at $T = 300$ K gives

$$\langle v \rangle = \sqrt{\frac{8\tau}{\pi \mu}} = \sqrt{\frac{8RT}{\pi \mu}} \approx 500 \text{ m/s} \tag{S.4.15.3}$$

and

$$n = \frac{P}{k_B T} \tag{S.4.15.4}$$

Thus,

$$P \leq \frac{16 \gamma k_B T}{\pi d^2 t \langle v \rangle} = \frac{16 \cdot 0.05 \cdot 1.4 \cdot 10^{-23} \cdot 300}{\pi \cdot (3 \cdot 10^{-10})^2 \cdot 3600 \cdot 5 \cdot 500} \tag{S.4.15.5}$$

$$\approx 10^{-9} \text{ Pa} \approx 10^{-11} \text{ Torr}$$

So, we will have to maintain a pressure better than 10^{-11} Torr, which can be quite a technical challenge. In fact, at such low pressures the residual gas composition is somewhat different from room air, since it may be more difficult to pump gases such as H_2 and He. Therefore, (S.4.15.3) and (S.4.15.5) are only order-of-magnitude estimates.

4.16 Bell Jar (Moscow Phys-Tech)

The pressure inside the vessel

$$P_1 = n_1 k_B T_1 = 4 n_1 k_B T_0 = 4 P_0 \frac{n_1}{n_0} \tag{S.4.16.1}$$

where n_1 is the concentration of the molecules inside the vessel and n_0 is the concentration of the molecules in the chamber. Disregarding the thickness of the walls of the vessel, we can write the condition that the number of molecules entering the vessel per second is equal to the number of molecules leaving it:

$$\frac{1}{4} n_0 \langle v_0 \rangle A = \frac{1}{4} n_1 \langle v_1 \rangle A \qquad (S.4.16.2)$$

where A is the area of the hole, and we used the result of Problem 4.14 for the number of molecules striking a unit area per second. Actually, the only important point here is that this number is proportional to the product of concentration and average velocity. Therefore,

$$n_1 = \frac{\langle v_0 \rangle}{\langle v_1 \rangle} \qquad (S.4.16.3)$$

The average velocity $\langle v \rangle \propto \sqrt{T}$. So, from (S.4.16.3), we have

$$n_1 = n_0 \sqrt{\frac{T_0}{T_1}} = \frac{n_0}{2} \qquad (S.4.16.4)$$

Substituting (S.4.16.4) into (S.4.16.1), we obtain

$$P_1 = 2P_0 \qquad (S.4.16.5)$$

4.17 Hole in Wall (Princeton)

a) If the diameter of the hole is large compared to the mean free path in both gases, we have regular hydrodynamic flow of molecules in which the pressures are the same in both parts. If the diameter of the hole (and thickness of the partition) is small compared to the mean free path, there are practically no collisions within the hole and the molecules thermalize far from the hole (usually through collisions with the walls).

b) In case $d \ll \lambda_1$, $d \ll \lambda_2$, there are two independent effusive flows from I to II and from II to I. The number of particles N_1 and N_2 going through the hole from parts I and II are, respectively (see Problem 4.14),

$$N_1 = \frac{1}{4} n_1 \langle v_1 \rangle A \qquad (S.4.17.1)$$

$$N_2 = \frac{1}{4} n_2 \langle v_2 \rangle A$$

where $A = \pi d^2/4$ is the area of the hole. At equilibrium, $N_1 = N_2$, so we have

$$n_1\langle v_1 \rangle = n_2 \langle v_2 \rangle \qquad (\text{S.4.17.2})$$

The mean free path is related to the volume concentration of the molecules n by the formula

$$\lambda = \frac{1}{\sqrt{2}n\sigma} \qquad (\text{S.4.17.3})$$

where σ is the effective cross section of the molecule, which depends only on the type of molecule, helium in both halves. Substituting (S.4.17.3) into (S.4.17.2) gives

$$\frac{\langle v_1 \rangle}{\lambda_1} = \frac{\langle v_2 \rangle}{\lambda_2}$$

or

$$\frac{\lambda_1}{\lambda_2} = \frac{\langle v_1 \rangle}{\langle v_2 \rangle} = \sqrt{\frac{T_1}{T_2}} = \frac{1}{\sqrt{2}} \approx 0.7$$

c) When $d \gg \lambda_1$, $d \gg \lambda_2$, we have to satisfy the condition

$$P_1 = P_2$$

or

$$n_1 k_{\text{B}} T_1 = n_2 k_{\text{B}} T_2$$

which gives for the ratio of the mean free paths:

$$\frac{\lambda_1}{\lambda_2} = \frac{T_1}{T_2} = 0.5$$

4.18 Ballast Volume Pressure (Moscow Phys-Tech)

The number of molecules per second entering the volume B from the left container I is proportional to the density of the molecules in I, n_1, the average velocity, $\langle v_1 \rangle$, and the area of the opening, A. The constant of proportionality (see Problem 4.14) is unimportant for this problem. So, equating the rate of molecular flow in and out of volume B, we can write for flow rates $\tilde{R}_i$ in equilibrium (see Figure S.4.18)

$$\tilde{R}_1 + \tilde{R}_2 = \tilde{R}_b \qquad (\text{S.4.18.1})$$

or

$$n_1\langle v_1 \rangle + n_2 \langle v_2 \rangle = 2n_b \langle v_b \rangle \qquad (\text{S.4.18.2})$$

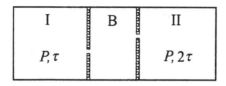

Figure S.4.18

The factor 2 appears for the flow rate $\tilde{R}_b$ since there are two portals from region B. On the other hand, for an ideal gas, $P = n\tau$, and therefore

$$n \propto \frac{P}{\tau} \qquad \langle v \rangle \propto \sqrt{\tau} \qquad \text{(S.4.18.3)}$$

We can rewrite (S.4.18.1) as

$$\frac{P_1}{\sqrt{\tau_1}} + \frac{P_2}{\sqrt{\tau_2}} = 2\frac{P_b}{\sqrt{\tau_b}} \qquad \text{(S.4.18.4)}$$

yielding

$$\frac{P}{\sqrt{\tau}} + \frac{P}{\sqrt{2\tau}} = 2\frac{P_b}{\sqrt{\tau_b}} \qquad \text{(S.4.18.5)}$$

For the state of equilibrium, the energy in the volume B is constant. This means that the total rate of energy transfer out of volumes I and II should be equal to the rate of energy transfer out of volume B :

$$\tilde{E}_1 + \tilde{E}_2 = \tilde{E}_b \qquad \text{(S.4.18.6)}$$

The average energy per particle is proportional to the temperature, so (S.4.18.6) becomes

$$\tilde{R}_1\tau_1 + \tilde{R}_2\tau_2 = \tilde{R}_b\tau_b \qquad \text{(S.4.18.7)}$$

We then have

$$P\sqrt{\tau} + P\sqrt{2\tau} = 2P_b\sqrt{\tau_b} \qquad \text{(S.4.18.8)}$$

Dividing (S.4.18.8) by (S.4.18.5), we can obtain

$$\tau_b = \frac{1 + \sqrt{2}}{1 + 1/\sqrt{2}}\tau = \sqrt{2}\,\tau \qquad \text{(S.4.18.9)}$$

and

$$P_b = \frac{1 + \sqrt{2}}{2\sqrt[4]{2}}P \qquad \text{(S.4.18.10)}$$

4.19 Rocket in Drag (Princeton)

a) Use dimensional analysis to derive the drag force F on the rocket:

$$F \propto \rho^\alpha v^\beta A^\gamma \qquad (S.4.19.1)$$

$$\frac{[\text{kg}]\,[\text{m}]}{[\text{s}]^2} = \left[\frac{\text{kg}}{\text{m}^3}\right]^\alpha \left[\frac{\text{m}}{\text{s}}\right]^\beta \left[\text{m}^2\right]^\gamma$$

We then have

$$\alpha = 1$$

$$-3\alpha + \beta + 2\gamma = 1 \qquad (S.4.19.2)$$

$$-\beta = -2$$

So $\alpha = \gamma = 1$, $\beta = 2$, and

$$F \propto \rho v^2 A \qquad (S.4.19.3)$$

This formula is generally correct for high Reynolds numbers; for low Reynolds numbers we have Stokes' law:

$$F \propto \eta r v \qquad (S.4.19.4)$$

where η is the viscosity and r is the radius.

b) For an isothermal atmosphere, take a column of air of height dz and area A. The pressure difference between top and bottom should compensate the weight of the column:

$$[P(z) - P(z + dz)]\, A = \rho(z) g A \, dz \qquad (S.4.19.5)$$

or

$$-\frac{dP}{dz} = \rho g$$

Using

$$P = \frac{\rho}{\mu} RT \qquad (S.4.19.6)$$

where μ is the molar weight of the air, and substituting (S.4.19.6) into (S.4.19.5), we obtain

$$\frac{d\rho}{dz} = -\rho \frac{g\mu}{RT} \qquad (S.4.19.7)$$

Therefore,

$$\rho = \rho_0 e^{-g\mu z/RT} \tag{S.4.19.8}$$

c) At a height h we have, from (S.4.19.3),

$$F \propto \rho_0 e^{-g\mu h/RT} 2ahA \propto he^{-g\mu h/RT} \tag{S.4.19.9}$$

where we used $v^2 = 2ah$ for uniform acceleration. Now, the maximum force is defined by

$$\left(\frac{dF}{dh}\right)_{h=h_0} = e^{-g\mu h_0/RT} - h_0 \frac{g\mu}{RT} e^{-g\mu h_0/RT} = 0 \tag{S.4.19.10}$$

So, assuming that the average temperature for the isothermal atmosphere $T = 250$ K, we find

$$h_0 = \frac{RT}{g\mu} = \frac{8.3 \cdot 10^3 \cdot 250}{10 \cdot 29} \approx 7000 \text{ m} = 7 \text{ km} \tag{S.4.19.11}$$

4.20 Adiabatic Atmosphere (Boston, Maryland)

a) Starting from the ideal gas law, we can express the temperature T as a function of pressure P and the mass density ρ:

$$T(P, \rho) = \frac{PV\mu}{mR} = \frac{\mu}{R}\frac{P}{\rho} \tag{S.4.20.1}$$

where P and ρ are functions of the height z above the surface of the Earth: $P \equiv P(z), \rho \equiv \rho(z)$. Taking the derivative of T with respect to z, we have

$$\frac{dT}{dz} = \frac{\mu}{R}\frac{1}{\rho}\frac{dP}{dz} - \frac{\mu}{R}\frac{P}{\rho^2}\frac{d\rho}{dz} \tag{S.4.20.2}$$

We need to express $d\rho$ in terms of dP. The fact that PV^γ is independent of altitude allows us to write

$$P = B\rho^\gamma$$

$$dP = B\gamma\rho^{\gamma-1}\, d\rho = \gamma\frac{P}{\rho}\, d\rho$$

where B is some constant. So

$$d\rho = \frac{1}{\gamma}\frac{\rho}{P}\, dP \tag{S.4.20.3}$$

Substituting (S.4.20.3) into (S.4.20.2), we obtain

$$\frac{dT}{dz} = \frac{\mu}{R}\frac{1}{\rho}\frac{dP}{dz} - \frac{\mu}{R}\frac{1}{\gamma}\frac{1}{\rho}\frac{dP}{dz} = \left(\frac{\gamma-1}{\gamma}\right)\frac{\mu}{\rho R}\frac{dP}{dz} \qquad (S.4.20.4)$$

Assuming that the acceleration of gravity is constant, using the hydrostatic pressure formula

$$dP = -\rho g\ dz \qquad (S.4.20.5)$$

and substituting (S.4.20.5) into (S.4.20.4), we can write

$$\frac{dT}{dz} = -\left(\frac{\gamma-1}{\gamma}\right)\frac{\mu g}{R} \qquad (S.4.20.6)$$

b) For the atmosphere, using diatomic molecules with $C_V = 5R/2$ and $C_P = 7R/2$, we have from (S.4.20.6),

$$\frac{dT}{dz} = -\frac{7/5-1}{7/5}\frac{29\cdot 10}{8.3\cdot 10^3} = -\frac{2}{7}\frac{10\cdot 0.029}{8.3} \approx -10^{-2}\ \text{K/m} = -10\ \text{K/km}$$

This value of $|dT/dz|$ is about a factor of 2 larger than that for the actual atmosphere.

4.21 Atmospheric Energy (Rutgers)

a) Again starting with the ideal gas law

$$PV = nRT = \frac{m}{\mu}RT$$

we have

$$P(z) = \frac{\rho(z)}{\mu}RT(z) \qquad (S.4.21.1)$$

b) The gravitational energy of a slice of atmosphere of cross section A and thickness dz at a height z is simply

$$\mathcal{E}_g = Ag\rho(z)z\ dz \qquad (S.4.21.2)$$

while the internal energy of the same slice is

$$\mathcal{E}_i = \frac{AC_V}{\mu}\rho(z)T(z)\ dz \qquad (S.4.21.3)$$

The total internal energy E_i is given by the integral of (S.4.21.3):

$$E_i = \frac{AC_V}{\mu} \int_0^{P_0} \rho(z)T(z)\,dz \qquad (S.4.21.4)$$

We wish to change the integral over z into an integral over P. To do so, first consider the forces on the slice of atmosphere:

$$AP(z+dz) + Ag\rho(z)\,dz = AP(z) \qquad (S.4.21.5)$$

Rearranging (S.4.21.5), we have

$$\frac{dP(z)}{dz} = -g\rho(z) \qquad (S.4.21.6)$$

$$dz = -\frac{dP}{g\rho(z)}$$

Substituting (S.4.21.6) into (S.4.21.4), we obtain

$$E_i = -\frac{AC_V}{\mu} \int_0^{P_0} \rho(z)T(z)\,\frac{dP}{g\rho(z)} \qquad (S.4.21.7)$$

$$= -\frac{AC_V}{\mu g} \int_0^{P_0} T(z)\,dP = \frac{AC_V}{\mu g} \int_0^{P_0} T(P)\,dP$$

The total gravitational energy E_g may be found by integrating (S.4.21.2):

$$E_g = Ag \int_0^{z_0} \rho(z)z\,dz \qquad (S.4.21.8)$$

Integrating by parts gives

$$E_g = Ag \left[z \int^z \rho(z')\,dz' \Big|_0^{z_0} - \int_0^{z_0} dz \int_0^z \rho(z')\,dz' \right] \qquad (S.4.21.9)$$

The first term on the RHS of (S.4.21.9) is zero since at the limits of evaluation either $z = 0$ or $\int \rho(z') = 0$, so we have

$$E_g = -Ag \int_0^{z_0} dz \int^z \rho(z')\,dz' = A \int_0^{z_0} dz \int^z \frac{dP(z')}{dz'}\,dz' \qquad (S.4.21.10)$$

$$= A \int_{0}^{z_0} P(z)dz = \frac{AR}{\mu} \int_{0}^{z_0} T(z)\rho(z)\, dz = \frac{AR}{\mu g} \int_{0}^{P_0} T(P)\, dP$$

The ratio of energies E_g/E_i, from (S.4.21.10) and (S.4.21.7) is

$$\frac{E_g}{E_i} = \frac{R}{C_V} = \frac{C_P - C_V}{C_V} = \frac{C_P}{C_V} - 1 = \gamma - 1 \qquad \text{(S.4.21.11)}$$

Finally,

$$E \equiv E_i + E_g = \frac{A\,(C_V + R)}{\mu g} \int_{0}^{P_0} T(P)\, dP = \frac{AC_P}{\mu g} \int_{0}^{P_0} T(P)\, dP \quad \text{(S.4.21.12)}$$

4.22 Puncture (Moscow Phys-Tech)

a) Use Bernoulli's equation (see, for instance, Landau and Lifshitz, *Fluid Mechanics*, Chapter 5) for an arbitrary flow line with one point inside the tire and another just outside it. We then have

$$w_0 + \frac{v_0^2}{2} = w_1 + \frac{v_1^2}{2} \qquad \text{(S.4.22.1)}$$

where w_0 and w_1 are the enthalpy per unit mass inside and outside the vessel, respectively, and v_0 and v_1 are the velocities of the gas. The velocity v_0 is very small and can be disregarded. Then the velocity of the gas outside is

$$v_1 \equiv v = \sqrt{2\,(w_0 - w_1)} \qquad \text{(S.4.22.2)}$$

For an ideal gas the heat capacity does not depend on temperature, so we may write for the enthalpy

$$w = E + \frac{p}{\rho} = \frac{C_V T}{\mu} + \frac{RT}{\mu} \qquad \text{(S.4.22.3)}$$

$$= \frac{T\,(C_V + R)}{\mu} = \frac{T C_P}{\mu}$$

Therefore, the velocity is

$$v = \sqrt{\frac{2C_P\,(T_0 - T_1)}{\mu}} \qquad \text{(S.4.22.4)}$$

The temperature T_1 may be found from the equation for adiabats and the ideal gas law:

$$\frac{P_0^{\gamma-1}}{T_0^{\gamma}} = \frac{P_1^{\gamma-1}}{T_1^{\gamma}} \qquad \text{(S.4.22.5)}$$

Rewriting gives

$$T_1 = T_0 \left(\frac{P_1}{P_0}\right)^{(\gamma-1)/\gamma} \qquad \text{(S.4.22.6)}$$

Substituting into (S.4.22.4) gives

$$v = \sqrt{\frac{2C_P T_0}{\mu}\left[1 - \left(\frac{P_1}{P_0}\right)^{(\gamma-1)/\gamma}\right]} \qquad \text{(S.4.22.7)}$$

The maximum velocity v_{m} will be reached when $P_1 = 0$, flow into vacuum.

$$v_{\mathrm{m}} = \sqrt{\frac{2C_P T_0}{\mu}} \qquad \text{(S.4.22.8)}$$

b) For one mole of an ideal gas

$$C_P - C_V = R \qquad \text{(S.4.22.9)}$$

and, by definition,

$$\gamma = \frac{C_P}{C_V} \qquad \text{(S.4.22.10)}$$

From (S.4.22.9) and (S.4.22.10), we may express C_V and C_P through R and γ:

$$C_V = \frac{R}{\gamma - 1} \qquad \text{(S.4.22.11)}$$

$$C_P = \frac{\gamma R}{\gamma - 1}$$

Then (S.4.22.8) becomes

$$v_{\mathrm{m}} = \sqrt{\frac{2\gamma RT}{\mu(\gamma - 1)}} \qquad \text{(S.4.22.12)}$$

For molecular hydrogen ($\gamma = 7/5$) we have

$$v_{\mathrm{m}} = \sqrt{\frac{2\,7/5}{2\,2/5} \cdot 8.3 \cdot 10^3 \cdot 10^3} \approx 5400 \text{ m/s}$$

Note that this estimate implies that $T_1 = 0$, i.e., that the gas would cool to absolute zero. This is, of course, not true; several assumptions would break down long before that. The flow during expansion into vacuum is always turbulent; the gas would condense and phase-separate and therefore would cease to be ideal. The velocity of sound inside the vessel

$$s_0 = \sqrt{\gamma \frac{P_0}{\rho_0}} \qquad (S.4.22.13)$$

or

$$s_0^2 = \gamma \frac{P_0}{\rho_0} = \gamma \frac{RT_0}{\mu} \qquad (S.4.22.14)$$

Substituting (S.4.22.14) into (S.4.22.12) yields

$$v_m = \sqrt{\frac{2}{\gamma - 1}} s_0 \approx 2.2 s_0 \qquad (S.4.22.15)$$

Heat and Work

4.23 Cylinder with Massive Piston (Rutgers, Moscow Phys-Tech)

When the piston is released, it will move in some, as yet unknown, direction. The gas will obey the ideal gas law at equilibrium, so

$$PV = nRT \qquad (S.4.23.1)$$

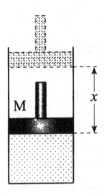

Figure S.4.23

On the other hand, at equilibrium, there is no net force acting on the piston (see Figure S.4.23), and we have

$$PA = P_a A + Mg \qquad (S.4.23.2)$$

Substituting (S.4.23.2) into (S.4.23.1) gives

$$\left(P_a + \frac{Mg}{A}\right) V = nRT \qquad (S.4.23.3)$$

We can also use energy conservation in this thermally insulated system. Then the work done to the gas equals its energy change $\Delta \mathcal{E}$. For an ideal gas

$$\Delta \mathcal{E} = nC_v \left(T - T_0\right) \qquad (S.4.23.4)$$

where C_v is the heat capacity of one mole of the gas (for a monatomic gas, $C_v = 3R/2$). The work done to the gas

$$W = (P_a A + Mg) \, x \qquad (S.4.23.5)$$

where $x = \left(V_0 - V\right)/A$ is the distance the piston moves, where downward is positive. From (S.4.23.4) and (S.4.23.5), we have

$$\left(P_a + \frac{Mg}{A}\right) (V_0 - V) = \frac{3}{2} nR \left(T - T_0\right) \qquad (S.4.23.6)$$

Solving (S.4.23.3) and (S.4.23.6) yields

$$T = \frac{3}{5} T_0 + \frac{2}{5} \frac{\left(P_a + Mg/A\right) V_0}{nR} \qquad (S.4.23.7)$$

$$V = \frac{2}{5} V_0 + \frac{3}{5} \frac{nRT_0}{P_a + Mg/A}$$

We may check that if $P_0 = P_a + Mg/A$, i.e., that the piston was initially balanced, (S.4.23.7) gives $T = T_0$ and $V = V_0$.

4.24 Spring Cylinder (Moscow Phys-Tech)

For a thermally insulated system (no heat transfer), the energy change $\Delta \mathcal{E}$ is given by

$$\Delta \mathcal{E} = \int_0^1 P \, dV \qquad (S.4.24.1)$$

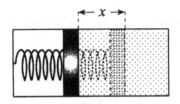

Figure S.4.24

where 0 and 1 correspond to the initial and final equilibrium states of the system, with sets of parameters T_0, P_0, V_0 and T_1, P_1, V_1, respectively. In this case, the gas is expanding, therefore some positive work is done by the gas, which indicates that the energy change is negative, and the temperature decreases. $\Delta \mathcal{E}$ for an ideal gas depends only on the change in temperature:

$$\Delta \mathcal{E} = C_v \Delta T = C_v (T_1 - T_0) \qquad \text{(S.4.24.2)}$$

where C_v is the heat capacity of one mole of the gas at constant volume (for a monatomic gas $C_v = 3R/2$). The work done by the gas goes into compressing the spring:

$$U = \frac{Kx^2}{2} = -\Delta \mathcal{E} \qquad \text{(S.4.24.3)}$$

where K is the spring constant and x is the change of the piston position (see Figure S.4.24). On the other hand, when equilibrium is reached, the compression force of the spring

$$F = Kx = P_1 A \qquad \text{(S.4.24.4)}$$

where A is the cross section of the piston. So

$$K = \frac{P_1 A}{x} = \frac{RT_1 A}{xV_1} \qquad \text{(S.4.24.5)}$$

where we used the ideal gas law for one mole of gas. Substituting (S.4.24.5) into (S.4.24.3), we have

$$U = \frac{RT_1 A}{2xV_1} x^2 = \frac{RT_1 Ax}{2V_1} \qquad \text{(S.4.24.6)}$$

Notice that Ax is the volume change of the gas:

$$Ax = V_1 - V_0 = 2V_0 - V_0 = V_0 \qquad \text{(S.4.24.7)}$$

Therefore,

$$U = \frac{RT_1 V_0}{4V_0} = \frac{RT_1}{4} \qquad \text{(S.4.24.8)}$$

Substituting (S.4.24.8) and (S.4.24.2) into (S.4.24.3), we obtain

$$\Delta \mathcal{E} = C_v (T_1 - T_0) = -\frac{RT_1}{4} \qquad \text{(S.4.24.9)}$$

and

$$T_1 = \frac{C_v T_0}{C_v + R/4} = \frac{T_0}{1 + R/4C_v} = \frac{300}{1 + 1/6} \approx 257 \text{ K} \qquad \text{(S.4.24.10)}$$

The temperature indeed has decreased. As for the pressure, we have for the initial state

$$P_0 V_0 = RT_0 \qquad \text{(S.4.24.11)}$$

Now $V_1 = 2V_0$, so

$$P_1 V_1 = 2P_1 V_0 = RT_1 \qquad \text{(S.4.24.12)}$$

and

$$P_1 = \frac{1}{2}\frac{T_1}{T_0} P_0 = \frac{P_0}{2 + R/2C_v} = \frac{3}{7} P_0 \approx 0.43 \text{ atm}$$

4.25 Isothermal Compression and Adiabatic Expansion of Ideal Gas (Michigan)

a) We can calculate the work as an integral, using the ideal gas law:

$$W = -\int_1^2 P \, dV = -\int_1^2 \frac{N\tau}{V} \, dV = N\tau \ln \frac{V_1}{V_2} \qquad \text{(S.4.25.1)}$$

where τ is, as usual, the absolute temperature. Graphically, it is simply the area under the curve (see Figure S.4.25). Alternatively, we can say that the work done is equal to the change of free energy F of the system (see Problem 4.38):

$$W = \Delta F = F_2 - F_1 \qquad \text{(S.4.25.2)}$$

$$= -N\tau \ln \frac{eV_2}{N} + Nf(\tau) + N\tau \ln \frac{eV_1}{N} - Nf(\tau) = N\tau \ln \frac{V_1}{V_2}$$

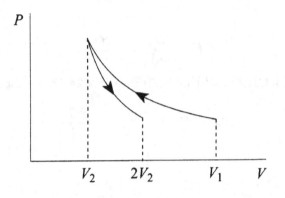

Figure S.4.25

The total energy $\mathcal{E}$ of the ideal gas depends only on the temperature, which is constant, so the heat absorbed by the gas is

$$Q = \Delta\mathcal{E} - W = -N\tau\ln\frac{V_1}{V_2} \qquad (S.4.25.3)$$

i.e., heat is rejected from the gas into the reservoir. Alternatively, since $\delta Q = \tau\,dS$,

$$Q = \int_1^2 \tau\,dS = \tau\,(S_2 - S_1)$$

$$= \tau\left(-\frac{\partial F_2}{\partial\tau} + \frac{\partial F_1}{\partial\tau}\right) = \tau\left(N\ln\frac{eV_2}{\tau} - N\ln\frac{eV_1}{\tau}\right) \qquad (S.4.25.4)$$

$$= -N\tau\ln\frac{V_1}{V_2}$$

the same result as in (S.4.25.3).

b) For an adiabatic expansion the entropy is conserved, so

$$dE = -P\,dV \qquad (S.4.25.5)$$

On the other hand,

$$dE = C_V\,d\tau \qquad (S.4.25.6)$$

where C_V is the specific heat for an ideal gas at constant volume. From (S.4.25.5) and (S.4.25.6), and using the ideal gas law, we obtain

$$-\frac{dV}{V} = \frac{C_V}{N}\frac{d\tau}{\tau} = \frac{c_v\,d\tau}{\tau} \qquad (S.4.25.7)$$

where $C_V/N \equiv c_v$, the specific heat per one molecule. Integrating (S.4.25.7) yields

$$\tau_f = \left(\frac{V_i}{V_f}\right)^{1/c_v} \tau_i = \left(\frac{1}{2}\right)^{1/c_v} \tau_i$$

c) For air we may take $c_v = 5/2$ (in regular units, $c_v = 5k_B/2$; it is mostly diatomic). Therefore,

$$T_f = \left(\frac{1}{2}\right)^{2/5} T_i \approx 227 \text{ K}$$

4.26 Isochoric Cooling and Isobaric Expansion (Moscow Phys-Tech)

The process diagram is shown in Figure S.4.26. The work W done by the gas occurs only during the $2 \to 3$ leg since there is no work done during the $1 \to 2$ leg. The work is given by

$$W_{2 \to 3} = \int_2^3 P \, dV = P_2 \, (V_2 - V_1) = P_2 V_2 \left(1 - \frac{V_1}{V_2}\right) \qquad \text{(S.4.26.1)}$$

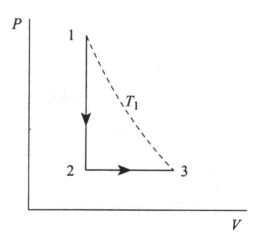

Figure S.4.26

Using the ideal gas law, we may relate V_1 and V_2:

$$P_1 V_1 = \frac{m}{\mu} R T_1 \qquad \text{(S.4.26.2)}$$

$$P_2 V_2 = \frac{m}{\mu} R T_2 = \frac{m}{\mu} R T_1$$

since the initial and final temperatures are the same. Substituting into (S.4.26.1) we find

$$W = \frac{m}{\mu} R T_1 \left(1 - \frac{P_2}{P_1}\right) = \frac{m}{\mu} R T_1 \left(1 - \frac{1}{n}\right) \qquad \text{(S.4.26.3)}$$

4.27 Venting (Moscow Phys-Tech)

The air surrounding the chamber may be thought of as a very large reservoir of gas at a constant pressure P_0 and temperature T_0. The process of venting is adiabatic, so we can assume that there is no energy dissipation. We then find that the energy of the gas admitted to the chamber equals the sum of its energy E_0 in the reservoir plus the work done by the gas of the reservoir at P_0 to expel the gas into the chamber. This may be calculated by considering the process of filling a cylinder by pushing a piston back, where the piston offers a resistant force of $P_0 A$, A being the cross section of the cylinder. The total energy E is then given by

$$E = E_0 + P_0 V_0 \qquad \text{(S.4.27.1)}$$

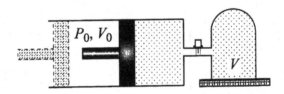

Figure S.4.27

where V_0 is the volume of the gas needed to fill the volume of the chamber V (note that V does not coincide with V_0 because the temperature of the gas in the chamber T presumably is not the same as T_0; see Figure S.4.27). On the other hand,

$$\Delta E = E - E_0 = C_V \, \Delta T = C_V \left(T - T_0\right) = N_0 c_v \left(T - T_0\right) \qquad \text{(S.4.27.2)}$$

where C_V is the heat capacity of the gas, c_v is the heat capacity per molecule, and N_0 is the number of molecules. From (S.4.27.1) and (S.4.27.2), we have

$$P_0 V_0 = N_0 c_v \left(T - T_0\right) \qquad \text{(S.4.27.3)}$$

Using the ideal gas law

$$P_0 V_0 = N_0 k_B T_0 \qquad \text{(S.4.27.4)}$$

we have

$$k_B T_0 = c_v (T - T_0) \qquad \text{(S.4.27.5)}$$

So

$$T = \frac{k_B + c_v}{c_v} T_0 = \frac{c_p}{c_v} T_0 = \gamma T_0 \qquad \text{(S.4.27.6)}$$

The air is mostly nitrogen and oxygen (78% nitrogen and 21% oxygen = 99%), diatomic gases, so that

$$c_v = \frac{5}{2} k_B \qquad \gamma = \frac{7}{5}$$

and therefore $T = 420$ K. Thus, the temperature of the gas in the chamber will increase. Note that the result does not depend on the outside pressure P_0, the volume of the chamber V, or whether it is filled to P_0.

4.28 Cylinder and Heat Bath (Stony Brook)

a) Since the process takes place at constant temperature, PV is constant for each side of the piston. When the piston is released, we can write

$$\frac{P_{0l} V_0}{2} = P V_l \qquad\qquad \frac{P_{0r} V_0}{2} = P V_r \qquad \text{(S.4.28.1)}$$

where P_{0l} and P_{0r} are the initial pressures on the left and right sides of the cylinder, respectively, P is the final pressure on both sides of the cylinder, and V_l and V_r are the final volumes. From S.4.28.1 we have

$$\frac{P_{0l}}{P_{0r}} = \frac{V_l}{V_r} \qquad \text{(S.4.28.2)}$$

or

$$\frac{P_{0l} + P_{0r}}{P_{0r}} = \frac{V_l + V_r}{V_r} \qquad \text{(S.4.28.3)}$$

Therefore,

$$V_r = \frac{V_0}{1 + P_{0l}/P_{0r}} = \frac{V_0}{5} \qquad \text{(S.4.28.4)}$$

So, the piston winds up 20 cm from the right wall of the cylinder.

b) The energy of the ideal gas does not change in the isothermal process, so all the work done by the gas goes into heating the reservoir. Denoting by n_l and n_r the number of moles on the left and right sides of the cylinder, respectively, and using $n_r = 1/4$ mole, we obtain the total work and, hence, heat given by the integral

$$\int \frac{nRT}{V} \, dV = n_l RT \ln \left(\frac{V_l}{V_0/2} \right) + n_r RT \ln \left(\frac{V_r}{V_0/2} \right)$$

$$= n_l RT \ln \left(\frac{4V_0/5}{V_0/2} \right) + n_r RT \ln \left(\frac{V_0/5}{V_0/2} \right)$$

$$= RT \ln \left(\frac{8}{5} \right) + \left(\frac{1}{4} \right) RT \ln \left(\frac{2}{5} \right) = 600 \text{ J}$$

4.29 Heat Extraction (MIT, Wisconsin-Madison)

a) For a mass of fixed volume we have

$$d\varepsilon = T \, dS \tag{S.4.29.1}$$

So, by the definition of C,

$$\frac{\partial \varepsilon}{\partial T} \equiv MC = T \frac{\partial S}{\partial T} \tag{S.4.29.2}$$

Since C is independent of T, we may rewrite (S.4.29.2) and integrate:

$$\frac{dS}{dT} = \frac{MC}{T} \tag{S.4.29.3}$$

$$S = MC \ln T$$

The change in entropy is then

$$\Delta S = MC \ln \frac{T_f}{T_i} \tag{S.4.29.4}$$

b) The maximum heat may be extracted when the entropy remains constant. Equating the initial and final entropies yields the final temperature of the two bodies:

$$MC \ln T_H + MC \ln T_C = MC \ln T + MC \ln T$$

$$T_H T_C = T^2 \tag{S.4.29.5}$$

$$T = \sqrt{T_H T_C}$$

The heat extracted, Q, is then equal to the difference in initial and final internal energies of the bodies:

$$Q = MC\left(T_H + T_C\right) - 2MC\sqrt{T_H T_C}$$

$$= MC\left(T_H + T_C - 2\sqrt{T_H T_C}\right) = MC\left(\sqrt{T_H} - \sqrt{T_C}\right)^2 \quad \text{(S.4.29.6)}$$

$$= MC\left(\sqrt{400} - \sqrt{100}\right)^2 = 100MC$$

c) Here we may calculate the maximum useful work by using the Carnot efficiency of a reversible heat engine operating between two reservoirs, one starting at a high temperature (100°C) and the other fixed (the lake) at 10°C. The efficiency may be written for a small heat transfer as

$$\varepsilon = \frac{dW}{dQ} = \left(1 - \frac{T_C}{T_H}\right) \quad \text{(S.4.29.7)}$$

where the heat dQ transferred from the hot reservoir equals its change in internal energy $-MC\,dT$. We may then find the total work by integrating dW as follows:

$$W = \int_{T_H}^{T_C} dW = -MC\int_{T_H}^{T_C}\left(1 - \frac{T_C}{T}\right)\,dT = -MC\left(T - T_C\ln T\right)|_{T_H}^{T_C}$$

$$= MC\left[T_H - T_C - T_C\ln\left(\frac{T_H}{T_C}\right)\right] \quad \text{(S.4.29.8)}$$

$$= 4.2\cdot 10^3\cdot 10^6\left(90 - 10\ln 10\right) \approx 2.8\cdot 10^{11}\text{ J}$$

We may also use the method of part (b) and the fact that the entropy is conserved. Denote the mass of the hot water M and the lake M_L. Equating the initial and final entropies gives

$$S = MC\ln T_H + M_L C\ln T_C = (M + M_L)C\ln T \quad \text{(S.4.29.9)}$$

Writing the final temperature T as $T_C + \delta$, where δ is small (it's a big lake), and expanding the logarithm, we obtain

$$(M + M_L)C\ln(T_C + \delta) \approx (M + M_L)C\ln T_C + \frac{(M + M_L)C\delta}{T_C} \quad \text{(S.4.29.10)}$$

Substituting (S.4.29.10) back into (S.4.29.9) gives

$$\delta = \frac{M}{M + M_L}T_C\ln\frac{T_H}{T_C} \quad \text{(S.4.29.11)}$$

As before, the work extracted equals the change in internal energy of the bodies, so

$$W = \varepsilon_i - \varepsilon_f = MCT_H + M_LCT_C - (M + M_L)C\,(T_C + \delta)$$

$$= MCT_H + M_LCT_C - (M + M_L)C\left(T_C + \frac{M}{M + M_L}T_C\ln\frac{T_H}{T_C}\right)$$

$$= MC\left(T_H - T_C - T_C\ln\frac{T_H}{T_C}\right) \qquad\qquad \text{(S.4.29.12)}$$

which is the same as above.

4.30 Heat Capacity Ratio (Moscow Phys-Tech)

If the gas is heated at constant volume, then the amount of heat Q transferred to the gas is

$$Q = C_V\,(T - T_0) = c_v m\,(T - T_0) \qquad\qquad \text{(S.4.30.1)}$$

where c_v is the heat capacity by weight of the gas, m is the mass, and T is the temperature at pressure P_1. Using the ideal gas law at the beginning and end of heating gives

$$P_0 V_0 = nRT_0 \qquad\qquad \text{(S.4.30.2)}$$

$$P_1 V_0 = nRT$$

where n is the number of moles of the gas. From (S.4.30.1) and (S.4.30.2),

$$T - T_0 = \frac{(P_1 - P_0)\,V_0}{nR} \qquad\qquad \text{(S.4.30.3)}$$

and

$$Q = c_v m\frac{(P_1 - P_0)\,V_0}{nR} \qquad\qquad \text{(S.4.30.4)}$$

For heating at constant pressure,

$$Q' = C_P\,(T' - T_0) = c_p m\,(T' - T_0) \qquad\qquad \text{(S.4.30.5)}$$

Similarly,

$$P_0 V_0 = nRT_0 \qquad\qquad \text{(S.4.30.6)}$$

$$P_0 V_1 = nRT'$$

So

$$T' - T_0 = \frac{P_0 (V_1 - V_0)}{nR} \tag{S.4.30.7}$$

and

$$Q' = c_p m \frac{P_0 (V_1 - V_0)}{nR} \tag{S.4.30.8}$$

Since the time t during which the current flows through the wire is the same in both experiments, the amount of heat transferred to the gas is also the same: $Q' = Q$. Equating (S.4.30.4) and (S.4.30.8), we obtain

$$\frac{C_P}{C_V} = \frac{P_1 - P_0}{V_1 - V_0} \frac{V_0}{P_0} = \frac{P_1/P_0 - 1}{V_1/V_0 - 1} \tag{S.4.30.9}$$

4.31 Otto Cycle (Stony Brook)

a) The efficiency of the cycle is $\eta = W/Q_1$, where W is the work done by the cycle and Q_1 is the amount of heat absorbed by the gas. Because the working medium returns to its initial state $W = Q_1 - Q_2$, where Q_2 is the amount of heat transferred from the gas, therefore

$$\eta = \frac{Q_1 - Q_2}{Q_1} = 1 - \frac{Q_2}{Q_1} \tag{S.4.31.1}$$

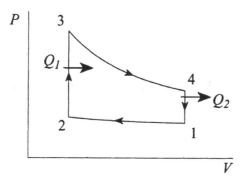

Figure S.4.31

Let us calculate Q_1 and Q_2. Since both processes are at constant volume (see Figure S.4.31), we may write

$$Q_1 = \int_2^3 dQ = c_v \int_2^3 dT = c_v(T_3 - T_2) \tag{S.4.31.2}$$

$$Q_2 = -c_v \int_4^1 \mathrm{d}T = -c_v(T_1 - T_4) = c_v(T_4 - T_1)$$

and

$$\frac{Q_2}{Q_1} = \frac{T_4 - T_1}{T_3 - T_2} \qquad\qquad \text{(S.4.31.3)}$$

We know that for an adiabatic process

$$pV^\gamma = \text{const} \quad \text{or} \quad TV^{\gamma-1} = \text{const}$$

So

$$T_1 V_i^{\gamma-1} = T_2 V_f^{\gamma-1} \qquad\qquad \text{(S.4.31.4)}$$

$$T_3 V_f^{\gamma-1} = T_4 V_i^{\gamma-1}$$

Using

$$\frac{T_4}{T_1} = \frac{T_3}{T_2} \quad \text{and} \quad \frac{T_4 - T_1}{T_1} = \frac{T_3 - T_2}{T_2}$$

we find

$$\frac{T_4 - T_1}{T_3 - T_2} = \frac{T_1}{T_2} = \left(\frac{V_f}{V_i}\right)^{\gamma-1} = \left(\frac{1}{r}\right)^{\gamma-1} \qquad\qquad \text{(S.4.31.5)}$$

and therefore the efficiency is

$$\eta = 1 - \left(\frac{1}{r}\right)^{\gamma-1} \qquad\qquad \text{(S.4.31.6)}$$

For $\gamma = 1.4$ and $r = 10$ the efficiency is

$$\eta = 1 - (0.1)^{0.4} = 0.6 \qquad\qquad \text{(S.4.31.7)}$$

b) The work done on the gas in the compression process is

$$W = -\int_1^2 P \, \mathrm{d}V = -\int_1^2 P_i V_i^\gamma V^{-\gamma} \, \mathrm{d}V = -P_i V_i^\gamma \int_{V_i}^{V_f} V^{-\gamma} \, \mathrm{d}V$$

$$= -P_i V_i^\gamma \left. \frac{V^{-\gamma+1}}{1-\gamma} \right|_{V_i}^{V_f} = \frac{P_i V_i^\gamma}{\gamma-1} \left(V_f^{-\gamma+1} - V_i^{-\gamma+1}\right) \qquad \text{(S.4.31.8)}$$

$$= \frac{P_i V_i}{\gamma-1} \left(\left(\frac{V_i}{V_f}\right)^{\gamma-1} - 1\right) = \frac{P_i V_i}{\gamma-1} \left(r^{\gamma-1} - 1\right)$$

For $V_i = 2$ L and $P_i = 1$ atm,

$$W = \frac{10^5 \cdot 2 \cdot 10^{-3}}{0.4} (10^{0.4} - 1) \approx 755 \, \text{J}$$

4.32 Joule Cycle (Stony Brook)

The efficiency η of the cycle is given by the work W during the cycle divided by the heat Q absorbed in path $2 \to 3$ (see Figure S.4.32). W is defined by the area enclosed by the four paths of the P-V plot. The integral $\int P \, dV$ along the paths of constant pressure $2 \to 3$ and $4 \to 1$ is simply the difference in volume of the ends times the pressure, and the work along the adiabats, where there is no heat transfer $\delta Q = 0$, is given by the change in internal energy $C_V \delta T$:

$$W = C_V \left[(T_3 - T_4) - (T_2 - T_1) \right] + P_2 \left(V_3 - V_2 \right) - P_1 \left(V_4 - V_1 \right) \quad \text{(S.4.32.1)}$$

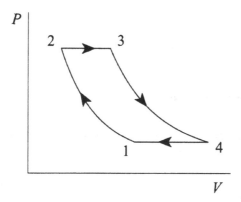

Figure S.4.32

Substituting the ideal gas law $PV = nRT$ into (S.4.32.1) and rearranging, we find

$$W = (C_V + nR) \left[(T_3 - T_4) - (T_2 - T_1) \right] \qquad \text{(S.4.32.2)}$$

$$= C_P \left[(T_3 - T_2) - (T_4 - T_1) \right]$$

where we used $C_P = C_V + nR$. In the process $2 \to 3$, the gas absorbs the heat Q:

$$Q = C_P \left(T_3 - T_2 \right) \qquad \text{(S.4.32.3)}$$

What remains is to write W and Q in terms of P and form the quotient. Using the equation for an adiabatic process in an alternative form,

$$PV^\gamma = \text{const}$$

$$TP^{(1-\gamma)/\gamma} = \text{const}$$

we have

$$\frac{T_1}{T_2} = \left(\frac{P_1}{P_2}\right)^{(\gamma-1)/\gamma} \tag{S.4.32.4}$$

$$\frac{T_4}{T_3} = \left(\frac{P_1}{P_2}\right)^{(\gamma-1)/\gamma}$$

Substituting for T_1 and T_4 by putting (S.4.32.4) into (S.4.32.1) yields

$$W = C_P \left[(T_3 - T_2) - (T_3 - T_2)\left(\frac{P_1}{P_2}\right)^{(\gamma-1)/\gamma} \right] \tag{S.4.32.5}$$

The efficiency η is then

$$\eta = \frac{W}{Q} = 1 - \left(\frac{P_1}{P_2}\right)^{(\gamma-1)/\gamma}$$

4.33   Diesel Cycle (Stony Brook)

We calculate the efficiency $\eta = W/Q$ as in Problem 4.32. The work W in the cycle (see Figure S.4.33) is

$$W = C_V \left[(T_3 - T_4) - (T_2 - T_1) \right] + P_2 (V_3 - V_2)$$

$$= C_V \left[(T_3 - T_4) - T_2 + T_1 + (\gamma - 1)(T_3 - T_2) \right] \tag{S.4.33.1}$$

$$= C_V \left[\gamma (T_3 - T_2) - (T_4 - T_1) \right]$$

where we have again used the ideal gas law $PV = nRT$ and $nR = C_P - C_V$. The heat Q absorbed by the gas during $2 \rightarrow 3$ is

$$Q = C_P (T_3 - T_2) \tag{S.4.33.2}$$

The efficiency η is

$$\eta = \frac{W}{Q} = 1 - \frac{1}{\gamma}\frac{T_4 - T_1}{T_3 - T_2} \tag{S.4.33.3}$$

Using the equation for the adiabats gives

$$T_1 V_1^{\gamma-1} = T_2 V_2^{\gamma-1} \tag{S.4.33.4}$$

$$T_3 V_3^{\gamma-1} = T_4 V_1^{\gamma-1}$$

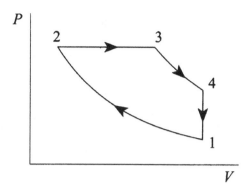

Figure S.4.33

The ideal gas law gives

$$\frac{T_3}{T_2} = \frac{V_3}{V_2} \tag{S.4.33.5}$$

Substituting (S.4.33.4) and (S.4.33.5) into (S.4.33.3) gives

$$\eta = 1 - \frac{1}{\gamma} \cdot \frac{(V_3/V_1)^{\gamma-1}\, T_3 - (V_2/V_1)^{\gamma-1}\, (V_2/V_3)\, T_3}{T_3 - (V_2/V_3)\, T_3} \tag{S.4.33.6}$$

$$= 1 - \frac{1}{\gamma} \cdot \frac{(V_3/V_1)^{\gamma} - (V_2/V_1)^{\gamma}}{V_3/V_1 - V_2/V_1}$$

4.34 Modified Joule–Thomson (Boston)

The work done by the piston goes into changing the internal energy of the part of the gas of volume dV that enters the plug and into the work done by the gas to enter container B occupying volume dV'. So we may write

$$P\,dV = c_v\,dN\,(\tau' - \tau) + (P - P')\,dV' \tag{S.4.34.1}$$

where c_v is the constant-volume heat capacity for one molecule and dN is the number of molecules in the volume dV. On the other hand, before and after the plug, we have, respectively,

$$P\,dV = dN\tau \tag{S.4.34.2}$$

$$P'\,dV' = dN\tau' \tag{S.4.34.3}$$

Substituting dV from (S.4.34.2) and dV' from (S.4.34.3) into (S.4.34.1), we have

$$\tau = c_v (\tau' - \tau) + (P - P') \frac{\tau'}{P'} \tag{S.4.34.4}$$

So,

$$\tau' = \frac{(1 + c_v)\tau}{c_v + P/P' - 1} = \frac{c_p \tau}{c_v + P/P' - 1} \tag{S.4.34.5}$$

$$= \frac{C_P \tau}{C_V + (P/P' - 1) N_A}$$

When $P = P'$, (S.4.34.5) becomes

$$\tau_1 = \frac{C_P}{C_V} \tau \tag{S.4.34.6}$$

Ideal Gas and Classical Statistics

4.35 Poisson Distribution in Ideal Gas (Colorado)

The probability ω_1 of finding a particular molecule in a volume V is

$$\omega_1 = \frac{V}{\tilde{V}}$$

The probability ω_N of finding N marked molecules in a volume V is

$$\omega_N = \left(\frac{V}{\tilde{V}}\right)^N \tag{S.4.35.1}$$

Similarly, the probability of finding one particular molecule outside of the volume V is

$$\overline{\omega_1} = \frac{\tilde{V} - V}{\tilde{V}}$$

and for $\tilde{N} - N$ particular molecules outside V,

$$\overline{\omega_{\tilde{N}-N}} = \left(\frac{\tilde{V} - V}{\tilde{V}}\right)^{\tilde{N}-N} \tag{S.4.35.2}$$

Therefore, the probability P_N of finding any N molecules in a volume V is the product of the two probabilities (S.4.35.1) and (S.4.35.2) weighted by the number of combinations for such a configuration:

$$P_N = \frac{\tilde{N}!}{N! \left(\tilde{N} - N\right)!} \left(\frac{V}{\tilde{V}}\right)^N \left(1 - \frac{V}{\tilde{V}}\right)^{\tilde{N}-N} \tag{S.4.35.3}$$

The condition $V \ll \tilde{V}$ also implies that $N \ll \tilde{N}$. Then we may approximate

$$\tilde{N}! \approx \left(\tilde{N} - N\right)! \, \tilde{N}^N \tag{S.4.35.4}$$

$$\tilde{N} - N \approx \tilde{N}$$

So, (S.4.35.3) becomes

$$P_N = \frac{\left(\tilde{N} - N\right)! \tilde{N}^N}{N! \left(\tilde{N} - N\right)!} \left(\frac{V}{\tilde{V}}\right)^N \left(1 - \frac{V}{\tilde{V}}\right)^{\tilde{N}}$$

$$= \frac{1}{N!} \left(\frac{\tilde{N}V}{\tilde{V}}\right)^N \left(1 - \frac{V}{\tilde{V}}\right)^{\tilde{N}} \tag{S.4.35.5}$$

$$= \frac{\langle N \rangle^N}{N!} \left(1 - \frac{\langle N \rangle}{\tilde{N}}\right)^{\tilde{N}}$$

where we used the average number of molecules in V:

$$\langle N \rangle = \frac{\tilde{N}}{\tilde{V}} V$$

Noticing that, for large $\tilde{N}$,

$$\left(1 - \frac{x}{\tilde{N}}\right)^{\tilde{N}} \approx e^{-x}$$

we obtain

$$P_N = \frac{\langle N \rangle^N}{N!} e^{-\langle N \rangle} \tag{S.4.35.6}$$

where we used

$$\frac{\langle N \rangle}{\tilde{N}} \ll 1 \qquad \tilde{N} \gg 1$$

(S.4.35.6) can be applied to find the mean square fluctuation in an ideal gas (see Problem 4.94) when the fluctuations are not necessarily small (i.e., it is possible to have $(N - \langle N \rangle) / \langle N \rangle \sim 1$, although N is always much smaller than the total number of particles in the gas $\tilde{N}$).

4.36 Polarization of Ideal Gas (Moscow Phys-Tech)

The potential energy of a dipole in an electric field $\mathbf{E}$ is

$$U = -\mathbf{p} \cdot \mathbf{E} = -pE \cos \theta$$

where the angle θ is between the direction of the electric field (which we choose to be along the $\hat{\mathbf{z}}$ axis) and the direction of a the dipole moment. The center of the spherical coordinate system is placed at the center of the dipole. The probability dw that the direction of the dipole is within a solid angle $d\omega = \sin \theta \, d\varphi \, d\theta$ is

$$dw = Ae^{-U/\tau} \, d\Omega = \frac{e^{pE \cos \theta/\tau} \sin \theta \, d\varphi \, d\theta}{\int\limits_0^{2\pi} \int\limits_0^{\pi} e^{pE \cos \theta/\tau} \sin \theta \, d\varphi \, d\theta} \tag{S.4.36.1}$$

The total electric dipole moment per unit volume of the gas is

$$P = N \langle p \rangle = N \langle p_z \rangle = N \int p \cos \theta \, d\Omega \tag{S.4.36.2}$$

$$= \frac{N \int\limits_0^{2\pi} \int\limits_0^{2\pi} p \cos \theta \, e^{pE \cos \theta/\tau} \sin \theta \, d\varphi \, d\theta}{\int\limits_0^{2\pi} \int\limits_0^{\pi} e^{pE \cos \theta/\tau} \sin \theta \, d\varphi \, d\theta}$$

Introducing a new variable $x \equiv \cos \theta$ and denoting $\alpha \equiv pE/\tau$, we obtain

$$P = \frac{Np \int\limits_{-1}^{1} x e^{\alpha x} \, dx}{\int\limits_{-1}^{1} e^{\alpha x} \, dx} = Np \frac{d}{d\alpha} \ln \left[\int\limits_{-1}^{1} e^{\alpha x} \, dx \right]$$

$$= Np \frac{d}{d\alpha} \ln \left(\frac{2 \sinh \alpha}{\alpha} \right)$$

$$= Np \left[\frac{\alpha}{\sinh \alpha} \left(\frac{\cosh \alpha}{\alpha} - \frac{\sinh \alpha}{\alpha^2} \right) \right] \tag{S.4.36.3}$$

$$= Np \left(\coth \alpha - \frac{1}{\alpha} \right) = Np \left[\coth \frac{pE}{\tau} - \frac{\tau}{pE} \right]$$

$$= Np \, \mathcal{L} \left(\frac{pE}{\tau} \right)$$

where $\mathcal{L}(pE/\tau)$ is the Langevin function. For $pE \ll \tau$ $(\alpha \ll 1)$, we can expand (S.4.36.3) to obtain

$$P = \frac{p^2 N}{3\tau} E \qquad (S.4.36.4)$$

Since $\mathbf{D} = \varepsilon \mathbf{E}$ and

$$\mathbf{D} = \mathbf{E} + 4\pi \mathbf{P} = \mathbf{E}\left(1 + 4\pi \frac{p^2 N}{3\tau}\right)$$

we have for the dielectric constant

$$\varepsilon = 1 + \frac{4\pi}{3}\frac{p^2 N}{\tau} \qquad (S.4.36.5)$$

4.37 Two-Dipole Interaction (Princeton)

Introduce spherical coordinates with the $\hat{\mathbf{z}}$ axis along the line of the separation between the dipoles. Then the partition function reads

$$Z = \int \int d\Omega_1 \, d\Omega_2 \, e^{-U/\tau} \qquad (S.4.37.1)$$

The potential energy of the interaction can be rewritten in the form

$$U = \frac{\mu_{1x}\mu_{2x} + \mu_{1y}\mu_{2y} + \mu_{1z}\mu_{2z}}{r^3} - 3\frac{\mu_{1z}\mu_{2z}}{r^3} \qquad (S.4.37.2)$$

$$= \frac{\mu_{1x}\mu_{2x} + \mu_{1y}\mu_{2y} - 2\mu_{1z}\mu_{2z}}{r^3}$$

Since

$$\mu_{ix} = \mu_i \sin\theta_i \, \cos\varphi_i$$

$$\mu_{iy} = \mu_i \sin\theta_i \, \sin\varphi_i \qquad (S.4.37.3)$$

$$\mu_{iz} = \mu_i \cos\theta_i$$

(S.4.37.2) becomes

$$U = \frac{\mu_1\mu_2}{r^3}\left[\sin\theta_1 \, \sin\theta_2 \, \cos(\varphi_1 - \varphi_2) - 2\cos\theta_1 \, \cos\theta_2\right] \qquad (S.4.37.4)$$

$$= \frac{\mu_1\mu_2}{r^3}f(\theta_1, \theta_2)$$

and

$$Z = \int_0^\pi \int_0^{2\pi} \int_0^\pi \int_0^{2\pi} \sin\theta_1 \, d\theta_1 \, d\varphi_1 \, \sin\theta_2 \, d\theta_2 \, d\varphi_2 \qquad \text{(S.4.37.5)}$$

$$\cdot \exp\left[-\frac{\mu_1\mu_2}{\tau r^3} f(\theta_1, \theta_2)\right]$$

We can expand the exponential at high temperatures $\mu_1\mu_2/\tau r^3 \ll 1$ so that

$$\exp\left[-\frac{\mu_1\mu_2}{\tau r^3} f(\theta_1, \theta_2)\right] \qquad \text{(S.4.37.6)}$$

$$\approx 1 - \frac{\mu_1\mu_2}{\tau r^3} f(\theta_1, \theta_2) + \frac{1}{2!} A f^2(\theta_1, \theta_2) + \cdots$$

where $A \equiv \mu_1^2\mu_2^2/\tau^2 r^6$. The first-order terms are all zero upon integration, and we have

$$Z = (4\pi)^2 + \frac{A}{2!} \int_0^\pi \int_0^{2\pi} \int_0^\pi \int_0^{2\pi} d\cos\theta_1 \, d\cos\theta_2 \, d\varphi_1 \, d\varphi_2 \qquad \text{(S.4.37.7)}$$

$$\cdot \left[\sin^2\theta_1 \, \sin^2\theta_2 \, \cos^2(\varphi_2 - \varphi_1) + 4\cos^2\theta_1 \, \cos^2\theta^2\right]$$

where the cross term also vanishes, and we find

$$Z = (4\pi)^2 + 2\pi^2 A \int_0^\pi \int_0^\pi d\cos\theta_1 \, d\cos\theta_2$$

$$\cdot \left(\frac{1}{2} \sin^2\theta_1 \, \sin^2\theta_2 + 4\cos^2\theta_1 \, \cos^2\theta_2\right)$$

$$= (4\pi)^2 + 2\pi^2 A \int_{-1}^1 \int_{-1}^1 dz_1 \, dz_2 \left[\frac{1}{2}(1 - z_1^2)(1 - z_2^2) + 4z_1^2 z_2^2\right]$$

$$= (4\pi)^2 + 2\pi^2 A \int_{-1}^1 \int_{-1}^1 dz_1 \, dz_2 \left[\frac{1}{2} - \frac{1}{2}(z_1^2 + z_2^2) + \frac{9}{2} z_1^2 z_2^2\right] \qquad \text{(S.4.37.8)}$$

$$= (4\pi)^2 + 2\pi^2 A \left[2 - \frac{1}{2} \cdot 2 \cdot \frac{4}{3} + \frac{9}{2} \cdot \frac{4}{9}\right]$$

$$= (4\pi)^2 + \frac{(4\pi)^2}{3} \frac{\mu_1^2\mu_2^2}{\tau^2 r^6} = (4\pi)^2 \left(1 + \frac{\mu_1^2\mu_2^2}{3\tau^2 r^6}\right)$$

The average force is given by

$$\langle \mathbf{f} \rangle = -\left\langle \frac{\partial U}{\partial r} \right\rangle = -\left(\frac{\partial F}{\partial r} \right)_{\tau,V}$$

where F is the free energy. So,

$$\langle \mathbf{f} \rangle = \tau \frac{\partial \ln Z}{\partial \mathbf{r}} \approx -2 \frac{\mu_1^2 \mu_2^2}{\tau r^7} \frac{\mathbf{r}}{r} \qquad \text{(S.4.37.9)}$$

The minus sign indicates an average attraction between the dipoles.

4.38   Entropy of Ideal Gas (Princeton)

a) For an ideal gas the partition function factors; however, we must take the sum of N identical molecules divided by the number of interchanges $N!$ to account for the fact that one microscopic quantum state corresponds to a number of different points in phase space. So

$$Z = \frac{1}{N!} \left(\sum_k e^{-E_k/\tau} \right)^N \qquad \text{(S.4.38.1)}$$

Now, the Helmholtz free energy, F, is given by

$$F = -\tau \ln Z = -\tau \ln \frac{1}{N!} \left(\sum_k e^{-E_k/\tau} \right)^N \qquad \text{(S.4.38.2)}$$

$$= -N\tau \ln \left(\sum_k e^{-E_k/\tau} \right) + \tau \ln N!$$

Using Stirling's formula, $\ln N! \approx \int\limits_1^N \ln x \cdot dx \approx N \ln(N/e)$, we obtain

$$F = -N\tau \ln \sum_k e^{-E_k/\tau} + N\tau \ln \frac{N}{e} = -N\tau \ln \left[\frac{e}{N} \sum_k e^{-E_k/\tau} \right] \qquad \text{(S.4.38.3)}$$

Using the explicit expression for the molecular energy E_k, we can rewrite (S.4.38.3) in the form

$$F = -N\tau \ln \left[\frac{e}{N} \int\limits_0^\infty \int\limits_{-\infty}^\infty \int\limits_{-\infty}^\infty \int\limits_{-\infty}^\infty \frac{dp_x\, dp_y\, dp_z\, dV}{(2\pi\hbar)^3} \right.$$

$$\cdot \, e^{-(p_x^2+p_y^2+p_z^2)/2m\tau} \sum_k e^{-\varepsilon_k/\tau} \Bigg] \qquad (\text{S.4.38.4})$$

$$= -N\tau \ln \left[\frac{eV_1}{N} \left(\frac{m\tau}{2\pi\hbar^2} \right)^{3/2} \sum_k e^{-\varepsilon_k/\tau} \right]$$

$$= -N\tau \ln \frac{eV_1}{N} + N f(\tau)$$

Here we used the fact that the sum depends only on temperature, so we can define $f(\tau)$:

$$f(\tau) \equiv -\tau \ln \left[\left(\frac{m\tau}{2\pi\hbar^2} \right)^{3/2} \sum_k e^{-\varepsilon_k/\tau} \right] \qquad (\text{S.4.38.5})$$

b) Now we can calculate the total entropy S of the two gases (it is important that the gases be identical so that $f(\tau)$ is the same for both vessels):

$$S = -\left(\frac{\partial F}{\partial \tau} \right)_V \qquad (\text{S.4.38.6})$$

where F is defined by (S.4.38.4).

$$S_1 = N \ln \frac{eV_1}{N} - N f'(\tau) \qquad (\text{S.4.38.7})$$

$$S_2 = N \ln \frac{eV_2}{N} - N f'(\tau)$$

We have for total entropy

$$S = S_1 + S_2 = N \ln \frac{eV_1}{N} - N f'(\tau) + N \ln \frac{eV_2}{N} - N f'(\tau) \quad (\text{S.4.38.8})$$

$$= N \ln \frac{\tau}{P_1} + N \ln \frac{\tau}{P_2} - 2N f'(\tau) = -N \ln P_1 P_2 + 2N \ln \tau - 2N f'(\tau)$$

c) After the vessels are connected their volume becomes $V = V_1 + V_2$, the number of particles becomes $2N$, and the temperature remains the same (no work is done in mixing the two gases). So now

$$\tilde{S} = 2N \ln \frac{e(V_1 + V_2)}{2N} - 2N f'(\tau) \qquad (\text{S.4.38.9})$$

$$= -2N \ln P + 2N \ln \tau - 2N f'(\tau)$$

It can be easily seen that the pressure becomes $\tilde{P} = 2P_1P_2/(P_1 + P_2)$, so

$$\tilde{S} = -2N \ln \frac{2P_1P_2}{P_1 + P_2} + 2N \ln \tau - 2N f'(\tau) \qquad \text{(S.4.38.10)}$$

$$= -N \ln \left(\frac{2P_1P_2}{P_1 + P_2}\right)^2 + 2N \ln \tau - 2N f'(\tau)$$

and

$$\Delta S = \tilde{S} - S = -N \ln \frac{4P_1P_2}{(P_1 + P_2)^2} = N \ln \frac{(P_1 + P_2)^2}{4P_1P_2} \qquad \text{(S.4.38.11)}$$

Let us show that ΔS is always nonnegative. This is equivalent to the condition

$$\frac{(P_1 + P_2)^2}{4P_1P_2} \geq 1 \qquad (P_1 - P_2)^2 \geq 0 \qquad \text{(S.4.38.12)}$$

which is always true. At $P_1 = P_2$ ($V_1 = V_2$), $\Delta S = 0$, which makes perfect sense.

4.39   Chemical Potential of Ideal Gas (Stony Brook)

The expression for the Helmholtz free energy was derived in Problem 4.38:

$$F = -N\tau \ln \left[\frac{eV}{N} \left(\frac{m\tau}{2\pi\hbar^2}\right)^{3/2} \sum_k e^{-\varepsilon_k/\tau} \right] \qquad \text{(S.4.39.1)}$$

Since all the molecules are in the ground state, the sum only includes one term, which we can take as an energy zero, $\varepsilon_0 = 0$. Then (S.4.39.1) becomes

$$F = -N\tau \ln \left[\frac{eV}{N} \left(\frac{m\tau}{2\pi\hbar^2}\right)^{3/2} g \right] \qquad \text{(S.4.39.2)}$$

where we took into account a degeneracy of the ground state g. The Gibbs free energy G is then

$$G = F + PV = -N\tau \ln \left[\frac{eV}{N} \left(\frac{m\tau}{2\pi\hbar^2}\right)^{3/2} g \right] + PV$$

$$= -N\tau \ln \left[\frac{eV}{N} \left(\frac{m\tau}{2\pi\hbar^2}\right)^{3/2} g \right] + N\tau \qquad \text{(S.4.39.3)}$$

$$= -N\tau \ln \left[\frac{gV}{N} \left(\frac{m\tau}{2\pi\hbar^2}\right)^{3/2} \right] = -N\tau \ln \left[\frac{g\tau}{P} \left(\frac{m\tau}{2\pi\hbar^2}\right)^{3/2} \right]$$

where we have expressed G as a function of τ, P. The chemical potential $\mu = G/N$, so we obtain, from (S.4.39.3),

$$\mu = \tau \ln \left[\frac{g\tau^{5/2}}{P} \left(\frac{m}{2\pi\hbar^2} \right)^{3/2} \right] \qquad \text{(S.4.39.4)}$$

$$= \tau \ln \left[\frac{P}{g\tau^{5/2}} \left(\frac{2\pi\hbar^2}{m} \right)^{3/2} \right] = \tau \ln \left[\frac{N}{gV} \left(\frac{2\pi\hbar^2}{m\tau} \right)^{3/2} \right]$$

This approximation is valid when the temperature is much lower than the energy difference ΔE between the electronic ground state and the first excited state; since this ΔE is comparable to the ionization energy ε_{ion}, this condition is equivalent to $\tau \ll \varepsilon_{\text{ion}}$. However, even at temperatures $\tau \sim \varepsilon_{\text{ion}}$, the gas is almost completely ionized (see Landau and Lifshitz, *Statistical Physics*, Sect. 106). Therefore (S.4.39.4) is always valid for a nonionized gas.

4.40 Gas in Harmonic Well (Boston)

a) The partition function is given by a standard integral (compare with 4.38, where the molecules are indistinguishable):

$$Z = \int e^{-\epsilon/\tau} \frac{d^3p \, d^3x}{(2\pi\hbar)^3}$$

$$= \frac{1}{(2\pi\hbar)^3} \int_0^\infty e^{-p^2/2m\tau} 4\pi p^2 \, dp \int_{-\infty}^\infty e^{-(x^2+y^2+z^2)/2V^{2/3}\tau} \, dx \, dy \, dz$$

$$= \frac{1}{2\pi^2\hbar^3} \frac{\sqrt{\pi}}{4} (2m\tau)^{3/2} \, 2^{3/2} V \tau^{3/2} \pi^{3/2} \qquad \text{(S.4.40.1)}$$

$$= \frac{\tau^3 m^{3/2} V}{\hbar^3}$$

The Helmholtz free energy F follows directly from the partition function:

$$F = -N\tau \ln Z = -N\tau \ln V - 3N\tau \ln \tau - N\tau \ln \frac{m^{3/2}}{\hbar^3} \qquad \text{(S.4.40.2)}$$

$$= -N\tau \ln V + Nf(\tau)$$

b) We may find the force from F:

$$\tilde{P} = -\left(\frac{\partial F}{\partial V} \right)_\tau = \frac{N\tau}{V} \qquad \text{(S.4.40.3)}$$

The equation of state is therefore analogous to the gas in a container with rigid walls, where

$$\tilde{P}V = N\tau$$

c) The entropy, energy, and heat capacity all follow in quick succession from F:

$$S = -\left(\frac{\partial F}{\partial \tau}\right)_V = N\ln V + 3N\ln\tau + 3N + N\ln\frac{m^{3/2}}{\hbar^3} \quad \text{(S.4.40.4)}$$

$$E = F + \tau S = 3N\tau \quad \text{(S.4.40.5)}$$

$$C_V = \left(\frac{\partial E}{\partial \tau}\right)_V = 3N \quad \text{(S.4.40.6)}$$

4.41 Ideal Gas in One-Dimensional Potential (Rutgers)

a) The coordinate- and momentum-dependent parts of the partition function can be separated. The coordinate-dependent part of the partition function

$$Z_q = \int_0^\infty e^{-U(x)/\tau}\,dx \quad \text{(S.4.41.1)}$$

For the potential in this case we have

$$Z_q = \int_0^\infty e^{-Ax^n/\tau}\,dx = \frac{1}{n}\left(\frac{\tau}{A}\right)^{1/n}\int_0^\infty e^{-z}z^{1/n-1}\,dz \quad \text{(S.4.41.2)}$$

$$= \frac{\tau^{1/n}}{nA^{1/n}}\Gamma\left(\frac{1}{n}\right)$$

where we substituted

$$x = \left(\frac{\tau}{A}\right)^{1/n}z^{1/n}$$

$$dx = \frac{1}{n}\left(\frac{\tau}{A}\right)^{1/n}z^{1/n-1}\,dz$$

and

$$\int_0^\infty z^{1/n-1}e^{-z}\,dz \equiv \Gamma\left(\frac{1}{n}\right)$$

The free energy associated with the coordinate-dependent part of the partition function is

$$F_q = -\tau \ln Z = -\tau \frac{1}{n} \ln \tau - \tau \ln \left[\frac{\Gamma(1/n)}{nA^{1/n}} \right] \qquad (S.4.41.3)$$

The average potential energy is given by

$$\langle U \rangle = F_q - \tau \left(\frac{\partial F_q}{\partial \tau} \right)_V = -\tau^2 \left[\frac{\partial}{\partial \tau} \left(\frac{F_q}{\tau} \right) \right] = \frac{\tau}{n} \qquad (S.4.41.4)$$

For $n = 2$ we have a harmonic oscillator, and in agreement with the equipartition theorem (see Problem 4.42)

$$\langle U \rangle = \langle T \rangle = \frac{\tau}{2}$$

b) For $n = 1$, $U = mgx$, and the average potential energy per particle

$$\langle U \rangle = \tau$$

which also agrees with the generalized equipartition theorem.

4.42 Equipartition Theorem (Columbia, Boston)

a) For both of these averages the method is identical, since the Hamiltonian depends on the same power of either p or q. Compose the first average as follows:

$$\frac{\langle p_i^2 \rangle}{2m_i} = \frac{\int (p_i^2/2m_i) e^{-p_i^2/2m_i\tau} \, dp_i \int e^{-E'/\tau} \, dq_1 \cdots dp_N}{\int e^{-p_i^2/2m_i\tau} \, dp_i \int e^{-E'/\tau} \, dq_1 \cdots dp_N} \qquad (S.4.42.1)$$

where the energy is broken into the p_i-dependent term and E', the rest of the sum. The second integrals in the numerator and denominator cancel, so the remaining expression may be written

$$\frac{\langle p_i^2 \rangle}{2m_i} = -\frac{\partial}{\partial \beta} \ln \int_{-\infty}^{\infty} e^{-\beta p_i^2/2m_i} \, dp_i \qquad (S.4.42.2)$$

where, as usual, $\beta \equiv 1/\tau$. A change of variables produces a piece dependent on β and an integral that is not:

$$p_i \equiv y\sqrt{2m_i/\beta}$$

$$\frac{\langle p_i^2 \rangle}{2m_i} = -\frac{\partial}{\partial \beta} \ln \left[\sqrt{\frac{2m_i}{\beta}} \int_{-\infty}^{\infty} e^{-y^2} \, dy \right] \qquad \text{(S.4.42.3)}$$

$$= -\frac{\partial}{\partial \beta} \ln \sqrt{\beta} = \frac{1}{2\beta} = \frac{\tau}{2}$$

The $(k_i/2)\langle q_i^2 \rangle$ average proceeds in precisely the same way, yielding

$$\frac{k_i}{2} \langle q_i^2 \rangle = \frac{\tau}{2} \qquad \text{(S.4.42.4)}$$

b) The heat capacity, C_V, at constant volume is equal to $\partial E/\partial \tau$. From part (a), we have

$$E = \langle H \rangle = \left\langle \sum_{i=1}^{3N} \frac{p_i^2}{2m_i} + \sum_{i=1}^{3N} \frac{k_i}{2} q_i^2 \right\rangle = 3N\frac{\tau}{2} + 3N\frac{\tau}{2} = 3N\tau \qquad \text{(S.4.42.5)}$$

where we now sum over the 3-space and momentum degrees of freedom per atom. The heat capacity,

$$C_V = \frac{\partial E}{\partial \tau} = 3N \qquad \text{(S.4.42.6)}$$

is the law of Dulong and Petit.

c) Now take the average:

$$\left\langle x_i \frac{\partial H}{\partial x_j} \right\rangle = \frac{\int e^{-E/\tau} x_i (\partial H/\partial x_j) \prod_k dx_k}{\int e^{-E/\tau} \prod_k dx_k} \qquad \text{(S.4.42.7)}$$

$$= \frac{-\int \tau \left(\partial e^{-E/\tau}/\partial x_j \right) x_i \prod_k dx_k}{\int e^{-E/\tau} \prod_k dx_k}$$

Integration by parts yields

$$= \frac{-\tau \int \left(e^{-E/\tau} x_i \right) \big|_{x_j=-\infty}^{\infty} \prod_k{}' dx_k + \tau \int e^{-E/\tau} \left(\partial x_i/\partial x_j \right) \prod_k dx_k}{\int e^{-E/\tau} \prod_k dx_k}$$

$$\text{(S.4.42.8)}$$

where the prime on the product sign in the first term indicates that we integrate over all x_i except $i = j$. If $i \neq j$, then the first term in the numerator equals zero. If x_j is one of the q's, then by the assumption of U infinite, the term still equals zero. Finally, if $i = j > N$, then by l'Hôpital's rule the first term again gives zero. In the second term, $\partial x_i / \partial x_j = \delta_{ij}$, so the expression reduces to

$$\frac{\tau \int e^{-E/\tau} \delta_{ij} \prod_k dx_k}{\int e^{-E/\tau} \prod_k dx_k} \tag{S.4.42.9}$$

Finally,

$$\left\langle x_i \frac{\partial H}{\partial x_j} \right\rangle = \tau \delta_{ij} \tag{S.4.42.10}$$

d) By definition,

$$\left\langle q \frac{\partial \varepsilon(q)}{\partial q} \right\rangle = \frac{\int\limits_{-\infty}^{\infty} q \frac{\partial \varepsilon(q)}{\partial q} e^{-\varepsilon(q)/\tau} \, dq}{\int\limits_{-\infty}^{\infty} e^{-\varepsilon(q)/\tau} \, dq}$$

$$= \frac{-\tau \int\limits_{-\infty}^{\infty} q \frac{\partial}{\partial q} e^{-\varepsilon(q)/\tau} \, dq}{\int\limits_{-\infty}^{\infty} e^{-\varepsilon(q)/\tau} \, dq} \tag{S.4.42.11}$$

$$= -\tau \frac{q e^{-\varepsilon(q)/\tau} \big|_{-\infty}^{\infty}}{\int\limits_{-\infty}^{\infty} e^{-\varepsilon(q)/\tau} \, dq} - \tau \frac{-\int\limits_{-\infty}^{\infty} e^{-\varepsilon(q)/\tau} \, dq}{\int\limits_{-\infty}^{\infty} e^{-\varepsilon(q)/\tau} \, dq} = \tau$$

Given a polynomial dependence of the energy on the generalized coordinate:

$$\varepsilon(q) = \alpha q^n \tag{S.4.42.12}$$

(S.4.42.11) yields

$$\left\langle q \frac{\partial \varepsilon(q)}{\partial q} \right\rangle = \langle n \alpha q^n \rangle = n \langle \alpha q^n \rangle \tag{S.4.42.13}$$

$$= n \langle \varepsilon(q) \rangle = \tau$$

To satisfy the equipartition theorem:

$$\langle \varepsilon(q) \rangle = \frac{\tau}{n} = \frac{\tau}{2} \tag{S.4.42.14}$$

Thus, we should have $n = 2$.

4.43 Diatomic Molecules in Two Dimensions (Columbia)

a) The partition function Z_{rot} may be calculated in the usual way by multiplying the individual Boltzmann factors by their degeneracies and summing:

$$Z_{rot} = 1 + 2\sum_{J=1}^{\infty} e^{-hcBJ^2/\tau} = 2\sum_{J=0}^{\infty} e^{-hcBJ^2/\tau} - 1 \qquad (S.4.43.1)$$

This is difficult to sum, but we may consider the integral instead, given the assumption that $\tau \gg hcB$:

$$Z_{rot} = 2\int_0^{\infty} e^{-hcBx^2/\tau}\,dx - 1 = \sqrt{\frac{\pi\tau}{hcB}} - 1 \qquad (S.4.43.2)$$

b) The energy and heat capacity of the set of diatomic molecules described above may be determined from the partition function for the set:

$$Z = \frac{1}{N!}(Z_{rot})^N = \frac{1}{N!}\left(\sqrt{\frac{\pi\tau}{hcB}} - 1\right)^N \approx \frac{1}{N!}\left(\frac{\pi\tau}{hcB}\right)^{N/2} \qquad (S.4.43.3)$$

where the N-fold product has been divided by the number of permutations of the N indistinguishable molecules. Recall that

$$E = \tau^2 \frac{\partial}{\partial\tau}\ln Z$$

We then find that

$$E = \tau^{3/2}\frac{N}{2}\sqrt{\frac{\pi}{hcB}}\frac{1}{\sqrt{\pi\tau/hcB} - 1} \approx \frac{N\tau}{2} \qquad (S.4.43.4)$$

Again, for $\tau \gg hcB$, the heat capacity is

$$C_V = \frac{\partial E}{\partial\tau} \approx \frac{N}{2} \qquad (S.4.43.5)$$

A diatomic rotor in three dimensions would have contributions to the energy of $(1/2)\tau$ per degree of freedom. Three degrees of translation and two degrees of rotation (assuming negligible inertia perpendicular to its length) gives for one molecule

$$E = \frac{5}{2}\tau \qquad (S.4.43.6)$$

$$E = \frac{5}{2}N\tau \qquad (S.4.43.7)$$

A diatomic rotor confined to a plane would have three degrees of freedom, two translational and one rotational. Hence,

$$E = \frac{3}{2}\tau \qquad (S.4.43.8)$$

$$c_v = \frac{3}{2} \qquad (S.4.43.9)$$

The quantization of energy is not apparent since we have assumed $\tau \gg hcB$.

4.44 Diatomic Molecules in Three Dimensions (Stony Brook, Michigan State)

a) We first transform the expression of the kinetic energy ε_k:

$$\varepsilon_k = \frac{1}{2}\sum_{i=1}^{2} m_i \left(\dot{x}_i^2 + \dot{y}_i^2 + \dot{z}_i^2\right) \qquad (S.4.44.1)$$

where x_i, y_i, z_i are the Cartesian coordinates of the molecule in the frame with the c.m. at the origin to spherical coordinates:

$$x_i = r_i \sin\theta_i \cos\varphi_i$$

$$y_i = r_i \sin\theta_i \sin\varphi_i \qquad (S.4.44.2)$$

$$z_i = r_i \cos\theta_i$$

For the rigid diatom,

$$\theta_1 = \pi - \theta_2 \equiv \theta$$

$$\varphi_1 = \pi + \varphi_2 \equiv \varphi$$

We may substitute (S.4.44.2) into (S.4.44.1), obtaining

$$\varepsilon_k = \frac{1}{2}\left(m_1 r_1^2 + m_2 r_2^2\right)\left(\dot{\theta}^2 + \sin^2\theta\,\dot{\varphi}^2\right) \qquad (S.4.44.3)$$

Using the definition of c.m., we may write

$$m_1 r_1 = m_2 r_2 \qquad (S.4.44.4)$$

$$|r_1| + |r_2| = a$$

yielding

$$r_1 = \frac{m_2}{m_1 + m_2} a \qquad\qquad (S.4.44.5)$$

$$r_2 = \frac{m_1}{m_1 + m_2} a$$

Then (S.4.44.3) becomes

$$\varepsilon_k = \frac{1}{2} \frac{m_1 m_2}{m_1 + m_2} a^2 \left(\dot\theta^2 + \sin^2\theta \, \dot\varphi^2 \right) = \frac{1}{2} I \left(\dot\theta^2 + \sin^2\theta \, \dot\varphi^2 \right) \qquad (S.4.44.6)$$

with

$$I = \frac{m_1 m_2}{m_1 + m_2} a^2$$

b) In order to find the conjugate momenta p_θ, p_φ, we must compute the Lagrangian $\mathcal{L}$:

$$\mathcal{L} = \varepsilon_k - V = \varepsilon_k = \frac{1}{2} I \left(\dot\theta^2 + \sin^2\theta \, \dot\varphi^2 \right)$$

$$p_\varphi = \frac{\partial \mathcal{L}}{\partial \dot\varphi} = I \sin^2\theta \, \dot\varphi \qquad\qquad (S.4.44.7)$$

$$p_\theta = \frac{\partial \mathcal{L}}{\partial \dot\theta} = I \dot\theta$$

Expressing $\dot\varphi$, $\dot\theta$ through p_φ, p_θ,

$$\dot\theta = \frac{p_\theta}{I} \qquad\qquad \dot\varphi = \frac{p_\varphi}{I \sin^2\theta}$$

we may rewrite the Hamiltonian as

$$H = \frac{1}{2} I \left(\frac{p_\theta^2}{I^2} + \frac{p_\varphi^2}{I^2 \sin^2\theta} \right) = \frac{p_\theta^2}{2I} + \frac{p_\varphi^2}{2I \sin^2\theta} \qquad (S.4.44.8)$$

c) The single-diatom partition function may be computed as follows:

$$Z_{cl} = \frac{1}{\hbar^2} \int_{-\infty}^{\infty} \int_{-\infty}^{\infty} \int_{0}^{2\pi} \int_{0}^{\pi} e^{-H/\tau} \, d\theta \, d\varphi \, dp_\theta \, dp_\varphi$$

$$= \frac{2\pi}{\hbar^2} \int_{-\infty}^{\infty} dp_\theta \, e^{-p_\theta^2/2I\tau} \int_{0}^{\pi} d\theta \int_{-\infty}^{\infty} dp_\varphi \, e^{-p_\varphi^2/2I \sin^2\theta} \qquad (S.4.44.9)$$

$$= \frac{2\pi}{\hbar^2} \sqrt{2I\tau\pi} \int_{0}^{\pi} d\theta \, \sqrt{2I\tau\pi} \, \sin\theta = \frac{4\pi^2 I\tau}{\hbar^2} \int_{0}^{\pi} d\theta \, \sin\theta = \frac{8\pi^2 I}{\hbar^2} \tau$$

Now the free energy F for N such classical molecules may be found from

$$F = -N\tau \ln Z = -N\tau \ln \frac{8\pi^2 I\tau}{\hbar^2} \qquad \text{(S.4.44.10)}$$

The entropy S is then

$$S = -\frac{\partial F}{\partial \tau} = N \ln \frac{8\pi^2 I\tau}{\hbar^2} + N \qquad \text{(S.4.44.11)}$$

and the energy E and heat capacity C are

$$E = F + \tau S = -N\tau \ln \frac{8\pi^2 I\tau}{\hbar^2} + N\tau \ln \frac{8\pi^2 I\tau}{\hbar^2} + N\tau = N\tau$$

$$C = \frac{\partial E}{\partial \tau} = N \qquad \text{(S.4.44.12)}$$

d) For the quantum case the Schrödinger equation for a rigid rotator

$$E\psi = \frac{J^2}{2I}\psi = -\frac{\hbar^2}{2I}\frac{1}{\sin^2\theta}\left(\sin\theta\frac{\partial}{\partial\theta}\sin\theta\frac{\partial}{\partial\theta} + \frac{\partial^2}{\partial\varphi^2}\right)\psi \qquad \text{(S.4.44.13)}$$

admits the standard solution

$$\psi_{jm} = e^{im\varphi} Y_{jm}(\theta) \qquad \text{(S.4.44.14)}$$

$$E_{jm} = \frac{\hbar^2}{2I}j(j+1)$$

where each of the energy states is $(2j+1)$-degenerate. The partition function Z_q is given by

$$Z_q = \sum_{j=0}^{\infty} (2j+1)\, e^{-\hbar^2 j(j+1)/2I\tau} \qquad \text{(S.4.44.15)}$$

For low temperatures we may neglect high-order terms and write

$$Z_q \approx 1 + 3e^{-\hbar^2/I\tau}$$

where we left only terms with $j = 0$ and $j = 1$. For N molecules we find for the free energy that

$$F = -N\tau \ln Z_q = -N\tau \ln\left(1 + 3e^{-\hbar^2/I\tau}\right) \approx -3N\tau e^{-\hbar^2/I\tau} \qquad \text{(S.4.44.16)}$$

The energy E and heat capacity C are then

$$E = -\tau^2 \frac{\partial}{\partial \tau} \left(\frac{F}{\tau} \right) = \frac{3N\tau^2 \hbar^2}{I\tau^2} e^{-\hbar^2/I\tau} = \frac{3N\hbar^2}{I} e^{-\hbar^2/I\tau} \text{(S.4.44.17)}$$

$$C = \frac{\partial E}{\partial \tau} = \frac{3N\hbar^2}{I} \frac{\hbar^2}{I\tau^2} e^{-\hbar^2/I\tau} = \frac{3N\hbar^4}{I^2\tau^2} e^{-\hbar^2/I\tau}$$

So, at low temperatures the heat capacity corresponding to the rotational degrees of freedom is exponentially small. This implies that there would be no difference, in this limit, between the heat capacity for monatomic and diatomic molecules. In the opposite case, at high temperatures, $\hbar^2/2I \ll \tau$, the sum may be replaced by an integral:

$$Z_q = \sum_{j=0}^{\infty} (2j+1) e^{-\alpha j(j+1)} \tag{S.4.44.18}$$

where $\alpha \equiv \hbar^2/2I\tau$. Proceeding from (S.4.44.18), we have

$$Z_q = 2 \sum_{j=0}^{\infty} \left(j + \frac{1}{2} \right) e^{-\alpha[(j+1/2)^2 - 1/4]} \tag{S.4.44.19}$$

$$= 2e^{\alpha/4} \sum_{j=0}^{\infty} \left(j + \frac{1}{2} \right) e^{-\alpha(j+1/2)^2}$$

Replacing the sum by an integral, we obtain

$$Z_q \approx 2e^{\alpha/4} \int_{1/2}^{\infty} dx\, x e^{-\alpha x^2} = e^{\alpha/4} \int_{1/4}^{\infty} dx^2\, e^{-\alpha x^2} \tag{S.4.44.20}$$

$$= \frac{1}{\alpha} = \frac{2I\tau}{\hbar^2}$$

Therefore, in the classical limit (high temperatures),

$$F = -N\tau \ln Z_q \approx -N\tau \ln \frac{2I\tau}{\hbar^2} \tag{S.4.44.21}$$

The energy E and heat capacity C are given by

$$E = -\tau^2 \frac{\partial}{\partial \tau} \left(\frac{F}{\tau} \right) = \tau^2 N \frac{1}{\tau} = N\tau \tag{S.4.44.22}$$

$$C = \frac{\partial E}{\partial \tau} \approx N$$

We see that this is the same as found in (S.4.44.12). Since we expect a heat capacity per degree of freedom of $1/2$, we see that there are two degrees of freedom for each molecule since

$$c = \frac{C}{N} = 1$$

They correspond to the two rotational degrees of freedom of a classical rod. (There are no spatial degrees of freedom since the molecule is considered fixed.)

4.45 Two-Level System (Princeton)

a) There is nothing to prevent giving each atom its larger energy ε; hence, $\eta = E/N\varepsilon$ has a maximum of 1 with $E/N = \varepsilon$. Clearly, the system would not be in thermal equilibrium. To compute the problem in equilibrium, we need to determine the partition function, Z. For distinguishable non-interacting particles, the partition function factors, so for identical energy spectra

$$Z = Z_1^N = \left(1 + e^{-\varepsilon/\tau}\right)^N \tag{S.4.45.1}$$

The free energy would be

$$F = -\tau \ln Z = -N\tau \ln Z_1 = -N\tau \ln(1 + e^{-\varepsilon/\tau}) \tag{S.4.45.2}$$

The energy is then

$$E = F + \tau S = F - \tau \frac{\partial F}{\partial \tau}$$

$$= -N\tau \ln Z_1 + N\tau^2 \frac{\partial}{\partial \tau} \ln Z_1 + N\tau \ln Z_1 \tag{S.4.45.3}$$

$$= N\tau^2 \frac{\partial}{\partial \tau} \ln Z_1$$

or

$$\frac{E}{N} = \tau^2 \frac{\partial}{\partial \tau} \ln Z_1 = \tau^2 \frac{\partial}{\partial \tau} \ln \left(1 + e^{-\varepsilon/\tau}\right) \tag{S.4.45.4}$$

$$= \frac{\varepsilon e^{-\varepsilon/\tau}}{1 + e^{-\varepsilon/\tau}} = \varepsilon \frac{x}{1+x} = \varepsilon f(x)$$

where $x = e^{-\varepsilon/\tau}$. Obviously, since both ε and τ are positive, x cannot be larger than 1. On the other hand, $x/(1+x)$ is a monotonic function which goes to $1/2$ when τ goes to infinity; hence, $\max\{E/N\varepsilon\} = f(1) = 1/2$ at $\tau \to \infty$.

b) The entropy S may be found from (S.4.45.2)–(S.4.45.4):

$$S = \frac{E - F}{\tau} = \frac{E}{\tau} - \frac{F}{\tau} = \frac{N}{\tau}\frac{\varepsilon x}{1 + x} + N \ln(1 + x) \qquad \text{(S.4.45.5)}$$

The entropy per particle, $s = S/N$, is given by

$$s = \frac{\varepsilon}{\tau}\frac{x}{1 + x} + \ln(1 + x) \qquad \text{(S.4.45.6)}$$

Writing

$$\eta \equiv \frac{x}{1 + x} \qquad x = \frac{\eta}{1 - \eta} \qquad 1 + x = \frac{1}{1 - \eta}$$

$$s = \frac{\varepsilon}{\tau}\eta - \ln(1 - \eta) \qquad \text{(S.4.45.7)}$$

$$= \eta[\ln(1 - \eta) - \ln \eta] - \ln(1 - \eta) = -[\eta \ln \eta + (1 - \eta) \ln(1 - \eta)]$$

We can check that

$$s \to \begin{cases} 0 & \text{as } \tau \to 0, \ \eta \to 0 \\ \ln 2 & \text{as } \tau \to \infty, \ \eta \to 1/2 \end{cases} \qquad \text{(S.4.45.8)}$$

as it should.

4.46 Zipper (Boston)

a) A partition function may be written as

$$Z = \sum_{n=0}^{N} e^{-n\varepsilon/\tau} \qquad \text{(S.4.46.1)}$$

where we have used the fact that a state with n open links has an energy $\varepsilon_n = n\varepsilon$. So

$$Z = \sum_{n=0}^{N} \eta^n = \frac{1 - \eta^{N+1}}{1 - \eta} \qquad \text{(S.4.46.2)}$$

where $\eta \equiv e^{-\varepsilon/\tau}$.

b) The average number of open links $\langle n \rangle$ is given by

$$\langle n \rangle = \frac{\sum\limits_{n=0}^{N} n e^{-n\varepsilon/\tau}}{Z} = \frac{\sum\limits_{n=0}^{N} n \eta^n}{\sum\limits_{n=0}^{N} \eta^n}$$

$$= \eta \frac{\partial}{\partial \eta} \ln Z = \eta \frac{\partial}{\partial \eta} \ln \left(\frac{1 - \eta^{N+1}}{1 - \eta} \right) \qquad (\text{S.4.46.3})$$

$$= \frac{\eta}{1 - \eta} - \frac{(N+1)\eta^{N+1}}{1 - \eta^{N+1}}$$

If $\tau \ll \varepsilon$ then $\eta \ll 1$ and

$$\langle n \rangle \approx \eta = e^{-\varepsilon/\tau}$$

which does not depend on N. It is also zipped up tight!

4.47 Hanging Chain (Boston)

a) Let the number of links with major axis vertical be n; the number of horizontal major axis links will then be $N - n$. The total length of the chain is then

$$L(n) = (l + a)\, n + (l - a)\, (N - n) \qquad (\text{S.4.47.1})$$

The energy of the system is also a function of n since

$$E(n) = -MgL(n) = -Mg(l + a)n - Mg(l - a)(N - n) \qquad (\text{S.4.47.2})$$

$$= -E_1 n - E_2(N - n)$$

where we let $E_{1,2} = Mg(l \pm a)$. The partition function

$$Z = \sum_n g_n e^{-E_n/\tau} = \sum_{n=0}^{N} \frac{N!}{n!(N-n)!} e^{E_1 n/\tau + E_2(N-n)/\tau} \qquad (\text{S.4.47.3})$$

$$= \left(e^{E_1/\tau} + e^{E_2/\tau} \right)^N$$

where $g_n = N!/\left[n!(N - n)! \right]$ is the number of possible configurations with n major axis vertical links.

b) The average energy can be found from (S.4.47.3):

$$\langle E \rangle = \tau^2 \frac{\partial \ln Z}{\partial \tau} = -\frac{\partial}{\partial \beta} (\ln Z) = -N \frac{\partial}{\partial \beta} \ln \left(e^{\beta E_1} + e^{\beta E_2} \right) \qquad (\text{S.4.47.4})$$

where $\beta = 1/\tau$. Therefore,

$$\langle E \rangle = -N \frac{E_1 e^{\beta E_1} + E_2 e^{\beta E_2}}{e^{\beta E_1} + e^{\beta E_2}}$$

$$= -NMg \frac{(l+a)e^{\beta Mg(l+a)} + (l-a)e^{\beta Mg(l-a)}}{e^{\beta Mg(l+a)} + e^{\beta Mg(l-a)}} \quad \text{(S.4.47.5)}$$

$$= -NMg \frac{l\left(e^{\beta Mga} + e^{-\beta Mga}\right) + a\left(e^{\beta Mga} - e^{-\beta Mga}\right)}{e^{\beta Mga} + e^{-\beta Mga}}$$

$$= -NMg\left[l + a\tanh(\beta Mga)\right]$$

The average length is

$$\langle L \rangle = -\frac{\langle E \rangle}{Mg} \quad \text{(S.4.47.6)}$$

We can check that, if $M = 0$,

$$\langle L \rangle = Nl = L_0$$

At $\tau \to 0$, $\langle L \rangle \to N(l+a)$ (lowest energy state). At $\tau \to \infty$, $\langle L \rangle \to Nl + N\beta Mga^2 = L_0 + N\beta Mga^2$.

4.48 Molecular Chain (MIT, Princeton, Colorado)

a) Consider one link of the chain in its two configurations: α and β. The energy of the link is

$$E_\alpha = \varepsilon_\alpha - fa \qquad E_\beta = \varepsilon_\beta - fb \quad \text{(S.4.48.1)}$$

The partition function for the entire chain is given by

$$Z = \left(\sum_{\alpha,\beta} e^{-E_{\alpha,\beta}}\right)^N = \left(e^{(fa-\varepsilon_\alpha)/\tau} + e^{(fb-\varepsilon_\beta)/\tau}\right)^N \quad \text{(S.4.48.2)}$$

b) The average length of the chain may be found from the partition function:

$$\langle L \rangle = \tau \left(\frac{\partial \ln Z}{\partial f}\right)_{\tau,N} = \frac{N\left(ae^{(fa-\varepsilon_\alpha)/\tau} + be^{(fb-\varepsilon_\beta)/\tau}\right)}{e^{(fa-\varepsilon_\alpha)/\tau} + e^{(fb-\varepsilon_\beta)/\tau}} \quad \text{(S.4.48.3)}$$

c) If $f = 0$, (S.4.48.3) becomes

$$\langle L \rangle = N \frac{ae^{-\varepsilon_\alpha/\tau} + be^{-\varepsilon_\beta/\tau}}{e^{-\varepsilon_\alpha/\tau} + e^{-\varepsilon_\beta/\tau}} \quad \text{(S.4.48.4)}$$

$$= N \frac{a + be^{(\varepsilon_\alpha-\varepsilon_\beta)/\tau}}{1 + e^{(\varepsilon_\alpha-\varepsilon_\beta)/\tau}}$$

If $\varepsilon_{\alpha,\beta} \ll \tau$, high temperature,

$$\langle L \rangle \approx N \frac{a+b}{2} \tag{S.4.48.5}$$

If $\varepsilon_{\alpha,\beta} \gg \tau$,

$$\langle L \rangle \approx N \left(ae^{-\delta} + b \right) \tag{S.4.48.6}$$

where we let $\delta = (\varepsilon_\alpha - \varepsilon_\beta)/\tau$. The changeover temperature is obviously $\varepsilon_\alpha - \varepsilon_\beta$.

d) From (S.4.48.3),

$$\langle L \rangle = N \frac{ae^{f a/\tau} + be^{f b/\tau + \delta}}{e^{f a/\tau} + e^{f b/\tau + \delta}} \tag{S.4.48.7}$$

$$= N \frac{a + be^{f(b-a)/\tau + \delta}}{1 + e^{f(b-a)/\tau + \delta}}$$

At small f, (S.4.48.7) becomes

$$\langle L \rangle \approx N \frac{a + be^{\delta} + bf(b-a)/\tau \cdot e^{\delta}}{1 + e^{\delta} + f(b-a)/\tau \cdot e^{\delta}}$$

$$\approx N \left[\frac{a + be^{\delta}}{1 + e^{\delta}} + \frac{bf(b-a)/\tau \cdot e^{\delta}}{1 + e^{\delta}} - \frac{\left(a + be^{\delta} \right) f(b-a)/\tau \cdot e^{\delta}}{\left(1 + e^{\delta} \right)^2} \right]$$

$$= N \frac{a + be^{\delta}}{1 + e^{\delta}} + \frac{Nfe^{\delta}(b-a)}{\tau \left(1 + e^{\delta} \right)} \left[b - \frac{a + be^{\delta}}{1 + e^{\delta}} \right] \tag{S.4.48.8}$$

$$= N \frac{a + be^{\delta}}{1 + e^{\delta}} + \frac{Nfe^{\delta}(b-a)^2}{\tau \left(1 + e^{\delta} \right)^2}$$

Therefore,

$$\left. \frac{\partial \langle L \rangle}{\partial f} \right|_{f=0} = \frac{Ne^{\delta}}{\tau} \left(\frac{b-a}{1 + e^{\delta}} \right)^2 > 0 \tag{S.4.48.9}$$

as it should, since (for the specified direction of the tensile force f) it corresponds to a thermodynamic inequality for a system at equilibrium:

$$-\left(\frac{\partial V}{\partial P} \right) > 0$$

Nonideal Gas

4.49   Heat Capacities (Princeton)

From the definition of C_P for a gas,

$$C_P = \tau \left(\frac{\partial S}{\partial \tau} \right)_P \qquad \text{(S.4.49.1)}$$

Since we are interested in a relation between C_P and C_V, it is useful to transform to other variables than in (S.4.49.1), namely τ, V instead of τ, P. We will use the Jacobian transformation (see Landau and Lifshitz, *Statistical Physics*, Sect. 16):

$$
\begin{aligned}
C_P &= \tau \left(\frac{\partial S}{\partial \tau} \right)_P = \tau \frac{\partial (S,P)}{\partial (\tau,P)} = \tau \frac{\partial (S,P)/\partial (\tau,V)}{\partial (\tau,P)/\partial (\tau,V)} \\
&= \tau \frac{(\partial S/\partial \tau)_V (\partial P/\partial V)_\tau - (\partial S/\partial V)_\tau (\partial P/\partial \tau)_V}{(\partial P/\partial V)_\tau} \qquad \text{(S.4.49.2)} \\
&= \tau \left(\frac{\partial S}{\partial \tau} \right)_V - \tau \frac{(\partial S/\partial V)_\tau (\partial P/\partial \tau)_V}{(\partial P/\partial V)_\tau} \\
&= C_V - \tau \frac{(\partial S/\partial V)_\tau (\partial P/\partial \tau)_V}{(\partial P/\partial V)_\tau}
\end{aligned}
$$

A useful identity is obtained from

$$S = - \left(\frac{\partial F}{\partial \tau} \right)_V \qquad \text{(S.4.49.3)}$$

$$\left(\frac{\partial S}{\partial V} \right)_\tau = - \frac{\partial^2 F}{\partial \tau \partial V} = \left(\frac{\partial P}{\partial \tau} \right)_V \qquad \text{(S.4.49.4)}$$

So

$$C_P - C_V = -\tau \frac{(\partial P/\partial \tau)_V^2}{(\partial P/\partial V)_\tau} \qquad \text{(S.4.49.5)}$$

Since

$$\left(\frac{\partial P}{\partial \tau} \right)_V^2 > 0$$

and

$$\left(\frac{\partial P}{\partial V} \right)_\tau < 0 \qquad \tau < \tau_{\text{cr}}$$

$$C_P - C_V > 0$$

$$C_P > C_V$$

b) Let us write the van der Waals equation for one mole of the gas in the form

$$\left(P + \frac{N_A^2 a}{V^2}\right)(V - N_A b) = N_A \tau \qquad \text{(S.4.49.6)}$$

from which we obtain

$$P = \frac{N_A \tau}{V - N_A b} - \frac{N_A^2 a}{V^2} \qquad \text{(S.4.49.7)}$$

Substituting for P in (S.4.49.5) yields

$$\left(\frac{\partial P}{\partial \tau}\right)_V^2 = \left(\frac{N_A}{V - N_A b}\right)^2$$

$$\left(\frac{\partial P}{\partial V}\right)_\tau = -\frac{N_A \tau}{(V - N_A b)^2} + \frac{2 N_A^2 a}{V^3}$$

$$C_P - C_V = \tau \frac{(N_A/V - N_A b)^2}{\left[N_A \tau/(V - N_A b)^2\right] - (2 N_A^2 a/V^3)}$$

$$= \frac{N_A}{1 - (2 N_A a/\tau V^3)(V - N_A b)^2}$$

We can see that $C_P - C_V = N_A$ (in regular units $C_P - C_V = R$) for an ideal gas where $a = b = 0$.

4.50 Return of Heat Capacities (Michigan)

a) We will again use the Jacobian transformation to find $c_p - c_v$ as a function of τ, P.

$$c_v = \tau \left(\frac{\partial s}{\partial \tau}\right)_v = \tau \frac{\partial(s, v)}{\partial(\tau, v)} = \frac{\tau \partial(s, v)/\partial(\tau, P)}{\partial(\tau, v)/\partial(\tau, P)}$$

$$= \tau \frac{(\partial s/\partial \tau)_P (\partial v/\partial P)_\tau - (\partial s/\partial P)_\tau (\partial v/\partial \tau)_P}{(\partial v/\partial P)_\tau} \qquad \text{(S.4.50.1)}$$

$$= c_p - \tau \frac{(\partial s/\partial P)_\tau (\partial v/\partial \tau)_P}{(\partial v/\partial p)_\tau} = c_p + \tau \frac{(\partial v/\partial \tau)_P^2}{(\partial v/\partial P)_\tau}$$

where we used

$$\left(\frac{\partial s}{\partial P}\right)_\tau = -\frac{\partial}{\partial P}\left(\frac{\partial G}{\partial \tau}\right)_P = -\frac{\partial}{\partial \tau}\left(\frac{\partial G}{\partial P}\right)_\tau = -\left(\frac{\partial v}{\partial \tau}\right)_P$$

So, we obtain

$$c_p - c_v = -\tau\frac{(\partial v/\partial \tau)_P^2}{(\partial v/\partial P)_\tau} \qquad (S.4.50.2)$$

Substituting $v(\tau, P)$ into (S.4.50.2) yields

$$c_p - c_v = -\tau\frac{v^2/\tau_1^2}{-v/P_1} = \frac{P_1 v_1}{\tau_1^2}\tau\exp\left(\frac{\tau}{\tau_1} - \frac{P}{P_1}\right) \qquad (S.4.50.3)$$

b,c) We cannot determine the temperature dependence of c_p or c_v, but we can find $c_p(P)$ and $c_v(v)$, as follows:

$$\left(\frac{\partial c_p}{\partial P}\right)_\tau = \tau\frac{\partial^2 s}{\partial P \partial \tau} = -\tau\frac{\partial^3 G}{\partial P \partial \tau^2} = -\tau\left(\frac{\partial^2 v}{\partial \tau^2}\right)_P \qquad (S.4.50.4)$$

Similarly,

$$\left(\frac{\partial c_v}{\partial v}\right)_\tau = \tau\frac{\partial^2 s}{\partial v \partial \tau} = -\tau\frac{\partial^3 F}{\partial v \partial \tau^2} = -\tau\frac{\partial^2}{\partial \tau^2}\left(\frac{\partial F}{\partial v}\right)_\tau \qquad (S.4.50.5)$$

$$= \tau\left(\frac{\partial^2 P}{\partial \tau^2}\right)_v$$

where F is the Helmholtz free energy, and we used

$$-\left(\frac{\partial F}{\partial v}\right)_\tau = P$$

From (S.4.50.4) and the equation of state, we have

$$\left(\frac{\partial c_p}{\partial P}\right)_\tau = -\frac{\tau v}{\tau_1^2} \qquad (S.4.50.6)$$

and from (S.4.50.5),

$$\left(\frac{\partial c_v}{\partial v}\right)_\tau = 0 \qquad (S.4.50.7)$$

(since $v = \text{const}$ implies $\tau/\tau_1 = P/P_1$). Integrating (S.4.50.6) and (S.4.50.7), we obtain

$$c_p = -\frac{\tau}{\tau_1^2} \int v_1 \exp\left(\frac{\tau}{\tau_1} - \frac{P}{P_1}\right) dP = \frac{\tau P_1 v}{\tau_1^2} + f_1(\tau) \quad \text{(S.4.50.8)}$$

$$= \frac{P_1}{\tau_1^2} \tau v + f_1(\tau) = \frac{P_1 v_1}{\tau_1^2} \tau \exp\left(\frac{\tau}{\tau_1} - \frac{P}{P_1}\right) + f_1(\tau)$$

and

$$c_v = f_2(\tau) \quad \text{(S.4.50.9)}$$

where f_1 and f_2 are some functions of temperature. Since we know $c_p - c_v$ from (a), we infer that $f_1 = f_2 \equiv f$, and finally

$$c_p = \frac{P_1 v_1}{\tau_1^2} \tau \exp\left(\frac{\tau}{\tau_1} - \frac{P}{P_1}\right) + f(\tau) \quad \text{(S.4.50.10)}$$

$$c_v = f(\tau)$$

4.51 Nonideal Gas Expansion (Michigan State)

a) The work done in the expansion

$$W = -\int_{V_0}^{V_1} P \, dV = -\int_{V_0}^{2V_0} P \, dV \quad \text{(S.4.51.1)}$$

$$= -\int_{V_0}^{2V_0} \left(\frac{\tau}{V} + \frac{B(\tau)}{V^2}\right) dV = -\tau \ln 2 - \frac{B(\tau)}{2V_0}$$

b) To find the heat absorbed in the expansion use the Maxwell relations given in the problem:

$$\left(\frac{\partial S}{\partial V}\right)_\tau = \left(\frac{\partial P}{\partial \tau}\right)_V = \frac{1}{V} + \frac{B'(\tau)}{V^2} \quad \text{(S.4.51.2)}$$

where the prime indicates the derivative with respect to τ. Integrating (S.4.51.2), we obtain

$$S = \ln V - \frac{B'(\tau)}{V} + f(\tau) \quad \text{(S.4.51.3)}$$

where $f(\tau)$ is some function of τ. The heat absorbed in the expansion

$$\Delta Q = \tau \Delta S = \tau \left(S_1 - S_0\right) = \tau \ln 2 + \tau \frac{B'(\tau)}{2V_0} \quad \text{(S.4.51.4)}$$

4.52   van der Waals (MIT)

a) The heat capacity C_V is defined as

$$C_V \equiv \left.\frac{\partial \varepsilon}{\partial \tau}\right|_V = \tau \left.\frac{\partial S}{\partial \tau}\right|_V \qquad \text{(S.4.52.1)}$$

By using the Maxwell relation

$$\left.\frac{\partial S}{\partial V}\right|_\tau = \left.\frac{\partial P}{\partial \tau}\right|_V \qquad \text{(S.4.52.2)}$$

we may write

$$\left.\frac{\partial C_V}{\partial V}\right|_\tau = \tau \frac{\partial^2 S}{\partial V \partial \tau} = \tau \frac{\partial^2 S}{\partial \tau \partial V} = \tau \left.\frac{\partial^2 P}{\partial \tau^2}\right|_V \qquad \text{(S.4.52.3)}$$

Substituting the van der Waals equation of state

$$P = \frac{N\tau}{V - Nb} - \frac{N^2 a}{V^2} \qquad \text{(S.4.52.4)}$$

into (S.4.52.3) gives

$$\left.\frac{\partial C_V}{\partial V}\right|_\tau = \tau \frac{\partial}{\partial \tau} \left.\left(\frac{N}{V - Nb}\right)\right|_V = 0 \qquad \text{(S.4.52.5)}$$

b) The entropy $S(\tau, V)$ may be computed from

$$S(\tau, V) = \int \left.\frac{\partial S}{\partial \tau}\right|_V d\tau + \int \left.\frac{\partial S}{\partial V}\right|_\tau dV + \text{const} \qquad \text{(S.4.52.6)}$$

We were given that $C_V = 3N/2$ at $V \to \infty$; therefore, again using (S.4.52.2) and (S.4.52.4), we obtain

$$S(\tau, V) = \int \frac{3}{2}\frac{N}{\tau} d\tau + \int \frac{N}{V - Nb} dV + \text{const}$$

$$= \frac{3}{2} N \ln \tau + N \ln(V - Nb) + \text{const} \qquad \text{(S.4.52.7)}$$

$$= N \ln \left[(V - Nb)\, \tau^{3/2}\right]$$

c) The internal energy $\varepsilon(\tau, V)$ may be calculated in the same way from

$$\varepsilon(\tau, V) = \int \left.\frac{\partial \varepsilon}{\partial \tau}\right|_V d\tau + \int \left.\frac{\partial \varepsilon}{\partial V}\right|_\tau dV + \text{const} \qquad \text{(S.4.52.8)}$$

Now, from $d\varepsilon = \tau \, dS - P \, dV$, we have

$$\left.\frac{\partial \varepsilon}{\partial V}\right|_{\tau} = \tau \left.\frac{\partial S}{\partial V}\right|_{\tau} - P \tag{S.4.52.9}$$

and using (S.4.52.4) and (S.4.52.7), we get

$$\left.\frac{\partial \varepsilon}{\partial V}\right|_{\tau} = \frac{N\tau}{V - Nb} - \frac{N\tau}{V - Nb} + \frac{N^2 a}{V^2} = \frac{N^2 a}{V^2} \tag{S.4.52.10}$$

So, (S.4.52.8) becomes

$$\varepsilon(\tau, V) = \int \frac{3}{2} N \, d\tau + \int \frac{N^2 a}{V^2} \, dV + \text{const} \tag{S.4.52.11}$$

$$= \frac{3}{2} N\tau - \frac{N^2 a}{V} + \text{const}$$

d) During adiabatic compression, the entropy is constant, so from (S.4.52.7)

$$(V - Nb)\tau^{3/2} = \text{const} \tag{S.4.52.12}$$

and we have

$$\tau_2 = \left(\frac{V_1 - Nb}{V_2 - Nb}\right)^{2/3} \tau_1 \tag{S.4.52.13}$$

e) The work done is given by the change in internal energy ε since the entropy is constant:

$$\int d\varepsilon \equiv \int \tau \, dS - \int P \, dV = -\int P \, dV = W \tag{S.4.52.14}$$

From (S.4.52.11), we arrive at

$$W = \frac{3}{2} N (\tau_2 - \tau_1) - N^2 a \left(\frac{1}{V_2} - \frac{1}{V_1}\right) \tag{S.4.52.15}$$

$$= \frac{3}{2} N \left[\left(\frac{V_1 - Nb}{V_2 - Nb}\right)^{2/3} - 1\right] \tau_1 - N^2 a \left(\frac{V_1 - V_2}{V_1 V_2}\right)$$

4.53   Critical Parameters (Stony Brook)

At the critical point we have the conditions

$$\left(\frac{\partial P}{\partial V}\right)_{\tau} = 0 \tag{S.4.53.1}$$

$$\left(\frac{\partial^2 P}{\partial V^2}\right)_{\tau} = 0 \tag{S.4.53.2}$$

Substituting the Dietrici equation into (S.4.53.1) gives

$$\left(\frac{\partial P}{\partial V}\right)_\tau = \frac{nN_A\tau}{V-nB}e^{-nA/N_A\tau V}\left(-\frac{1}{V-nB}+\frac{nA}{N_A\tau V^2}\right) = 0 \quad \text{(S.4.53.3)}$$

so

$$V_c - nB = \frac{N_A T_c V_c^2}{nA} \quad \text{(S.4.53.4)}$$

Using the second criterion (S.4.53.2) gives

$$\left(\frac{\partial^2 P}{\partial V^2}\right)_\tau = \frac{nN_A\tau}{V-nB}e^{-nA/N_A\tau V}(D_1 + D_2) = 0 \quad \text{(S.4.53.5)}$$

$$D_1 + D_2 = 0 \quad \text{(S.4.53.6)}$$

where

$$D_1 = \left(-\frac{1}{V_c - nB} + \frac{nA}{N_A T_c V_c^2}\right)^2 = 0$$

by (S.4.53.3), so

$$D_2 = \frac{1}{(V_c - nB)^2} - \frac{2nA}{N_A T_c V_c^3} = 0 \quad \text{(S.4.53.7)}$$

by (S.4.53.6). (S.4.53.7) then yields

$$(V_c - nB)^2 = \frac{N_A T_c V_c^3}{2nA}$$

which combined with (S.4.53.4) gives

$$\frac{(N_A T_c)^2 V_c^4}{n^2 A^2} = \frac{N_A T_c V_c^3}{2nA} \quad \text{(S.4.53.8)}$$

$$N_A T_c V_c = \frac{nA}{2} \quad \text{(S.4.53.9)}$$

Substituting this result in the RHS of (S.4.53.4) finally yields

$$V_c = 2nB \quad \text{(S.4.53.10)}$$

$$T_c = \frac{A}{4N_A B} \quad \text{(S.4.53.11)}$$

Rearranging the original equation of state gives

$$P_c = \frac{nN_A T_c}{V_c - nB}e^{-nA/N_A\tau V} = \frac{A}{4B^2}e^{-2} \quad \text{(S.4.53.12)}$$

Mixtures and Phase Separation

4.54   Entropy of Mixing (Michigan, MIT)

a) The energy of the mixture of ideal gases is the sum of energies of the two gases (since we assume no interaction between them). Therefore the temperature will not change upon mixing. The pressure also remains unchanged. The entropy of the mixture is simply the sum of the entropies of each gas (as if there is no other gas) in the total volume. We may write the total entropy S (see Problem 4.38) as

$$S = N_1 \ln \frac{eV}{N_1} + N_2 \ln \frac{eV}{N_2} - N_1 f_1'(\tau) - N_2 f_2'(\tau) \qquad (S.4.54.1)$$

where N_1 and N_2 are the number of molecules of each gas in the mixture. V is the total volume of the mixture ($V = V_1 + V_2$). The entropy of the gases before they are allowed to mix is

$$S_1 + S_2 = N_1 \ln \frac{eV_1}{N_1} + N_2 \ln \frac{eV_2}{N_2} - N_1 f_1'(\tau) - N_2 f_2'(\tau) \qquad (S.4.54.2)$$

Therefore, the change in entropy, ΔS, is given by

$$\Delta S = S - S_1 - S_2 = N_1 \ln \frac{V}{V_1} + N_2 \ln \frac{V}{V_2} \qquad (S.4.54.3)$$

In our case $V_1 = V_2 = V/2$, and

$$N_1 = N_2 = \frac{PV_1}{\tau} = \frac{PV}{2\tau}$$

So, (S.4.54.3) becomes

$$\Delta S = \frac{PV}{\tau} \ln 2 \qquad (S.4.54.4)$$

In conventional units we find

$$\Delta S = \frac{PV}{T} \ln 2 \qquad (S.4.54.5)$$

$$\approx \frac{10^5 \cdot 2 \cdot 10^{-3}}{293} \cdot \ln 2 \approx 0.47 \text{ J/K}$$

The entropy increased as it should because the process is clearly irreversible.

b) If the gases are the same, then the entropy after mixing is given by

$$S = (N_1 + N_2) \ln \frac{V_1 + V_2}{N_1 + N_2} - (N_1 + N_2) f'(\tau) \qquad (S.4.54.6)$$

$$= 2N \ln \frac{V}{2N} - 2N f'(\tau)$$

and so $\Delta S = 0$. In the case of identical gases, reversing the process only requires the reinsertion of the partition, whereas in the case where two dissimilar gases are mixed, some additional work has to be done to separate them again.

c) The same arguments as in (a) apply for a mixture of two isotopes, ^{16}O and ^{18}O. The Gibbs free energy can be written in the form

$$G = N_1 \tau \ln \frac{N_1}{N} + N_2 \tau \ln \frac{N_2}{N} + N_1 \mu_{01} + N_2 \mu_{02} \qquad (S.4.54.7)$$

where μ_{01} and μ_{02} are the chemical potentials of pure isotopes. Therefore, the potential (S.4.54.7) has the same form as in the mixture of two different gases, and there is no correction to the result of (a). This is true as long as (S.4.54.7) can be written in this form, and it holds even after including quantum corrections to the order of h^2 (see, for further details, Landau and Lifshitz, *Statistical Physics*, Sect. 94).

4.55 Leaky Balloon (Moscow Phys-Tech)

Let us consider the bag as part of a very large system (the atmosphere) which initially has N molecules of air, which we consider as one gas, and N_1 molecules of helium. The bag has volume V_0, and the number of helium molecules is N_0. Using (S.4.38.7) from Problem 4.38 and omitting all the temperature-dependent terms, we may write for the initial entropy of the system

$$S = N_0 \ln \frac{eV_0}{N_0} + N_1 \ln \frac{eV_1}{N_1} + N \ln \frac{eV}{N} \qquad (S.4.55.1)$$

When the helium has diffused out, we have

$$\tilde{S} = (N_0 + N_1) \ln \frac{e(V_0 + V_1)}{N_0 + N_1} + N \ln \frac{e(V_0 + V)}{N} \qquad (S.4.55.2)$$

We wish to find $\Delta S = \tilde{S} - S$ in the limit where $V \gg V_0$. Then

$$\lim_{N,N_1,V \to \infty} \tilde{S} = (N_0 + N_1) \ln \frac{eV_1}{N_1} + N \ln \frac{eV}{N} \qquad (S.4.55.3)$$

We then obtain

$$\lim_{N,N_1,V \to \infty} \Delta S = N_0 \ln \frac{V_1/N_1}{V_0/N_0} = N_0 \ln \frac{n_0}{n_1} \qquad (S.4.55.4)$$

where $n_0/n_1 \approx 1/(5 \cdot 10^{-6})$ is the concentration ratio of helium molecules in the bag to their concentration in the air. In regular units

$$\Delta S = N_0 k_B \ln\left(\frac{n_0}{n_1}\right) = \frac{P_0 V_0}{T_0} \ln\left(\frac{n_0}{n_1}\right) \qquad (S.4.55.5)$$

Substituting the standard pressure and temperature into (S.4.55.5) gives

$$\Delta S = \frac{10^5 \cdot 40 \cdot 10^{-3}}{293} \cdot \ln\frac{1}{5 \cdot 10^{-6}} \approx 167 \text{ J/K} \qquad (S.4.55.6)$$

The minimum work necessary to separate the helium at constant temperature is (see Landau and Lifshitz, *Statistical Physics*, Sect. 20)

$$R_{\min} = -T_0 \, \Delta S_1 = T_0 \, \Delta S \qquad (S.4.55.7)$$

$\Delta S_1 = -\Delta S$, since after we separate the helium molecules from the rest of the air, the total entropy of that system would decrease. So

$$R_{\min} \approx 293 \cdot 167 \approx 50 \text{ kJ}$$

4.56 Osmotic Pressure (MIT)

a) The free energy for a one-component ideal gas is derived in Problem 4.38:

$$F = -N\tau \ln\frac{eV}{N} + N f(\tau) \qquad (S.4.56.1)$$

The Gibbs free energy

$$G = F + PV = -N\tau \ln\frac{eV}{N} + N f(\tau) + PV \qquad (S.4.56.2)$$

But $G = G(\tau, V, N)$, so (S.4.56.2) must be transformed:

$$G = -N\tau \ln\frac{\tau}{P} + N f(\tau)$$

$$= N\tau \ln P + N \left[f(\tau) - \tau \ln \tau\right] \qquad (S.4.56.3)$$

$$= N\tau \ln P + N\chi(\tau)$$

If we have a mixture of two types of molecules with N_0 and N_1 particles each, we find for the thermodynamic potential of the mixture:

$$\tilde{F} = -N_0\tau \ln\frac{eV}{N_0} - N_1\tau \ln\frac{eV}{N_1} + N_0 f_0(\tau) + N_1 f_1(\tau) \qquad (S.4.56.4)$$

Therefore $\tilde{F} = F_0 + F_1$. The Gibbs potential of the mixture

$$\tilde{G} = N_0\tau \ln P_0 + N_0\chi_0(\tau) + N_1\tau \ln P_1 + N_1\chi_1(\tau) \qquad (S.4.56.5)$$

where P_0, P_1 are partial pressures ($P = P_1 + P_2$) corresponding to particles A and B, respectively. So,

$$\tilde{G} = N_0\tau \ln \left(P\frac{P_0}{P} \right) + N_0\chi_0(\tau) + N_1\tau \ln \left(P\frac{P_1}{P} \right) + N_1\chi_1(\tau)$$

$$= N_0\tau \ln P + N_0\chi_0(\tau) + N_1\tau \ln P \qquad (S.4.56.6)$$

$$+ N_1\chi_1(\tau) + N_0\tau \ln \frac{N_0}{N} + N_1\tau \ln \frac{N_1}{N}$$

It can be seen that

$$\tilde{G} \neq G_0 + G_1$$

namely

$$\tilde{G} = G_0 + G_1 + N_0\tau \ln \frac{N_0}{N} + N_1\tau \ln \frac{N_1}{N} \qquad (S.4.56.7)$$

where $N = N_0 + N_1$ (see also Problem 4.54).

b) To derive the pressure difference, we notice that for the system with a semipermeable membrane, only the chemical potentials of the *solvent* are equal, whereas the chemical potentials of the solute do not have to be (since they cannot penetrate through the membrane). We will write first the Gibbs free energy on the left and right of the membrane, $\tilde{G}_L$ and $\tilde{G}_R$, respectively. $\tilde{G}_L$ will be defined by (S.4.56.6), with $P \to P_L$, whereas

$$\tilde{G}_R = N_0\tau \ln P_R + N_0\chi_0(\tau) + N_2\tau \ln P_R + N_2\chi_2(\tau) + N_0\tau \ln \frac{N_0}{N} + N_2\tau \ln \frac{N_2}{N} \qquad (S.4.56.8)$$

The chemical potentials of the solvent are given by

$$\mu_R = \frac{\partial \tilde{G}_R}{\partial N_0} \qquad (S.4.56.9)$$

$$\mu_L = \frac{\partial \tilde{G}_L}{\partial N_0}$$

Equating $\mu_R = \mu_L$, we obtain

$$\tau \ln P_L + \chi_0(\tau) + \tau \ln \frac{N_0}{N_0 + N_1} \qquad (S.4.56.10)$$

$$= \tau \ln P_R + \chi_0(\tau) + \tau \ln \frac{N_0}{N_0 + N_2}$$

or

$$\ln \frac{P_\mathrm{L}}{P_\mathrm{R}} \approx \frac{N_1}{N_0} - \frac{N_2}{N_0} \qquad (\text{S.4.56.11})$$

where we only take into account the first-order terms in the solute. If we also assume, which is usually the case, that the osmotic pressure is also small, i.e., $P_\mathrm{L} = P_\mathrm{R} + \Delta P \approx P + \Delta P$, we obtain, from (S.4.56.11),

$$\frac{\Delta P}{P} \approx \frac{N_1 - N_2}{N_0} = C_1 - C_2 \qquad (\text{S.4.56.12})$$

where C_1 and C_2 are the concentrations of the solutes: $C_1 \equiv N_1/N_0$; $C_2 \equiv N_2/N_0$. Therefore, with the same accuracy, we arrive at the final formula:

$$\Delta P = (C_1 - C_2)\, P = (C_1 - C_2)\,\frac{N_0 \tau}{V} \qquad (\text{S.4.56.13})$$

A different derivation of this formula may be found in Landau and Lifshitz, *Statistical Physics*, Sect. 88.

4.57 Clausius–Clapeyron (Stony Brook)

a) We know that, at equilibrium, the chemical potentials of two phases should be equal:

$$\mu_1\left[P(\tau), \tau\right] = \mu_2\left[P(\tau), \tau\right] \qquad (\text{S.4.57.1})$$

Here we write $P \equiv P(\tau)$ to emphasize the fact that the pressure depends on the temperature. By taking the derivative of (S.4.57.1) with respect to temperature, we obtain

$$\left(\frac{\partial \mu_1}{\partial \tau}\right)_P + \left(\frac{\partial \mu_1}{\partial P}\right)_\tau \frac{dP}{d\tau} = \left(\frac{\partial \mu_2}{\partial \tau}\right)_P + \left(\frac{\partial \mu_2}{\partial P}\right)_\tau \frac{dP}{d\tau} \qquad (\text{S.4.57.2})$$

Taking into account that $(\partial \mu/\partial \tau)_P = -s$ and $(\partial \mu/\partial P)_\tau = v$, where s and v are the entropy and volume per particle, and substituting into (S.4.57.2), we have

$$\frac{dP}{d\tau} = \frac{s_1 - s_2}{v_1 - v_2} \qquad (\text{S.4.57.3})$$

where subscripts 1 and 2 refer to the two phases. On the other hand, $q = \tau(s_2 - s_1)$, where q is the latent heat per particle, so we can rewrite (S.4.57.3) in the form

$$\frac{dP}{d\tau} = \frac{q}{\tau\, \Delta v} \qquad (\text{S.4.57.4})$$

which is the Clausius–Clapeyron equation.

b) Consider the particular case of equilibrium between liquid and vapor. The volume v_1 of the liquid is usually much smaller than that for the vapor v_2, so we can disregard v_1 in (S.4.57.4) and write

$$\frac{dP_v}{d\tau} = \frac{q}{\tau v_v}$$

Using the ideal gas law for vapor, $v_v = \tau/P_v$, we get

$$\frac{dP_v}{d\tau} = \frac{qP_v}{\tau^2} \qquad (S.4.57.5)$$

or

$$\ln P_v = A - \frac{q}{\tau} \qquad (S.4.57.6)$$

We can see that $B = q$. Rewriting (S.4.57.6) in usual units gives

$$\ln P_v = A - \frac{q}{k_B T} = A - \frac{qN_A}{k_B N_A T} = A - \frac{L}{RT}$$

where L is the latent heat per mole, N_A is Avogadro's number, and R is the gas constant.

4.58   Phase Transition (MIT)

For a system at equilibrium with an external reservoir, the Gibbs free energy $G = F + P_0 V$ is a minimum. Any deviation from equilibrium will raise G:

$$\delta G = \delta F + P_0 \, \delta V > 0 \qquad (S.4.58.1)$$

where P_0 is the pressure of the reservoir (see Landau and Lifshitz, *Statistical Physics*, Sect. 21). Expanding δF in δV, we have

$$\delta G \approx \left(\frac{\partial F}{\partial V}\right)_\tau \delta V + \frac{1}{2!}\left(\frac{\partial^2 F}{\partial V^2}\right)_\tau \delta V^2 \qquad (S.4.58.2)$$

$$+ \frac{1}{3!}\left(\frac{\partial^3 F}{\partial V^3}\right)_\tau \delta V^3 + \frac{1}{4!}\left(\frac{\partial^4 F}{\partial V^4}\right)_\tau \delta V^4 + P_0 \, \delta V > 0$$

Since $\partial F/\partial V = -P$, we may rewrite (S.4.58.2) as

$$-P_0 \, \delta V - \frac{1}{2}\left(\frac{\partial P}{\partial V}\right)_\tau \delta V^2 - \frac{1}{3!}\left(\frac{\partial^2 P}{\partial V^2}\right)_\tau \delta V^3$$

$$- \frac{1}{4!}\left(\frac{\partial^3 P}{\partial V^3}\right)_\tau \delta V^4 + P_0 \, \delta V > 0 \qquad (S.4.58.3)$$

$$\frac{1}{2}\left(\frac{\partial P}{\partial V}\right)_\tau \delta V^2 + \frac{1}{3!}\left(\frac{\partial^2 P}{\partial V^2}\right)_\tau \delta V^3 + \frac{1}{4!}\left(\frac{\partial^3 P}{\partial V^3}\right)_\tau \delta V^4 < 0$$

At the critical point, $\partial P/\partial V = 0$, so (S.4.58.3) becomes

$$\frac{1}{3!}\left(\frac{\partial^2 P}{\partial V^2}\right)_{\tau=\tau_c} \delta V^3 + \frac{1}{4!}\left(\frac{\partial^3 P}{\partial V^3}\right)_{\tau=\tau_c} \delta V^4 < 0 \qquad \text{(S.4.58.4)}$$

For (S.4.58.4) to hold for arbitrary δV, we have

$$\left(\frac{\partial^2 P}{\partial V^2}\right)_{\tau=\tau_c} = 0 \qquad \text{(S.4.58.5)}$$

$$\left(\frac{\partial^3 P}{\partial V^3}\right)_{\tau=\tau_c} < 0$$

See Landau and Lifshitz, *Statistical Physics*, Sect. 153 for further discussion.

4.59 Hydrogen Sublimation in Intergalactic Space (Princeton)

Using the Clausius–Clapeyron equation derived in Problem 4.57, we can estimate the vapor pressure P at $T = 3$ K. Namely,

$$P \approx P_t e^{-(\mu L/R)(1/T - 1/T_t)} \qquad \text{(S.4.59.1)}$$

where P_t is the pressure at the triple point and R is the gas constant. Here we disregard the volume per molecule of solid hydrogen compared to the one for its vapor. This formula is written under the assumption that the latent heat does not depend on the temperature, but for an order-of-magnitude estimate this is good enough.

Consider solid hydrogen at equilibrium with its vapor. Then the number of particles evaporating from the surface equals the number of particles striking the surface and sticking to it from the vapor. The rate $\tilde{R}$ of the particles striking the surface is given by

$$\tilde{R} = \frac{n\langle v\rangle}{4}\eta \qquad \text{(S.4.59.2)}$$

where ρ is the number density, $\langle v\rangle$ is the average speed, and η is a sticking coefficient, which for this estimate we take equal to 1. Here we used the result of Problem 4.14, where we calculated the rate of particles striking the surface. Now if the density is not too high, the number of particles leaving the surface does not depend on whether there is vapor outside, so this would be the sublimation rate. Taking the average velocity from Problem 4.13, we get

$$\langle v \rangle = \sqrt{\frac{8k_B T}{\pi m}}$$

where m is the mass of a hydrogen molecule, and substituting $n = P/k_B T$, we may rewrite (S.4.59.2) as

$$\tilde{R} = \frac{P}{4k_B T}\sqrt{\frac{8k_B T}{\pi m}} = \frac{P}{\sqrt{2\pi k_B m T}}$$

$$= \frac{P_t e^{-(\mu L/R)(1/T - 1/T_t)}}{\sqrt{2\pi k_B m T}} = \frac{P_t N_A e^{-(\mu L/R)(1/T - 1/T_t)}}{\sqrt{2\pi R\mu T}} \qquad (S.4.59.3)$$

$$= \frac{(50/760) \cdot 10^5 \cdot 6 \cdot 10^{23} \cdot e^{-(2\cdot 450/8.3)(1/3 - 1/15)}}{\sqrt{2\pi \cdot 8.3 \cdot 0.002 \cdot 3}} \approx 2 \cdot 10^{15}\ \mathrm{s}^{-1}\mathrm{m}^{-2}$$

4.60 Gas Mixture Condensation (Moscow Phys-Tech)

Consider three parts of the plot (see Figure S.4.60). At $V > V_2$ there is a regular gas mixture (no condensation). At $V_1 < V < V_2$, one of the gases is condensing; let us assume for now it is oxygen (it happens to be true). At $V < V_1$, they are both condensing, and there is no pressure change. Let us write

$$P_1 = P_O + P_A \qquad (S.4.60.1)$$

$$P_2 = P_O + P_N \qquad (S.4.60.2)$$

Here P_N is the partial nitrogen pressure at (V_2, P_2), P_O is the saturation vapor pressure of oxygen, and P_A is the saturated vapor pressure of nitrogen (1 atm) at $T = 77.4$ K. Between V_2 and V_1, nitrogen is a gas, and since the temperature is constant,

$$P_N = P_A \frac{V_1}{V_2} \qquad (S.4.60.3)$$

Using (S.4.60.3) and dividing (S.4.60.1) by (S.4.60.2), we have

$$\frac{7}{4} = \frac{P_1}{P_2} = \frac{P_O + P_A}{P_O + P_A V_1/V_2} = \frac{P_O + P_A}{P_O + P_A/2} \qquad (S.4.60.4)$$

yielding

$$P_O = P_A/6 \approx 1.7 \cdot 10^4\ \mathrm{Pa}$$

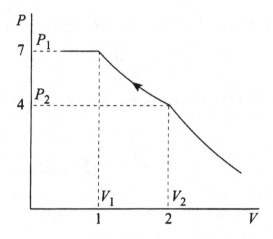

Figure **S.4.60**

Had we assumed that oxygen is condensing at (V_1, P_1) we would get $\tilde{P}_O = 6P_A$. This contradicts the fact that oxygen boils at a higher temperature. The saturated vapor pressure at $T = 77.4$ K should be less than P_A. To find the oxygen mass, we use the ideal gas law at (V_2, P_2) where the oxygen is just starting to condense (i.e., its pressure is P_O and it is all gas). So

$$P_O V_2 = \frac{m_O}{\mu_O} RT \tag{S.4.60.5}$$

where μ_O is the oxygen molar mass. For nitrogen a similar equation can be written for (V_1, P_1):

$$P_A V_1 = \frac{m_N}{\mu_N} RT \tag{S.4.60.6}$$

where μ_N is the molar mass of nitrogen. Dividing (S.4.60.5) by (S.4.60.6), we obtain

$$m_O = \frac{P_O}{P_A} \frac{\mu_O}{\mu_N} \frac{V_2}{V_1} m_N = \frac{1}{6} \frac{8}{7} 2 m_N = \frac{8}{21} m_N \approx 38 \text{ g}$$

4.61 Air Bubble Coalescence (Moscow Phys-Tech)

Writing the equilibrium conditions for the bubbles to exist, we find for the pressure inside each original bubble:

$$P_i = P_0 + \rho g h + \frac{2\sigma}{R_0} \tag{S.4.61.1}$$

where $\rho g h$ is the hydrostatic pressure (h is the height of the water). We disregard any effects due to the finite size of the bubble since they are small ($R_0 \ll h$). After merging, the pressure inside the new bubble will not change. This is due to the fact that the temperature is constant, and since the jar is closed and water is incompressible, the total volume also will not change. The new radius R_1 is given by

$$2 \cdot \frac{4}{3} \pi R_0^3 = \frac{4}{3} \pi R_1^3 \qquad \text{(S.4.61.2)}$$

$$R_1 = 2^{1/3} R_0$$

Writing (S.4.61.1) for the new bubble, we obtain

$$P_i = P_1 + \rho g h + \frac{2\sigma}{R_1} \qquad \text{(S.4.61.3)}$$

where we disregard any small change in hydrostatic pressure. From (S.4.61.1) and (S.4.61.3) we find that the change of pressure inside the jar is

$$\Delta P = P_1 - P_0 = 2\sigma \left(-\frac{1}{R_1} + \frac{1}{R_0} \right) \qquad \text{(S.4.61.4)}$$

$$= \frac{2\sigma}{R_0} \left(1 - \frac{1}{2^{1/3}} \right) \approx 0.4 \frac{\sigma}{R_0}$$

4.62 Soap Bubble Coalescence (Moscow Phys-Tech)

Assume that, during the coalescence, the total mass of air inside the bubbles and the temperature do not change. So,

$$m_0 = m_1 + m_2 \qquad \text{(S.4.62.1)}$$

where m_0, m_1, m_2 are the masses of air inside bubbles B_0, B_1, B_2, respectively. By the ideal gas law,

$$m_i = \frac{P_i V_i \mu}{RT} \qquad \text{(S.4.62.2)}$$

where m_i is the mass, P_i is the pressure, and $V_i = (4/3) \pi R_i^3$ is the volume in the ith bubble, and μ is the molar mass of the trapped air. The equilibrium condition for a bubble is

$$P_i = P_a + 2\frac{2\sigma}{R_i} \qquad \text{(S.4.62.3)}$$

The coefficient 2 in front of the second term results from the presence of two surfaces of the soap film enclosing the air (compare with Problem 4.61). From (S.4.62.2) and (S.4.62.3) we obtain

$$m_i = \left(P_a + \frac{4\sigma}{R_i} \right) \frac{4}{3}\pi R_i^3 \frac{\mu}{RT} \qquad \text{(S.4.62.4)}$$

Substituting (S.4.62.4) into (S.4.62.1), we obtain

$$\left(P_a + \frac{4\sigma}{R_0} \right) R_0^3 = \left(P_a + \frac{4\sigma}{R_1} \right) R_1^3 + \left(P_a + \frac{4\sigma}{R_2} \right) R_2^3 \qquad \text{(S.4.62.5)}$$

and so

$$\sigma = \frac{P_a}{4} \frac{R_0^3 - R_1^3 - R_2^3}{R_1^2 + R_2^2 - R_0^2} \qquad \text{(S.4.62.6)}$$

Note that if σ is very small the volume of the new bubble is close to the sum of the original volumes, whereas if it is very large the surface area of the new bubble is roughly the sum of the original surface areas.

4.63   Soap Bubbles in Equilibrium (Moscow Phys-Tech)

a) The equilibrium is unstable. It is obvious from purely mechanical considerations that if the radius of one bubble decreases and the other increases, the pressure in the first bubble (which is inversely proportional to R_0) will increase and that in the second bubble will decrease, leading to further changes in respective radii until the system becomes one bubble with radius R_1 (see Figure S.4.63). The same result can be obtained by considering the free energy of the system.

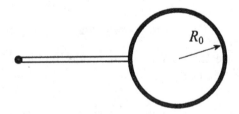

Figure S.4.63

b) The free energy of the bubble consists of two parts: a volume part, which is just the free energy of a gas (see Problem 4.38), and a surface part, which is associated with the surface tension:

$$F = F_V + F_S = -N\tau \ln \frac{eV}{N} + Nf(\tau) + \sigma A \qquad (S.4.63.1)$$

The Gibbs free energy

$$G = F + PV = N\tau \ln P + N\xi(\tau) + \sigma A \qquad (S.4.63.2)$$

The entropy change

$$\Delta S = -\frac{\partial}{\partial \tau}(G_1 - G_0)$$

where G_1 is the potential of the system with one bubble and G_0 is the potential of the initial configuration. We then find

$$\Delta S = -N \ln \frac{P_1}{P_0} - \frac{d\sigma}{d\tau}(A_1 - A_0) \qquad (S.4.63.3)$$

$$= N \ln \frac{P_0}{P_1} + \frac{q}{\tau}(A_1 - A_0)$$

where we used the fact that the number of particles did not change and q is the heat necessary to produce a unit area of the film:

$$q = -\tau \frac{d\sigma}{d\tau}$$

So

$$\Delta S = N \ln \left(\frac{P_a + 4\sigma/R_0}{P_a + 4\sigma/R_1} \right) + \frac{q}{\tau}(A_1 - A_0)$$

$$\approx N \left(\frac{4\sigma}{P_a R_0} - \frac{4\sigma}{P_a R_1} \right) + \frac{4\pi q}{\tau}(R_1^2 - 2R_0^2)$$

$$= \frac{4N\sigma}{P_a} \frac{R_1 - R_0}{R_0 R_1} + \frac{4\pi q}{\tau}(R_1^2 - 2R_0^2) \qquad (S.4.63.4)$$

$$\approx \frac{2 \cdot P_a \frac{4}{3}\pi R_0^3 \cdot 4\sigma}{P_a \tau} \frac{R_1 - R_0}{R_0 R_1} + \frac{4\pi q}{\tau}(R_1^2 - 2R_0^2)$$

$$= \frac{32\pi\sigma R_0^2(R_1 - R_0)}{3\tau R_1} + \frac{4\pi q}{\tau}(R_1^2 - 2R_0^2)$$

We can eliminate R_1 from the final result by using the following equations:

$$P_0 = P_a + 4\sigma/R_0 \qquad (S.4.63.5)$$

$$P_1 = P_a + 4\sigma/R_1 \qquad (S.4.63.6)$$

$$2P_0 R_0^3 = P_1 R_1^3 \qquad (S.4.63.7)$$

where the last equation represents the ideal gas law at constant temperature. This yields the equation

$$\left(1 + \frac{4\sigma}{P_a R_1}\right) R_1^3 = 2 \left(1 + \frac{4\sigma}{P_a R_0}\right) R_0^3 \qquad \text{(S.4.63.8)}$$

Solving this cubic equation in the small σ limit gives

$$R_1 = 2^{1/3} R_0 + \frac{4\sigma}{3P_a} \left(2^{1/3} - 1\right) \qquad \text{(S.4.63.9)}$$

Substituting (S.4.63.9) into (S.4.63.4) yields (in the same approximation)

$$\Delta S \approx \frac{8\pi R_0^2}{\tau} \left(\frac{4}{3}\sigma - q\right)\left(1 - \frac{1}{2^{1/3}}\right) \qquad \text{(S.4.63.10)}$$

Quantum Statistics

4.64 Fermi Energy of a 1D Electron Gas (Wisconsin-Madison)

For a one-dimensional gas the number of quantum states in the interval dp is

$$dN = g\frac{L\,dp}{2\pi\hbar}$$

where $g = 2s + 1 = 2$ and L is the "length" of the metal. The total number of electrons N (which in this case is equal to the number of atoms) is

$$N = \int_{-p_F}^{p_F} dN = g\frac{L}{2\pi\hbar} \int_{-p_F}^{p_F} dp = \frac{2Lp_F}{\pi\hbar} \qquad \text{(S.4.64.1)}$$

Therefore,

$$p_F = \pi\hbar\frac{N}{2L} = \frac{\pi\hbar}{2d}$$

where d is the atomic spacing. The Fermi energy

$$E_F = \frac{p_F^2}{2m} = \frac{(\pi\hbar)^2}{8md^2} \qquad \text{(S.4.64.2)}$$

where m is the electron mass.

$$E_F = \frac{\hbar^2\pi^2}{8md^2} = \frac{\left(1.05 \cdot 10^{-27}\right)^2 \pi^2}{8 \cdot 9 \cdot 10^{-28} \left(2.5 \cdot 10^{-8}\right)^2} \qquad \text{(S.4.64.3)}$$

$$\approx 2.5 \cdot 10^{-12} \text{ erg} \simeq 2 \cdot 10^4 \text{ K}$$

4.65 Two-Dimensional Fermi Gas (MIT, Wisconson-Madison)

a) At $T = 0$ the noninteracting fermions will be distributed among the available states so that the total energy is a minimum. The number of quantum states available to a fermion confined to a box of area A with momentum between p and $p + dp$ is given by

$$f(p)\ dp = g\frac{2\pi p\ dp \cdot A}{(2\pi\hbar)^2} = g\frac{Ap\ dp}{2\pi\hbar^2} \qquad (S.4.65.1)$$

where the multiplicity $g = 2s + 1$ and the spin $s = 1/2$. The N fermions at $T = 0$ fill all the states of momentum from 0 to p_F. We can therefore calculate this maximum momentum p_F from

$$N = \frac{gA}{2\pi\hbar^2} \int\limits_0^{p_F} p\ dp = \frac{gA}{2\pi\hbar^2}\frac{p_F^2}{2} \qquad (S.4.65.2)$$

The Fermi energy ε_F for this nonrelativistic gas is simply

$$\varepsilon_F = \frac{p_F^2}{2m} \qquad (S.4.65.3)$$

Using (S.4.65.2) and (S.4.65.3) we obtain

$$N = \frac{gmA\varepsilon_F}{2\pi\hbar^2} \qquad (S.4.65.4)$$

$$\varepsilon_F = \frac{2\pi\hbar^2}{gm}\frac{N}{A} \qquad (S.4.65.5)$$

For $s = 1/2$ and $g = 2$, (S.4.65.5) becomes

$$\varepsilon_F = \frac{\pi\hbar^2}{m}\frac{N}{A} \qquad (S.4.65.6)$$

b) The total energy of the gas

$$E = \int\limits_0^{p_F} dp\ f(p)\frac{p^2}{2m} = \frac{gA}{4\pi m\hbar^2} \int\limits_0^{p_F} p^3\ dp = \frac{gA}{4\pi m\hbar^2}\frac{p_F^4}{4} \qquad (S.4.65.7)$$

Substituting p_F from (S.4.65.2) into (S.4.65.7) gives

$$E = \frac{gA}{16\pi m\hbar^2} \cdot \frac{16\pi^2\hbar^4 N^2}{g^2 A^2} = \frac{\pi\hbar^2 N^2}{gmA} = \frac{\varepsilon_F N}{2} \qquad (S.4.65.8)$$

4.66   Nonrelativistic Electron Gas (Stony Brook, Wisconsin-Madison, Michigan State)

a) As $\tau \to 0$, the Fermi–Dirac distribution function

$$\langle n \rangle = \frac{1}{e^{(\varepsilon - \mu)/\tau} + 1} \qquad (S.4.66.1)$$

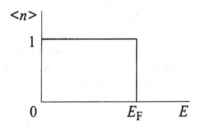

Figure S.4.66

becomes a step function. All the states above a certain energy, $\varepsilon > \mu$, are empty, and the states below, $\varepsilon < \mu$, are filled (see Figure S.4.66). This energy for an electron gas is called the Fermi energy. Physically, this results from the simple fact that the total energy of the gas should be a minimum. However, we have to reconcile this with the Pauli principle, which prohibits more than one electron per quantum state (i.e., same momentum and spin). This means that the states are filled gradually from zero energy to the limiting energy, ε_{F}. The number of states accessible to a free particle with absolute value of momentum between p and $p + \mathrm{d}p$ is

$$\mathrm{d}\Gamma = \frac{\mathrm{d}^3 p \, \mathrm{d}^3 q}{(2\pi\hbar)^3} = \frac{4\pi p^2 \, \mathrm{d}p \, \mathrm{d}V}{(2\pi\hbar)^3} \qquad (S.4.66.2)$$

In each of these states, we can put two electrons with opposite spin (up and down), so if we consider the total number of electrons, N, contained in a box of volume V, then N is given by

$$N = 2V \int_0^{p_{\mathrm{F}}} \frac{4\pi p^2 \, \mathrm{d}p}{(2\pi\hbar)^3} \qquad (S.4.66.3)$$

Substituting $p^2/2m = \varepsilon$, we obtain

$$N = \frac{2\sqrt{2}\, m^{3/2} \varepsilon_{\mathrm{F}}^{3/2}\, V}{3\pi^2 \hbar^3} \qquad (S.4.66.4)$$

and therefore

$$\varepsilon_{\mathrm{F}} = (3\pi^2)^{2/3} \frac{\hbar^2}{2m} \left(\frac{N}{V}\right)^{2/3} \qquad (S.4.66.5)$$

To calculate the total energy of the gas, we can write

$$E = 2 \int_0^{\varepsilon_{\mathrm{F}}} \varepsilon \, d\Gamma_p$$

where again $d\Gamma_p = 4\pi p^2 \, dp \, dV/(2\pi\hbar)^3$.

$$E = \frac{2\sqrt{2}}{5} \frac{Vm^{3/2}}{\pi^2 \hbar^3} (3\pi^2)^{5/3} \left(\frac{\hbar^2}{2m}\right)^{5/2} \left(\frac{N}{V}\right)^{5/2} \qquad (S.4.66.6)$$

$$= \frac{3}{10} (3\pi^2)^{2/3} \frac{\hbar^2}{m} \left(\frac{N}{V}\right)^{2/3} N$$

and therefore

$$P = -\left(\frac{\partial E}{\partial V}\right) = \frac{1}{5} (3\pi^2)^{2/3} \frac{\hbar^2}{m} \left(\frac{N}{V}\right)^{5/3} \qquad (S.4.66.7)$$

and we can check that

$$PV = \frac{2}{3} E \qquad (S.4.66.8)$$

b) The condition for strong degeneracy is that the temperature τ should be much smaller than the Fermi energy:

$$\tau \ll \frac{\hbar^2}{m} \left(\frac{N}{V}\right)^{2/3} \qquad (S.4.66.9)$$

For typical metals, if we assume that there is one free electron per atom and a typical interatomic distance $a \approx 2.5$ Å, we obtain an electron density $N/V \approx 5 \cdot 10^{22}$ e/cm^3, which indicates a Fermi energy of the order of 10^{-12} erg $\approx 10^4$ K. So, most of the metals are strongly degenerate, even at room temperature.

4.67 Ultrarelativistic Electron Gas (Stony Brook)

The fact that the gas is ultrarelativistic implies that the energy of the electron is large compared to its rest energy mc^2. In this case, the dispersion law is linear: $\varepsilon = cp$. The number of quantum states is the same as for the nonrelativistic case considered in Problem 4.66:

$$d\Gamma = \frac{d^3p\, d^3q}{(2\pi\hbar)^3} = \frac{4\pi p^2\, dp\, dV}{(2\pi\hbar)^3}$$

However, the Fermi energy now is different since $\varepsilon = cp$, and

$$N = 2V \int_0^{p_F} \frac{4\pi p^2\, dp}{(2\pi\hbar)^3} = \frac{V}{\pi^2\hbar^3 c^3} \int_0^{\varepsilon_F} \varepsilon^2\, d\varepsilon = \frac{\varepsilon_F^3 V}{3\pi^2\hbar^3 c^3}$$

Hence,

$$\varepsilon_F = (3\pi^2)^{1/3} \hbar c \left(\frac{N}{V}\right)^{1/3} \qquad (S.4.67.1)$$

The total energy is

$$E = 2V \int_0^{p_F} \varepsilon\, d\Gamma = 2V \int_0^{p_F} \varepsilon \frac{4\pi p^2\, dp}{(2\pi\hbar)^3} = \frac{V}{\pi^2\hbar^3 c^3} \int_0^{\varepsilon_F} \varepsilon^3\, d\varepsilon = \frac{\varepsilon_F^4 V}{4\pi^2\hbar^3 c^3}$$

After substituting ε_F from (S.4.67.1), we obtain

$$E = \frac{(3\pi^2)^{4/3}}{4\pi^2\hbar^3 c^3} \hbar^4 c^4 \left(\frac{N}{V}\right)^{4/3} V = \frac{3}{4}(3\pi^2)^{1/3} \hbar c \left(\frac{N}{V}\right)^{1/3} N$$

So, the pressure is

$$P = -\left(\frac{\partial E}{\partial V}\right)_S = \frac{1}{4}(3\pi^2)^{1/3} \hbar c \left(\frac{N}{V}\right)^{4/3} = \frac{1}{3}\frac{E}{V}$$

Hence, for an ultrarelativistic gas we have $PV = E/3$, the same as for massless particles (e.g., photons), which is not surprising since the dispersion law is the same.

4.68 Quantum Corrections to Equation of State (MIT, Princeton, Stony Brook)

a) Start with the particle distribution over the absolute value of momentum:

$$dN_p = g \frac{Vp^2\, dp}{2\pi^2\hbar^3 \left(e^{(\varepsilon-\mu)/\tau} \pm 1\right)} \qquad (S.4.68.1)$$

where the upper sign in (S.4.68.1) and below corresponds to Fermi statistics and the lower to Bose ($g = 2s + 1$). Using $\varepsilon = p^2/2m$, we obtain

$$dN_\varepsilon = g\frac{Vm^{3/2}}{\sqrt{2}\,\pi^2\hbar^3}\frac{\sqrt{\varepsilon}\,d\varepsilon}{e^{(\varepsilon-\mu)/\tau}\pm 1} \tag{S.4.68.2}$$

The total energy is given by

$$E = \int_0^\infty \varepsilon\,dN_\varepsilon = \frac{gVm^{3/2}}{\sqrt{2}\,\pi^2\hbar^3}\int_0^\infty \frac{\varepsilon^{3/2}\,d\varepsilon}{e^{(\varepsilon-\mu)/\tau}\pm 1} \tag{S.4.68.3}$$

On the other hand, using the grand canonical potential Ω, where

$$\Omega = \mp\tau\sum_k \ln\left(1\pm e^{(\mu-\varepsilon_k)/\tau}\right) \tag{S.4.68.4}$$

and replacing the sum by an integral, using (S.4.68.2), we obtain

$$\Omega = \mp\frac{Vg\tau m^{3/2}}{\sqrt{2}\,\pi^2\hbar^3}\int_0^\infty \sqrt{\varepsilon}\,\ln\left(1\pm e^{(\mu-\varepsilon)/\tau}\right)\,d\varepsilon \tag{S.4.68.5}$$

Integrating (S.4.68.5) by parts, we have

$$\Omega = -\frac{2}{3}\frac{gVm^{3/2}}{\sqrt{2}\pi^2\hbar^3}\int_0^\infty \frac{\varepsilon^{3/2}\,d\varepsilon}{e^{(\varepsilon-\mu)/\tau}\pm 1} \tag{S.4.68.6}$$

Comparing this expression with (S.4.68.3), we find that

$$\Omega = -\frac{2}{3}E \tag{S.4.68.7}$$

However, $\Omega = F - \mu N = F - G = -PV$. Therefore, we obtain the equation of state, which is valid both for Fermi and Bose gases (and is, of course, also true for a classical Boltzmann gas):

$$PV = \frac{2}{3}E \tag{S.4.68.8}$$

Note that (S.4.68.8) was derived under the assumption of a particular dispersion law $\varepsilon = p^2/2m$; for relativistic particles or photons with $\varepsilon = pc$, (S.4.68.8) becomes $PV = E/3$ (see Problem 4.67). From (S.4.68.8) and (S.4.68.3), we obtain

$$P = \frac{2E}{3V} = \frac{g\sqrt{2}\,m^{3/2}\tau^{5/2}}{3\pi^2\hbar^3}\int_0^\infty \frac{x^{3/2}\,dx}{e^{x-\mu/\tau}\pm 1} \tag{S.4.68.9}$$

where $x \equiv \varepsilon/\tau$. (S.4.68.9) defines the equation of state. To find quantum corrections to the classical equation of state (which corresponds to the case $e^{\mu/\tau} \ll 1$), expand the integral in (S.4.68.9), using $\exp(\mu/\tau - x)$ as a small parameter:

$$\int_0^\infty \frac{x^{3/2}\,dx}{e^{x-\mu/\tau} \pm 1} \approx \int_0^\infty x^{3/2} e^{\mu/\tau-x}\left(1 \mp e^{\mu/\tau-x}\right)\,dx \quad \text{(S.4.68.10)}$$

$$= \frac{3\sqrt{\pi}}{4} e^{\mu/\tau}\left(1 \mp \frac{1}{4\sqrt{2}} e^{\mu/\tau}\right)$$

Using $\Omega = -PV$ and substituting (S.4.68.10) into (S.4.68.9), we have

$$\Omega = -\frac{gVm^{3/2}\tau^{5/2}}{(2\pi)^{3/2}\hbar^3} e^{\mu/\tau} \pm \frac{gVm^{3/2}\tau^{5/2}}{16\pi^{3/2}\hbar^3} e^{2\mu/\tau} \quad \text{(S.4.68.11)}$$

The first term, which we may call Ω_B, corresponds to a Boltzmann gas with $g = 1$ (see Problem 4.39), and the second term gives the first correction

$$\Omega = \Omega_B \pm \frac{gVm^{3/2}\tau^{5/2}}{16\pi^{3/2}\hbar^3} e^{2\mu/\tau} \quad \text{(S.4.68.12)}$$

Using the fact that, for small corrections (see, for instance, Landau and Lifshitz, *Statistical Physics*, Sect. 24),

$$(\delta F)_{\tau,V,N} = (\delta \Omega)_{\tau,V,\mu}$$

we can write the first quantum correction to the free energy F. Using the classical expression for μ in terms of τ and V gives the result to the same accuracy:

$$F = F_B \pm \frac{\pi^{3/2}}{2g} \frac{N^2 \hbar^3}{v\sqrt{\tau}\, m^{3/2}} \quad \text{(S.4.68.13)}$$

Using

$$P = -\left(\frac{\partial F}{\partial V}\right)_\tau \qquad \text{and} \qquad -\left(\frac{\partial F_B}{\partial V}\right)_\tau = \frac{N\tau}{V}$$

we obtain, from (S.4.68.13),

$$P = \frac{N\tau}{V} \pm \frac{\pi^{3/2} N^2 \hbar^3}{2gV^2 \sqrt{\tau}\, m^{3/2}} \quad \text{(S.4.68.14)}$$

and

$$PV = N\tau\left(1 \pm \frac{\pi^{3/2}\hbar^3}{2gm^{3/2}} \frac{N}{V\tau^{3/2}}\right) \quad \text{(S.4.68.15)}$$

$$P = n\tau\left(1 \pm \frac{\pi^{3/2}\hbar^3}{2gm^{3/2}} \frac{n}{\tau^{3/2}}\right)$$

where $n \equiv N/V$.

b) The condition for validity of this approximation is that the first correction should be much less than unity:

$$\frac{\hbar^3 n}{(m\tau)^{3/2}} \ll 1 \tag{S.4.68.16}$$

This gives the condition on the density for which (S.4.68.15) is valid:

$$n \ll \frac{(m\tau)^{3/2}}{\hbar^3} \tag{S.4.68.17}$$

It is interesting to determine the de Broglie wavelength λ_{dB} at this temperature τ. We find that

$$\lambda_{dB} \sim \frac{\hbar}{p} \sim \frac{\hbar}{\sqrt{m\tau}}$$

$$\frac{1}{\lambda_{dB}^3} \sim \frac{(m\tau)^{3/2}}{\hbar^3}$$

We see that this approximation is valid when the separation between particles is much larger than the de Broglie wavelength. (S.4.68.16) expresses the same condition as for the applicability of Boltzmann statistics (which implies $\exp(\mu/\tau) \ll 1$). Since the chemical potential μ may be written (see Problem 4.39)

$$\mu = \tau \ln \left[n \left(\frac{2\pi}{m\tau} \right)^{3/2} \hbar^3 \right] \tag{S.4.68.18}$$

we see that

$$\frac{n\hbar^3}{(m\tau)^{3/2}} \ll 1$$

4.69 Speed of Sound in Quantum Gases (MIT)

a) The Gibbs free energy G is a function of (P, τ), which do not depend on the number of particles; i.e.,

$$G = Nf(P, \tau) \tag{S.4.69.1}$$

where $f(P, \tau)$ is some function of (P, τ). On the other hand,

$$\mu = \left(\frac{\partial G}{\partial N} \right)_{P,\tau} = f(P, \tau) \tag{S.4.69.2}$$

Therefore, $\mu = G/N$ for a system consisting of identical particles, and we may write for μ:

$$d\mu = -s\, d\tau + v\, dP \tag{S.4.69.3}$$

where $s = S/N$ and $v = V/N$. From (S.4.69.3) we have

$$v\left(\frac{\partial P}{\partial n}\right)_\tau = \left(\frac{\partial \mu}{\partial n}\right)_\tau$$

and we recover

$$\left(\frac{\partial P}{\partial \rho}\right)_\tau = \frac{1}{mv}\left(\frac{\partial \mu}{\partial n}\right)_\tau = \frac{n}{m}\left(\frac{\partial \mu}{\partial n}\right)_\tau \tag{S.4.69.4}$$

b) The number of quantum states in the interval between p and $p + dp$ for a Fermi gas is

$$dN_p = g\frac{4\pi p^2\, dp}{(2\pi\hbar)^3}V \tag{S.4.69.5}$$

where $g = 2s + 1$. At $\tau = 0$, electrons fill all the states with momentum from 0 to p_F, so the total number of electrons N is given by

$$N = \frac{gV}{2\pi^2\hbar^3}\int_0^{p_F} p^2\, dp = \frac{gVp_F^3}{6\pi^2\hbar^3} \tag{S.4.69.6}$$

For $s = 1/2$, $g = 2$, and

$$N = \frac{Vp_F^3}{3\pi^2\hbar^3} \tag{S.4.69.7}$$

or

$$p_F = \left(3\pi^2\right)^{1/3}\hbar n^{1/3} \tag{S.4.69.8}$$

The total energy of the gas

$$E = \int_0^{p_F} \frac{p^2}{2m}g\frac{4\pi p^2\, dp}{(2\pi\hbar)^3}V = \frac{gV}{4m\pi^2\hbar^3}\int_0^{p_F} p^4\, dp \tag{S.4.69.9}$$

$$= \frac{gVp_F^5}{20m\pi^2\hbar^3} = \frac{V}{10m\pi^2\hbar^3}p_F^5$$

Substituting p_F from (S.4.69.8), we obtain

$$E = \frac{V}{10m\pi^2\hbar^3}\left(3\pi^2\right)^{5/3}\hbar^5\left(\frac{N}{V}\right)^{5/3} \tag{S.4.69.10}$$

$$= \frac{3}{10}\left(3\pi^2\right)^{2/3}\frac{\hbar^2}{m}\left(\frac{N}{V}\right)^{2/3}N$$

Using the equation of state for a Fermi gas (see Problem 4.66),

$$PV = \frac{2}{3}E$$

we have

$$P = \frac{2}{3}\left(\frac{E}{V}\right) = \frac{1}{5}(3\pi^2)^{2/3}\frac{\hbar^2}{m}\left(\frac{N}{V}\right)^{5/3} \qquad (S.4.69.11)$$

$$= \frac{1}{5}(3\pi^2)^{2/3}\frac{\hbar^2}{m}n^{5/3}$$

Now, using (S.4.69.11), we can calculate u^2:

$$u^2 = \left(\frac{\partial P}{\partial \rho}\right)_{\tau=0} \qquad (S.4.69.12)$$

$$= \frac{1}{m}\left(\frac{\partial P}{\partial n}\right)_{\tau=0} = \frac{1}{3}(3\pi^2)^{2/3}\frac{\hbar^2}{m^2}n^{2/3}$$

Alternatively, we can use the expression obtained in (a) and the fact that, at $\tau = 0$, the chemical potential $\mu \equiv \varepsilon_{\text{F}}$. From (S.4.69.8),

$$\varepsilon_{\text{F}} = \frac{p_{\text{F}}^2}{2m} = (3\pi^2)^{2/3}\frac{\hbar^2}{2m}n^{2/3} \qquad (S.4.69.13)$$

and we again recover (S.4.69.12) in

$$u^2 = \frac{n}{m}\frac{\partial\mu}{\partial n} \qquad (S.4.69.14)$$

$$= \frac{n}{m}\frac{\partial\varepsilon_{\text{F}}}{\partial n} = \frac{1}{3}(3\pi^2)^{2/3}\frac{\hbar^2}{m^2}n^{2/3}$$

c) We can explicitly calculate the total energy of the Bose gas, which will be defined by the particles that are outside the condensate (since the condensed particles are in the ground state with $\varepsilon = 0$). At a temperature below the Bose–Einstein condensation $\tau < \tau_0$, the particles outside the condensate (with $\varepsilon > 0$) are distributed according to a regular Bose distribution with $\mu = 0$ (see Problem 4.70):

$$dN_\varepsilon = g\frac{Vm^{3/2}}{\sqrt{2}\,\pi^2\hbar^3}\frac{\sqrt{\varepsilon}\,d\varepsilon}{e^{\varepsilon/\tau}-1} \qquad (S.4.69.15)$$

The total number of particles outside the condensate $\tilde{N}$ is therefore

$$\tilde{N} = \int dN_\varepsilon = \frac{gV(m\tau)^{3/2}}{\sqrt{2}\,\pi^2\hbar^3}\int_0^\infty\frac{\sqrt{x}\,dx}{e^x-1} = N\left(\frac{\tau}{\tau_0}\right)^{3/2} \qquad (S.4.69.16)$$

The energy of the Bose gas at $\tau < \tau_0$ is

$$E = \frac{gV(m\tau)^{3/2}}{\sqrt{2}\pi^2\hbar^3}\tau\int_0^\infty \frac{x^{3/2}\,dx}{e^x - 1} \sim \frac{m^{3/2}\tau^{5/2}}{\hbar^3}V \qquad (S.4.69.17)$$

The free energy F is

$$F = G - PV = -PV = -\frac{2}{3}E$$

since $G = \mu N$ and $\mu = 0$. So the pressure

$$P = -\left(\frac{\partial F}{\partial V}\right)_\tau \sim \frac{m^{3/2}}{\hbar^3}\tau^{5/2} \qquad (S.4.69.18)$$

So we see that the pressure does not depend on the volume and

$$\left(\frac{\partial P}{\partial \rho}\right)_\tau = \frac{1}{m}\left(\frac{\partial P}{\partial n}\right)_\tau = 0$$

at $\tau < \tau_0$. We could have determined the result without the above calculations since $\mu = 0$ at $\tau \leq \tau_0$; the particles which are inside the condensate (with $\varepsilon = 0$) have no momentum and do not contribute to pressure.

4.70 Bose Condensation Critical Parameters (MIT)

a) The number of particles dN in an element of phase space is given by

$$dN = g\frac{d^3p\,d^3q}{(2\pi\hbar)^3}\frac{1}{e^{(\varepsilon-\mu)/\tau} - 1} \qquad (S.4.70.1)$$

where $g = 2S+1$. With the usual dispersion law for an ideal gas $\varepsilon = p^2/2m$ and integrating over d^3q, we find the particle distribution over energy:

$$dN_\varepsilon = g\frac{Vm^{3/2}}{\sqrt{2}\,\pi^2\hbar^3}\frac{\sqrt{\varepsilon}\,d\varepsilon}{e^{(\varepsilon-\mu)/\tau} - 1} \qquad (S.4.70.2)$$

Integrating (S.4.70.2), we obtain a formula for the total number N of particles in the gas:

$$N = g\frac{Vm^{3/2}}{\sqrt{2}\,\pi^2\hbar^3}\int_0^\infty \frac{\sqrt{\varepsilon}\,d\varepsilon}{e^{(\varepsilon-\mu)/\tau} - 1} \qquad (S.4.70.3)$$

Letting $\varepsilon/\tau = x$, we rewrite (S.4.70.3) as

$$\frac{N}{V} = g \frac{m^{3/2}\tau^{3/2}}{\sqrt{2}\,\pi^2\hbar^3} \int\limits_0^\infty \frac{\sqrt{x}\,dx}{e^{x-\mu/\tau} - 1} \qquad (S.4.70.4)$$

(S.4.70.4) defines a parametric equation for the chemical potential μ. The decrease of volume (or temperature) will increase the value of the integral, and therefore the value of μ (which is always negative in Bose statistics) will increase. The critical parameters V_c or τ_c correspond to the point where $\mu = 0$ (i.e., if you decrease the volume or temperature any further, μ should increase even further to provide a solution to (S.4.70.4), whereas it cannot become positive). So we can write at a certain temperature:

$$\frac{N}{V_c} = g \frac{m^{3/2}\tau^{3/2}}{\sqrt{2}\,\pi^2\hbar^3} \int\limits_0^\infty \frac{\sqrt{x}\,dx}{e^x - 1} \sim \frac{m^{3/2}\tau^{3/2}}{\hbar^3} \qquad (S.4.70.5)$$

Therefore,

$$V_c \sim \frac{N\hbar^3}{(m\tau)^{3/2}} \qquad (S.4.70.6)$$

b) In two dimensions the integral (S.4.70.3) becomes

$$N \sim \tau \int\limits_0^\infty \frac{dx}{e^{x-\mu/\tau} - 1}$$

and there is no Bose condensation (see Problem 4.71).

4.71 Bose Condensation (Princeton, Stony Brook)

For Bose particles,

$$\bar{n} \propto \frac{1}{e^{(\varepsilon-\mu)/\tau} - 1} \qquad (S.4.71.1)$$

where τ is the temperature in energy units. The total number of particles in a Bose distribution is

$$N = \sum_k \bar{n}_k = \frac{V}{(2\pi\hbar)^D} \int \frac{d^D p}{e^{(\varepsilon(p)-\mu)/\tau} - 1} \qquad (S.4.71.2)$$

Substituting $d^D p \propto p^{D-1}\,dp \propto \varepsilon^{(D-1)/\sigma}\varepsilon^{(1-\sigma)/\sigma}\,d\varepsilon \propto \varepsilon^{D/\sigma-1}\,d\varepsilon$ into the integral gives

$$N \propto \int\limits_0^\infty \frac{\varepsilon^{D/\sigma-1}\,d\varepsilon}{e^{(\varepsilon-\mu)/\tau} - 1} \propto \tau^{D/\sigma} \int\limits_0^\infty \frac{x^{D/\sigma-1}\,dx}{e^{x-\mu/\tau} - 1} \qquad (S.4.71.3)$$

The condition for Bose condensation to occur is that, at some particular temperature, the chemical potential goes to zero. Then the number of particles outside the Bose condensate will be determined by the integral

$$N \propto \tau_c^{D/\sigma} \int\limits_0^\infty \frac{x^{D/\sigma-1}\,\mathrm{d}x}{e^x - 1} \qquad \text{(S.4.71.4)}$$

This integral should converge since N is a given number. Expanding around $x = 0$ in order to determine conditions for convergence of the integral yields

$$\lim_{x \to 0} \int_0^{} \frac{x^{D/\sigma-1}\,\mathrm{d}x}{x} \propto \int x^{D/\sigma-2}\,\mathrm{d}x \qquad \text{(S.4.71.5)}$$

$$\propto \begin{cases} x^{D/\sigma-1}|_0 & D/\sigma \neq 1 \\ \ln|x||_0 & D/\sigma = 1 \end{cases}$$

So, this integral diverges at $D/\sigma \leq 1$, and there is no Bose condensation for this region. (For instance, in two dimensions, particles with ordinary dispersion law $E = p^2/2m$ would not Bose-condense.) In three dimensions, $D/\sigma = 3/2 > 1$, so that Bose condensation does occur.

4.72 How Hot the Sun? (Stony Brook)

(See Problem 2 of Chapter 4 in Kittel and Kroemer, *Thermal Physics*.) The distribution of photons over the quantum states k with energy $\varepsilon_k = \hbar\omega_k$ is given by Planck's distribution n_k (the Bose–Einstein distribution with chemical potential $\mu = 0$):

$$n_k = \frac{1}{e^{\hbar\omega_k/\tau} - 1} \qquad \text{(S.4.72.1)}$$

where τ is the temperature of the radiation which we consider equal to the temperature of the surface of the Sun. To find the total energy, we can replace the sum over modes by an integral over frequencies:

$$\mathrm{d}N_\omega = 2\int \frac{\mathrm{d}^3\mathbf{x}\,\mathrm{d}^3\mathbf{k}}{(2\pi)^3} n_k = 2\frac{V4\pi k^2\,\mathrm{d}k}{(2\pi)^3} n_k = \frac{V\omega^2\,\mathrm{d}\omega}{\pi^2 c^3} n_k \qquad \text{(S.4.72.2)}$$

where the factor 2 accounts for the two transverse photon polarizations. The energy of radiation $\mathrm{d}\varepsilon_\omega$ in an interval $\mathrm{d}\omega$ and unit volume is therefore

$$\mathrm{d}\varepsilon_\omega = \hbar\omega\frac{\mathrm{d}N_\omega}{V} = \frac{\hbar}{\pi^2 c^3}\frac{\omega^3\,\mathrm{d}\omega}{e^{\hbar\omega/\tau} - 1} \qquad \text{(S.4.72.3)}$$

The total radiation energy density $u = \varepsilon/V$ over all frequencies is

$$u = \frac{\hbar}{\pi^2 c^3} \cdot \int_0^\infty \frac{\omega^3 \, d\omega}{e^{\hbar \omega/\tau} - 1} = \frac{\tau^4}{\hbar^3 c^3} \frac{1}{\pi^2} \int_0^\infty \frac{x^3 \, dx}{e^x - 1} \qquad \text{(S.4.72.4)}$$

The integral with factor $1/\pi^2$ is just a number which we can take ≈ 1 (in fact it is $\pi^2/15$). So

$$u \approx \frac{\tau^4}{\hbar^3 c^3} \qquad \text{(S.4.72.5)}$$

The energy flux J_T per a unit solid angle is

$$J_T = \frac{cu}{4\pi} \approx \frac{\tau^4}{4\pi \hbar^3 c^2} \qquad \text{(S.4.72.6)}$$

The flux that illuminates the Earth is proportional to the solid angle $\Delta\Omega$ subtended by the Sun's surface at the Earth:

$$\Delta\Omega = \frac{\pi R_S^2}{d^2} \qquad \text{(S.4.72.7)}$$

The radiant energy flux at the Earth is therefore

$$J_E = \Delta\Omega J_T \approx \frac{c}{4} \frac{\tau^4}{\hbar^3 c^3} = \frac{k_B^4 T_S^4}{4\hbar^3 c^2} \frac{R_S^2}{d^2} \qquad \text{(S.4.72.8)}$$

where T_S is the temperature of the Sun's surface in K. Now we may estimate T_S:

$$T_S = \left(\frac{4 J_E \hbar^3 c^2}{k_B^4} \right)^{1/4} \left(\frac{d}{R_S} \right)^{1/2}$$

$$= \left(\frac{4 \cdot 0.14 \cdot 10^7 \cdot \left(10^{-27}\right)^3 \cdot 10^{21}}{(1.4)^4 \left(10^{-16}\right)^4} \right)^{1/4} \left(\frac{1.5 \cdot 10^{13}}{7 \cdot 10^{10}} \right)^{1/2} \qquad \text{(S.4.72.9)}$$

$$\approx 5000 \text{ K}$$

(The actual temperature is about 6000 K; see Problem 4.73.)

4.73 Radiation Force (Princeton, Moscow Phys-Tech, MIT)

a) The total radiation flux from the Sun is

$$J_S = \sigma T_S^4 4\pi R_S^2 \qquad \text{(S.4.73.1)}$$

where σ is the Stefan–Boltzmann constant. Only a fraction $\pi R_{\mathrm{E}}^2/4\pi d^2$ of this flux reaches the Earth. In equilibrium this fraction equals the total flux radiated from the Earth at temperature T_{E}. So

$$\sigma T_{\mathrm{S}}^4 4\pi R_{\mathrm{S}}^2 \frac{\pi R_{\mathrm{E}}^2}{4\pi d^2} = \sigma T_{\mathrm{E}}^4 4\pi R_{\mathrm{E}}^2 \tag{S.4.73.2}$$

From (S.4.73.2) we obtain

$$T_{\mathrm{E}} = \sqrt{\frac{R_{\mathrm{S}}}{2d}}\, T_{\mathrm{S}} \approx 290 \text{ K} \tag{S.4.73.3}$$

b) The radiation pressure on the Earth is given by

$$P_{\mathrm{r}} = \frac{4}{3c}\sigma T_{\mathrm{S}}^4 \frac{R_{\mathrm{S}}^2}{d^2}$$

$$= \frac{4}{3 \cdot 3 \cdot 10^8} 5.67 \cdot 10^{-8} \cdot \left(6 \cdot 10^3\right)^4 \left(\frac{7 \cdot 10^8}{1.5 \cdot 10^{11}}\right)^2 \tag{S.4.73.4}$$

$$= 7 \cdot 10^{-6} \text{ N/m}^2$$

where $(R_{\mathrm{S}}/d)^2$ is the ratio of the total flux from the Sun to the flux that reaches the Earth. The radiation force on the Earth

$$f_{\mathrm{E}} = P_{\mathrm{r}} A_{\mathrm{E}} = P_{\mathrm{r}} \pi R_{\mathrm{E}}^2 = 7 \cdot 10^{-6} \cdot \pi \left(6.4 \cdot 10^6\right)^2 \tag{S.4.73.5}$$

$$= 9 \cdot 10^8 \text{ N}$$

where A_{E} is the cross section of the Earth.

c) For the small "chondrule" the temperature will be the same because it depends only on the angle at which the Sun is seen and the radiation force:

$$f_{\mathrm{c}} = P_{\mathrm{r}} A_{\mathrm{c}} = P_{\mathrm{r}} \pi R^2 = 7 \cdot 10^{-6} \pi \left(10^{-3}\right)^2 \tag{S.4.73.6}$$

$$\approx 2 \cdot 10^{-11} \text{ N} \sim 2 \cdot 10^{-6} \text{ dyn}$$

d) Using (S.4.73.3) and denoting the melting temperature of the metallic particle T_{m} and the distance from the Sun d_{c}, we obtain

$$d_{\mathrm{c}} = \frac{1}{2} R_{\mathrm{S}} \left(\frac{T_{\mathrm{S}}}{T_{\mathrm{m}}}\right)^2 = \frac{1}{2} 7 \cdot 10^8 \left(\frac{6000}{1550}\right)^2 \tag{S.4.73.7}$$

$$\approx 5 \cdot 10^9 \text{ m} = 5 \cdot 10^6 \text{ km}$$

e) Let us estimate the radius r of a particle for which the radiation force will equal the gravitational force at the distance of the Earth's orbit d. Using (S.4.73.6), we have

$$P_r \pi r^2 = G \frac{M_S m_p}{d^2} \qquad \text{(S.4.73.8)}$$

where the particle mass $m_p = (4/3)\pi r^3 \rho$, and $\rho \sim 10^3$ kg/m^3

$$r = \frac{3}{4} \frac{P_r d^2}{G M_S \rho} \approx \frac{3 \cdot 7 \cdot 10^{-6} \left(1.5 \cdot 10^{11}\right)^2}{4 \cdot 7 \cdot 10^{-11} \cdot 2 \cdot 10^{30} \cdot 1 \cdot 10^3} \approx 1\ \mu\text{m} \qquad \text{(S.4.73.9)}$$

4.74   Hot Box and Particle Creation (Boston, MIT)

a) The number of photons is

$$N_{\text{ph}} = 2 \int\int_0^\infty \frac{dV\, d^3 p}{(2\pi\hbar)^3 \left(e^{\hbar\omega/\tau} - 1\right)} \qquad \text{(S.4.74.1)}$$

where the factor 2 comes from the two polarizations of photons; $p = \hbar k = \hbar\omega/c$. So,

$$N_{\text{ph}} = 2V \int_0^\infty \frac{4\pi p^2\, dp}{(2\pi\hbar)^3 \left(e^{\hbar\omega/\tau} - 1\right)} = 2V \int_0^\infty \frac{4\pi \hbar^3 \omega^2\, d\omega}{c^3 (2\pi\hbar)^3 \left(e^{\hbar\omega/\tau} - 1\right)}$$

$$= \frac{V}{\pi^2 c^3} \int_0^\infty \frac{\omega^2\, d\omega}{e^{\hbar\omega/\tau} - 1} = \frac{\tau^3 V}{\pi^2 c^3 \hbar^3} \int_0^\infty \frac{x^2\, dx}{e^x - 1} \qquad \text{(S.4.74.2)}$$

$$\approx 2.4 \frac{\tau^3 V}{\pi^2 c^3 \hbar^3} \approx 0.24 \left(\frac{\tau}{\hbar c}\right)^3 V$$

b) At low temperatures we can disregard any interaction between photons due to the creation of electron–positron pairs. We can therefore use the standard formula for energy flux:

$$J = \sigma \tau^4 \qquad \text{(S.4.74.3)}$$

where σ is the Stefan–Boltzmann constant. On the other hand, by analogy with molecular flow,

$$J = \frac{c}{4} \varepsilon \qquad \text{(S.4.74.4)}$$

where $\varepsilon = E/V$ is the energy density. So,

$$E = \frac{4V}{c}\sigma T^4 \qquad (S.4.74.5)$$

c) Using the equation of state for a photon gas

$$PV = \frac{E}{3} \qquad (S.4.74.6)$$

and

$$P = -\left(\frac{\partial F}{\partial V}\right)_\tau \qquad (S.4.74.7)$$

we have

$$F = -\frac{E}{3} \qquad (S.4.74.8)$$

The entropy S is then

$$S = -\left(\frac{\partial F}{\partial \tau}\right)_V = \frac{1}{3}\left(\frac{\partial E}{\partial \tau}\right)_V = \frac{16V}{3c}\sigma T^3 \qquad (S.4.74.9)$$

d) The energy E of the system of particles + photons is

$$E = Nmc^2 + \frac{3}{2}N\tau + \frac{4\sigma\tau^4 V}{c} \qquad (S.4.74.10)$$

The entropy of the system is the sum of the entropy of an ideal gas and radiation. The free energy of a single-particle ideal gas with $s = 0$ (see Problem 4.38) of created particles and the radiation is then

$$F = Nmc^2 - N\tau \ln\left[\frac{eV}{N}\left(\frac{m\tau}{2\pi\hbar^2}\right)^{3/2}\right] - \frac{4\sigma}{3c}\tau^4 V \qquad (S.4.74.11)$$

Minimizing the free energy with respect to the number of particles, we have

$$0 = \delta F = mc^2\,\delta N - \tau\,\delta N \ln\left[\frac{eV}{N}\left(\frac{m\tau}{2\pi\hbar^2}\right)^{3/2}\right] + \tau\,\delta N$$

$$\frac{mc^2}{\tau} = \ln\frac{V}{N} + \ln\left(\frac{m\tau}{2\pi\hbar^2}\right)^{3/2} \qquad (S.4.74.12)$$

From (S.4.74.12) we obtain

$$n = \left(\frac{m\tau}{2\pi\hbar^2}\right)^{3/2} e^{-mc^2/\tau} \qquad (S.4.74.13)$$

This result can be immediately obtained if we consider the process as a "chemical" reaction

$$\gamma = \chi$$

or

$$\gamma - \chi = 0$$

For chemical equilibrium

$$\mu_\chi - \mu_\gamma = 0$$

Since, for photons, $\mu = 0$, we have

$$\mu_\chi = \mu_{id} + mc^2 = 0$$

where μ_{id} is the chemical potential of an ideal gas (see part (e)). This result gives us (S.4.74.13).

e) Pair creation and annihilation can be written in the form

$$\gamma = e^- + e^+ \tag{S.4.74.14}$$

The chemical potential μ of the photon gas is zero (since the number of photons is not constant but is defined by equilibrium conditions). Therefore, we have for process (S.4.74.14) in equilibrium:

$$\mu_e + \mu_p = 0 \tag{S.4.74.15}$$

where μ_e and μ_p are the chemical potentials of electrons and positrons, respectively. If we disregard the walls of the box and assume that there are no electrons inside the box initially, then the number of electrons equals the number of positrons, and $\mu_e = \mu_p = 0$. We then find for the number of electrons (positrons)

$$N_e = N_p = 2V \int_0^\infty \frac{4\pi p^2 \, dp}{(2\pi\hbar)^3 \left[e^{\varepsilon/\tau} + 1\right]} = \frac{V}{\pi^2\hbar^3} \int_0^\infty \frac{p^2 \, dp}{e^{\varepsilon/\tau} + 1} \tag{S.4.74.16}$$

where the factor 2 comes from the double degeneracy of the electron gas $g = 2s + 1$, and we set $\mu = 0$. The energy may be written

$$\varepsilon = \sqrt{m_e^2 c^4 + p^2 c^2}$$

Disregarding the 1 in the denominator of (S.4.74.16) and expanding the square root with respect to the small parameter $p^2/m_e^2c^2$, for $\tau \ll m_e c^2$, we obtain, from (S.4.74.16),

$$N_e = N_p = \frac{V}{\pi^2\hbar^3} \int\limits_0^\infty p^2 \exp\left[-\frac{m_e c^2}{\tau}\left(1 + \frac{p^2}{2m_e^2 c^2}\right)\right] dp$$

$$= \frac{V}{\pi^2\hbar^3} e^{-m_e c^2/\tau} \left(-\frac{d}{d\alpha}\int\limits_0^\infty e^{-\alpha p^2}\, dp\right) \qquad \text{(S.4.74.17)}$$

$$= \frac{V}{\pi^2\hbar^3} e^{-m_e c^2/\tau}\frac{\sqrt{\pi}}{4\alpha^{3/2}} = \frac{V}{\sqrt{2}}\left(\frac{m_e\tau}{\pi\hbar^2}\right)^{3/2} e^{-m_e c^2/\tau}$$

where we set $\alpha \equiv 1/2m_e\tau$. We then find that the concentrations are

$$n_e = n_p = \frac{1}{\sqrt{2}}\left(\frac{m_e\tau}{\pi\hbar^2}\right)^{3/2} e^{-m_e c^2/\tau} \qquad \text{(S.4.74.18)}$$

and so

$$n = n_e + n_p = \sqrt{2}\left(\frac{m_e\tau}{\pi\hbar^2}\right)^{3/2} e^{-m_e c^2/\tau} \qquad \text{(S.4.74.19)}$$

Alternatively, we can take an approach similar to the one in (d). Using the formula for the chemical potential of an ideal gas (see Problem 4.39) gives

$$\mu_{\text{id}} = \tau \ln\left[\frac{P}{g\tau^{5/2}}\left(\frac{2\pi\hbar^2}{m}\right)^{3/2}\right] = \tau \ln\left[\frac{N}{gV}\left(\frac{2\pi\hbar^2}{m\tau}\right)^{3/2}\right] \qquad \text{(S.4.74.20)}$$

We can immediately write

$$\mu_e = \mu_p = \mu_{\text{id}} + m_e c^2 = 0$$

to obtain the same result in (S.4.74.18) and (S.4.74.19).

f) For $\tau \gg m_e c^2$ the electrons are highly relativistic, and we can write $\varepsilon = cp$ in (S.4.74.16). Then

$$N_e = N_p = \frac{V}{\pi^2\hbar^3} \int\limits_0^\infty \frac{p^2\, dp}{e^{cp/\tau}+1} = \frac{V}{\pi^2\hbar^3}\left(\frac{\tau}{c}\right)^3 \int\limits_0^\infty \frac{z^2\, dz}{e^z+1} \qquad \text{(S.4.74.21)}$$

$$\approx 1.8\frac{V}{\pi^2\hbar^3}\left(\frac{\tau}{c}\right)^3 \approx 0.18\left(\frac{\tau}{\hbar c}\right)^3 V$$

where we have used the integral given in the problem. Finally,

$$n_e = n_p = 0.18 \left(\frac{\tau}{\hbar c}\right)^3 \qquad \text{(S.4.74.22)}$$

$$n = n_e + n_p = 0.36 \left(\frac{\tau}{\hbar c}\right)^3$$

4.75 D-Dimensional Blackbody Cavity (MIT)

For a photon gas the average number of photons per mode ω is given by

$$\langle n \rangle = \frac{1}{e^{\hbar\omega/\tau} - 1} \qquad \text{(S.4.75.1)}$$

The energy

$$E = \frac{V}{(2\pi\hbar)^D} \int \frac{d^D p \, \hbar\omega}{e^{\hbar\omega/\tau} - 1} \qquad \text{(S.4.75.2)}$$

where V is the volume of the hypercube:

$$d^D p \propto p^{D-1} \, dp \propto \varepsilon^{D-1} \, d\varepsilon \propto \omega^{D-1} \, d\omega$$

Substituting $x = \hbar\omega/\tau$ into (S.4.75.2), we obtain

$$E \propto V \int_0^\infty \frac{\omega^D \, d\omega}{e^{\hbar\omega/\tau} - 1} \propto V\tau^{D+1} \int_0^\infty \frac{x^D \, dx}{e^x - 1} \qquad \text{(S.4.75.3)}$$

The energy density is simply

$$\frac{E}{V} \propto \tau^{D+1} \int_0^\infty \frac{x^D \, dx}{e^x - 1} \propto \tau^{D+1} \qquad \text{(S.4.75.4)}$$

For $D = 3$ we recover the Stefan–Boltzmann law:

$$\frac{E}{V} \propto T^4$$

4.76 Fermi and Bose Gas Pressure (Boston)

a) Using $F = E - \tau S$ and substituting the entropy from the problem, we obtain

$$F = E - \sum_i \frac{\hbar\omega_i}{e^{\hbar\omega_i/\tau} - 1} + \tau \sum_i \ln\left(1 - e^{-\hbar\omega_i/\tau}\right) \qquad \text{(S.4.76.1)}$$

But

$$\sum_i \frac{\hbar\omega_i}{e^{\hbar\omega_i/\tau} - 1} = E$$

Therefore,

$$F = \tau \sum_i \ln\left(1 - e^{-\hbar\omega_i/\tau}\right)$$

We may then find the pressure P of the gas:

$$P = -\left(\frac{\partial F}{\partial V}\right)_\tau = -\sum_i \frac{e^{-\hbar\omega_i/\tau}}{1 - e^{-\hbar\omega_i/\tau}} \hbar \frac{d\omega_i}{dV} \qquad \text{(S.4.76.2)}$$

$$= -\sum_i \frac{1}{e^{\hbar\omega_i/\tau} - 1} \hbar \frac{d\omega_i}{dV} = -\sum_i n_i \hbar \frac{d\omega_i}{dV}$$

The isothermal work done by the gas

$$dW = P\,dV = -\sum_i n_i \hbar \frac{d\omega_i}{dV}\,dV \qquad \text{(S.4.76.3)}$$

b) For a photon gas in a cuboid box

$$\omega_i \sim \sqrt{n_x^2 + n_y^2 + n_z^2}\,\frac{\pi c}{L} \propto \frac{\xi}{V^{1/3}} \qquad \text{(S.4.76.4)}$$

where ξ is a constant.

$$\frac{d\omega_i}{dV} = -\frac{1}{3}\frac{\xi}{V^{4/3}} = -\frac{1}{3}\frac{\omega_i}{V} \qquad \text{(S.4.76.5)}$$

So,

$$P = \frac{1}{3}\frac{1}{V}\sum_i n_i \hbar\omega_i = \frac{1}{3}\frac{E}{V} \qquad \text{(S.4.76.6)}$$

The same is true for a relativistic Fermi gas with dispersion law $E = cp$.

c) For a nonrelativistic Fermi gas the energy is

$$\hbar\omega_i = (n_x^2 + n_y^2 + n_z^2)\frac{\hbar^2}{2m}\frac{\pi^2}{L^2} \propto \frac{\eta}{V^{2/3}} \qquad \text{(S.4.76.7)}$$

where η is a constant. So,

$$\frac{d\omega_i}{dV} = -\frac{2}{3}\frac{\eta}{V^{5/3}} = -\frac{2}{3}\frac{\omega_i}{V} \qquad \text{(S.4.76.8)}$$

and

$$P = \frac{2}{3}\frac{E}{V} \qquad \text{(S.4.76.9)}$$

This result was already obtained directly in Problem 4.66 (see (S.4.66.8)).

4.77 Blackbody Radiation and Early Universe (Stony Brook)

a) By definition the free energy

$$F = -\tau \ln \sum_n e^{-\varepsilon_n/\tau}$$

$$= -\tau \ln \sum_n \left(e^{-\hbar\omega/\tau}\right)^n = -\tau \ln \left(\frac{1}{1 - e^{-\hbar\omega/\tau}}\right) \qquad \text{(S.4.77.1)}$$

$$= \tau \ln \left(1 - e^{-\hbar\omega/\tau}\right)$$

b) The entropy is then

$$S = -\left(\frac{\partial F}{\partial \tau}\right)_V = -\ln\left(1 - e^{-\hbar\omega/\tau}\right) + \frac{\hbar\omega\tau e^{-\hbar\omega/\tau}}{\tau^2 \left(1 - e^{-\hbar\omega/\tau}\right)} \text{(S.4.77.2)}$$

$$= -\ln\left(1 - e^{-\hbar\omega/\tau}\right) + \frac{\hbar\omega}{\tau}\frac{1}{e^{\hbar\omega/\tau} - 1}$$

The energy of the system

$$E = F + \tau S = \frac{\hbar\omega}{e^{\hbar\omega/\tau} - 1} \qquad \text{(S.4.77.3)}$$

Alternatively, the entropy can be found from

$$\tau = \left(\frac{\partial E}{\partial S}\right)_N$$

or

$$S = \int_0^E \frac{dE'}{\tau}$$

where τ can be expressed from (S.4.77.3) as

$$\frac{1}{\tau} = \frac{\ln(1 + \hbar\omega/E)}{\hbar\omega} \qquad \text{(S.4.77.4)}$$

So,

$$S = \frac{1}{\hbar\omega} \int_0^E dE' \ln\left(1 + \frac{\hbar\omega}{E'}\right) = \frac{1}{\xi} \int_0^E dE' \ln\left(1 + \frac{\xi}{E'}\right) \qquad \text{(S.4.77.5)}$$

where we let $\xi \equiv \hbar\omega$. Performing the integral gives

$$S = \frac{1}{\xi} \int_0^E dE' \left[\ln\left(1 + \frac{E'}{\xi}\right) - \ln\left(\frac{E'}{\xi}\right) \right]$$

$$= \int_0^E d\left(1 + \frac{E'}{\xi}\right) \ln\left(1 + \frac{E'}{\xi}\right) - \int_0^E d\left(\frac{E'}{\xi}\right) \ln\left(\frac{E'}{\xi}\right)$$

$$= \int_1^{1+E/\xi} dx \, \ln x - \int_0^{E/\xi} dx \, \ln x \qquad (S.4.77.6)$$

$$= x \ln x \big|_1^{1+E/\xi} - E/\xi - x \ln x \big|_0^{E/\xi} + E/\xi$$

$$= (1 + E/\xi) \ln(1 + E/\xi) - (E/\xi) \ln(E/\xi)$$

Substituting the energy for this mode $E = \hbar\omega \langle n \rangle$, we recover the entropy in the form

$$S = (1 + \langle n \rangle) \ln(1 + \langle n \rangle) - \langle n \rangle \ln \langle n \rangle \qquad (S.4.77.7)$$

4.78 Photon Gas (Stony Brook)

The photon gas is a Bose gas ($s = 0$) with zero chemical potential ($\mu = 0$), leading to Planck's distribution:

$$\langle n_k \rangle = \frac{1}{e^{\hbar\omega_k/\tau} - 1} \qquad (S.4.78.1)$$

Replacing the sum over different modes by an integral in spherical coordinates, we may write, for the number of quantum states in a volume V,

$$dn = \frac{V \, d^3p}{(2\pi\hbar)^3} = \frac{V 4\pi p^2 \, dp}{(2\pi\hbar)^3} \qquad (S.4.78.2)$$

Substituting $p = \hbar\omega/c$ into (S.4.78.2) and taking into account the two possible transverse polarizations of photons, we obtain

$$dn = \frac{V\omega^2 \, d\omega}{\pi^2 c^3} \qquad (S.4.78.3)$$

Let us calculate the Helmholtz free energy F. For a Bose gas with $\mu = 0$, the grand thermodynamic potential Ω is given by

$$\Omega = \tau \sum_k \ln \left(1 - e^{-\varepsilon_k/\tau}\right) \tag{S.4.78.4}$$

The free energy F would coincide with Ω (since $F = N\mu + \Omega = \Omega$). Again replacing the sum by an integral in (S.4.78.4) and substituting $\xi = \hbar\omega/\tau$ yield

$$F = \tau \int_0^\infty \ln \left(1 - e^{-\hbar\omega/\tau}\right) \frac{V\omega^2 \, d\omega}{\pi^2 c^3}$$

$$= \frac{\tau V}{\pi^2 c^3} \int_0^\infty \omega^2 \ln \left(1 - e^{-\hbar\omega/\tau}\right) \, d\omega \tag{S.4.78.5}$$

$$= \frac{V\tau^4}{\pi^2 c^3 \hbar^3} \int_0^\infty \xi^2 \ln \left(1 - e^{-\xi}\right) \, d\xi$$

Integrating by parts gives

$$F = \frac{V\tau^4}{3\pi^2 c^3 \hbar^3} \left[\xi^3 \ln \left(1 - e^{-\xi}\right)\Big|_0^\infty - \int_0^\infty \frac{\xi^3 e^{-\xi} \, d\xi}{1 - e^{-\xi}} \right] \tag{S.4.78.6}$$

$$= -\frac{V\tau^4}{3\pi^2 c^3 \hbar^3} \int_0^\infty \frac{\xi^3 \, d\xi}{e^\xi - 1} = -\frac{V\tau^4}{3\pi^2 c^3 \hbar^3} I$$

where

$$I = \int_0^\infty \frac{\xi^3 \, d\xi}{e^\xi - 1} = \frac{\pi^4}{15}$$

although we really do not need this, and so

$$F = -\alpha V \tau^4$$

with α a positive constant. The pressure P of the photon gas is

$$P = -\left(\frac{\partial F}{\partial V}\right)_\tau = \alpha \tau^4 \tag{S.4.78.7}$$

The entropy S of the gas is given by

$$S = -\left(\frac{\partial F}{\partial \tau}\right)_V = 4\alpha V \tau^3 \qquad \text{(S.4.78.8)}$$

The energy E may now be found from

$$E = F + \tau S = -\alpha V \tau^4 + 4\alpha V \tau^4 = 3\alpha V \tau^4 = -3F \qquad \text{(S.4.78.9)}$$

Comparing (S.4.78.7) and (S.4.78.9) gives

$$PV = \frac{E}{3} \qquad \text{(S.4.78.10)}$$

Note that this result is the same as for an ultrarelativistic electron gas (which has the same dispersion law $E = pc$; see Problem 4.67). The total number of photons is given by

$$N_{\text{ph}} = \frac{V}{\pi^2 c^3} \int_0^\infty \frac{\omega^2 \, d\omega}{e^{\hbar\omega/\tau} - 1} \qquad \text{(S.4.78.11)}$$

$$= \frac{V\tau^3}{\pi^2 c^3 \hbar^3} \int_0^\infty \frac{x^2 \, dx}{e^x - 1} = \beta V \tau^3$$

where we let

$$\beta \equiv \frac{1}{\pi^2 c^3 \hbar^3} \int_0^\infty \frac{x^2 \, dx}{e^x - 1}$$

Comparing (S.4.78.9)–(S.4.78.10) with (S.4.78.11), we can write

$$PV = \frac{\alpha}{\beta} N_{\text{ph}} \tau \qquad \text{(S.4.78.12)}$$

So, similar to the classical ideal gas, we have $PV/\tau = \text{const}$.

4.79   Dark Matter (Rutgers)

a) The virial theorem may be written relating the average kinetic energy, $\langle T \rangle$, and the forces between particles (see Sect. 3.4 of Goldstein, *Classical Mechanics*):

$$\langle T \rangle = -\frac{1}{2}\left\langle \sum_i \mathbf{F}_i \cdot \mathbf{r}_i \right\rangle$$

For an inverse square law force (gravitation), the average kinetic energy $\langle T \rangle$ and potential energy $\langle U \rangle$ are related as

$$\langle T \rangle = -\frac{1}{2} \langle U \rangle \qquad (S.4.79.1)$$

For a very rough estimate of the gravitational potential energy of Draco, consider the energy of a sphere of uniform density $\rho(0)$ and radius r_0:

$$\langle U \rangle = - \int_0^{r_0} \frac{G\frac{4\pi}{3}r^3\rho(0)4\pi r^2\rho(0) \; dr}{r} \qquad (S.4.79.2)$$

$$= -\frac{16\pi^2 G\rho(0)^2 r_0^5}{15}$$

The average kinetic energy may be approximated by

$$\langle T \rangle = \frac{\frac{4}{3}\pi r_0^3\rho(0)\sigma^2}{2} \qquad (S.4.79.3)$$

Substituting (S.4.79.2) and (S.4.79.3) into (S.4.79.1), we find

$$\frac{\frac{4}{3}\pi r_0^3\rho(0)\sigma^2}{2} = -\frac{1}{2}\left(-\frac{16\pi^2 G\rho(0)^2 r_0^5}{15}\right) \qquad (S.4.79.4)$$

$$\sigma^2 \sim G\rho(0)r_0^2$$

b) If most of the mass of Draco is in massive neutrinos, we may estimate the energy by considering the energy of a uniform distribution of fermions in a box of volume V. The energy of such a fermionic gas has been found in Problem 4.66:

$$E = \frac{3}{10}\left(3\pi^2\right)^{2/3}\frac{\hbar^2}{m}\left(\frac{N}{V}\right)^{2/3} N \qquad (S.4.79.5)$$

Rewriting (S.4.79.5) in terms of density and volume gives

$$E \sim \frac{\hbar^2 \rho^{2/3}\rho V}{m^{8/3}} = \frac{\hbar^2 \rho^{5/3}V}{m^{8/3}} \qquad (S.4.79.6)$$

If the mass of the neutrino is too low, in order to maintain the observed density, the number density would increase, and the Pauli principle would require the kinetic energy to increase. So, in (S.4.79.6), the energy increases as the mass of the neutrino decreases. Equating the kinetic energy from (a) with (S.4.79.6), we see

$$V\rho(0)\sigma^2 \sim \frac{\hbar^3 \rho^{5/3} V}{m^{8/3}} \qquad (S.4.79.7)$$

$$m > \left(\frac{\hbar^2 \rho(0)}{\sigma^3}\right)^{1/4}$$

c) Substituting (S.4.79.4) into (S.4.79.7), we determine that

$$m > \left(\frac{\hbar^3 \sigma^2}{Gr_0^2 \sigma^3}\right)^{1/4} = \left(\frac{\hbar^3}{Gr_0^2 \sigma}\right)^{1/4}$$

$$\approx \left(\frac{(1.06)^3 \cdot 10^{-102}}{6.7 \cdot 10^{-11}(150 \cdot 3.3 \cdot 9.5)^2 \cdot 10^{30} \cdot 10 \cdot 10^3}\right)^{1/4} \frac{1}{1.8 \cdot 10^{-36}} \text{ eV/c}^2$$

$$\approx 300 \text{ eV/c}^2$$

This value is at least an order of magnitude larger than any experimental results for neutrino masses, implying that the model does not explain the manner in which Draco is held together (see also D. W. Sciama, *Modern Cosmology and the Dark Matter Problem*).

4.80 Einstein Coefficients (Stony Brook)

a) At equilibrium the rates of excitation out of and back to state 1 should be equal, so

$$\left(\frac{dN_1}{dt}\right)_{abs} = \left(\frac{dN_2}{dt}\right)_{spon} + \left(\frac{dN_2}{dt}\right)_{abs} \qquad (S.4.80.1)$$

Substituting (P.4.80.1), (P.4.80.2), and (P.4.80.3) into (S.4.80.1) gives

$$-B_{12} N_1 \rho(\nu) = -[A_{21} + B_{21}\rho(\nu)] N_2 \qquad (S.4.80.2)$$

We may find the ratio of the populations from (S.4.80.2) to be

$$\frac{N_2}{N_1} = \frac{B_{12}\rho(\nu)}{A_{21} + B_{21}\rho(\nu)} \qquad (S.4.80.3)$$

b) At thermal equilibrium the population of the upper state should be smaller than that of the lower state by the Boltzmann factor $e^{-h\nu/\tau}$, so (S.4.80.3) gives

$$e^{-h\nu/\tau} = \frac{B_{12}\rho(\nu)}{A_{21} + B_{21}\rho(\nu)} \qquad (S.4.80.4)$$

Substituting the radiation density $\rho(\nu)$ into (S.4.80.4) gives

$$A_{21}e^{-h\nu/\tau} + B_{21}\frac{8\pi h\nu^3}{c^3}\frac{e^{-h\nu/\tau}}{e^{h\nu/\tau}-1} = B_{12}\frac{8\pi h\nu^3}{c^3}\frac{1}{e^{h\nu/\tau}-1} \qquad (S.4.80.5)$$

or

$$\frac{A_{21}}{B_{21}} + \frac{8\pi h\nu^3}{c^3}\frac{1}{e^{h\nu/\tau}-1} = \frac{B_{12}}{B_{21}}\frac{8\pi h\nu^3}{c^3}\frac{e^{h\nu/\tau}}{e^{h\nu/\tau}-1} \qquad (S.4.80.6)$$

The ratios of coefficients may be found by considering (S.4.80.6) for extreme values of ν since it must be true for all values of ν. For very large values of ν, we have

$$\frac{A_{21}}{B_{21}} = \frac{B_{12}}{B_{21}}\frac{8\pi h\nu^3}{c^3} \qquad (S.4.80.7)$$

Substituting (S.4.80.7) back into (S.4.80.6) yields

$$\frac{B_{12}}{B_{21}}\frac{8\pi h\nu^3}{c^3} + \frac{8\pi h\nu^3}{c^3}\frac{1}{e^{h\nu/\tau}-1} \qquad (S.4.80.8)$$

$$= \frac{B_{12}}{B_{21}}\frac{8\pi h\nu^3}{c^3}\frac{e^{h\nu/\tau}}{e^{h\nu/\tau}-1}$$

or

$$\frac{B_{12}}{B_{21}}\left(1 - \frac{e^{h\nu/\tau}}{e^{h\nu/\tau}-1}\right) = -\frac{1}{e^{h\nu/\tau}-1} \qquad (S.4.80.9)$$

which immediately yields

$$\frac{B_{12}}{B_{21}} = 1 \qquad (S.4.80.10)$$

and so, from (S.4.80.7),

$$\frac{A_{21}}{B_{21}} = \frac{8\pi h\nu^3}{c^3} \qquad (S.4.80.11)$$

c) Inspection of (S.4.80.11) shows that the ratio of the spontaneous emission rate to the stimulated emission rate grows as the cube of the frequency, which makes it more difficult to create the population inversion necessary for laser action. The pump power would therefore scale as $(1/\lambda)^3$.

4.81   Atomic Paramagnetism (Rutgers, Boston)

a) The energy associated with the magnetic field is

$$\varepsilon_{\mathrm{m}} = -\hat{\mu}\cdot\mathbf{H} = g\mu_{\mathrm{B}}\mathbf{J}\cdot H\hat{z} = g\mu_{\mathrm{B}}Hm \qquad (S.4.81.1)$$

where m is an integer varying in the range $-J \le m \le J$.

b) From (S.4.81.1) we may find the partition function Z:

$$Z = \sum_{m=-J}^{J} e^{-\varepsilon_m/\tau} = \sum_{m=-J}^{J} e^{-mg\mu_B H/\tau} = \sum_{m=-J}^{J} e^{-mx} \qquad \text{(S.4.81.2)}$$

where we define $x \equiv g\mu_B H/\tau$. The sum (S.4.81.2) may be easily calculated:

$$Z = \sum_{m=-J}^{J} e^{-mx} = e^{-Jx}\left(1 + e^x + \cdots + e^{(2J-1)x} + e^{2Jx}\right)$$

$$= e^{-Jx}\frac{1 - e^{(2J+1)x}}{1 - e^x} = \frac{e^{(J+1/2)x} - e^{-(J+1/2)x}}{e^{x/2} - e^{-x/2}} \qquad \text{(S.4.81.3)}$$

$$= \frac{\sinh\left(J + \frac{1}{2}\right)x}{\sinh \frac{x}{2}}$$

The mean magnetic moment per dipole $\langle\mu\rangle$ is given by

$$\langle\mu\rangle = \tau\frac{\partial \ln Z}{\partial H} = \frac{\tau}{Z}\frac{\partial Z}{\partial H} = \frac{\tau}{Z}\frac{\partial x}{\partial H}\frac{\partial Z}{\partial x} \qquad \text{(S.4.81.4)}$$

$$= \frac{\tau}{Z}\frac{g\mu_B}{\tau}\frac{\left(J + \frac{1}{2}\right)\cosh\left(J + \frac{1}{2}\right)x\sinh\frac{x}{2} - \frac{1}{2}\sinh\left(J + \frac{1}{2}\right)x\cosh\frac{x}{2}}{\sinh^2\frac{x}{2}}$$

$$= g\mu_B\left[\left(J + \frac{1}{2}\right)\coth\left(J + \frac{1}{2}\right)x - \frac{1}{2}\coth\frac{x}{2}\right]$$

Since the atoms do not interact,

$$M = N\langle\mu\rangle \qquad \text{(S.4.81.5)}$$

$$= Ng\mu_B\left[\left(J + \frac{1}{2}\right)\coth\left(J + \frac{1}{2}\right)x - \frac{1}{2}\coth\frac{x}{2}\right]$$

For $J = 1/2$,

$$M = Ng\mu_B\left[\coth\frac{g\mu_B H}{\tau} - \frac{1}{2}\coth\frac{g\mu_B H}{2\tau}\right] \qquad \text{(S.4.81.6)}$$

$$= \frac{1}{2}Ng\mu_B\tanh\frac{g\mu_B H}{2\tau}$$

This result can be obtained directly from (S.4.81.3) and (S.4.81.4):

$$Z_{1/2} = \sum_{m=-1/2}^{1/2} e^{-mx} = 2\cosh\frac{x}{2} \qquad \text{(S.4.81.7)}$$

For $J = 1$,

$$Z = 1 + 2\cosh x \qquad\qquad (S.4.81.8)$$

$$M = Ng\mu_{\mathrm{B}}\frac{2\sinh x}{1 + 2\cosh x} \qquad\qquad (S.4.81.9)$$

c) For large H the magnetization saturates ($\coth x \to 1$, $x \to \infty$):

$$M_{\mathrm{sat}} = Ng\mu_{\mathrm{B}}\left[\left(J + \frac{1}{2}\right)\coth\left(J + \frac{1}{2}\right)\frac{g\mu_{\mathrm{B}}H}{\tau} - \frac{1}{2}\coth\frac{1}{2}\frac{g\mu_{\mathrm{B}}H}{\tau}\right]$$

$$\approx Ng\mu_{\mathrm{B}}J \qquad\qquad (S.4.81.10)$$

It is convenient to define the so-called Brillouin function $B_J(x)$ [Brillouin, *Journal de Physique* **8**, 74 (1927)] in such a way that

$$M = M_{\mathrm{sat}}B_J(y) = Ng\mu_{\mathrm{B}}JB_J(y)$$

So,

$$B_J(y) \equiv \left(1 + \frac{1}{2J}\right)\coth\left(J + \frac{1}{2}\right)\frac{g\mu_{\mathrm{B}}H}{\tau}y - \frac{1}{2J}\coth\frac{y}{2J}$$

$$y = Jx$$

For small H, $H \to 0$, we can expand $\coth x$:

$$\coth x = \frac{1}{x} + \frac{x}{3} - \frac{x^3}{45} + \cdots$$

So,

$$M \approx Ng\mu_{\mathrm{B}}\left\{\left(J + \frac{1}{2}\right)\left[\frac{1}{(J + \frac{1}{2})x} + \frac{(J + \frac{1}{2})x}{3}\right] - \frac{1}{2}\left[\frac{2}{x} + \frac{x}{6}\right]\right\}$$

$$= Ng\mu_{\mathrm{B}}\left[\left(J + \frac{1}{2}\right)^2 - \frac{1}{4}\right]\frac{x}{3} \qquad\qquad (S.4.81.11)$$

$$= \frac{Ng^2\mu_{\mathrm{B}}^2 H}{3\tau}(J^2 + J) = N\frac{J(J+1)g^2\mu_{\mathrm{B}}^2 H}{3\tau}$$

The saturation value (S.4.81.10) corresponds to a classical dipole $g\mu_{\mathrm{B}}J$ per atom, where all the dipoles are aligned along the direction of H, whereas the value at small magnetic field H (S.4.81.11) reflects a competition between order (H) and disorder (τ).

4.82 Paramagnetism at High Temperature (Boston)

a) The specific heat c of a system that has N energy states is given by

$$c = \frac{\partial E}{\partial \tau} = \frac{\partial}{\partial \tau} \left(\frac{\sum\limits_{n=1}^{N} E_n e^{-E_n/\tau}}{\sum\limits_{n=1}^{N} e^{-E_n/\tau}} \right) \qquad \text{(S.4.82.1)}$$

Using $1/\tau = \beta$, we may rewrite c:

$$c = -\beta^2 \frac{\partial}{\partial \beta} \frac{\sum\limits_{n=1}^{N} E_n e^{-\beta E_n}}{\sum\limits_{n=1}^{N} e^{-\beta E_n}}$$

$$= -\beta^2 \frac{\partial}{\partial \beta} \left[-\frac{\partial}{\partial \beta} \ln \left(\sum_{n=1}^{N} e^{-\beta E_n} \right) \right] = \beta^2 \frac{\partial^2}{\partial \beta^2} \ln \left(\sum_{n=1}^{N} e^{-\beta E_n} \right)$$

$$\approx \beta^2 \frac{\partial^2}{\partial \beta^2} \ln \left(\sum_{n=1}^{N} 1 - \beta E_n + \frac{1}{2!} \beta^2 E_n^2 \right) \qquad \text{(S.4.82.2)}$$

$$= \beta^2 \frac{\partial^2}{\partial \beta^2} \ln \left(N - \beta N \langle \varepsilon \rangle + \frac{N\beta^2}{2} \langle \varepsilon^2 \rangle \right)$$

$$= \beta^2 \frac{\partial^2}{\partial \beta^2} \ln \left(1 - \beta \langle \varepsilon \rangle + \frac{\beta^2}{2} \langle \varepsilon^2 \rangle \right)$$

$$\approx \beta^2 \frac{\partial^2}{\partial \beta^2} \left(-\beta \langle \varepsilon \rangle + \frac{\beta^2}{2} \langle \varepsilon^2 \rangle - \frac{\beta^2}{2} \langle \varepsilon \rangle^2 \right)$$

$$= \beta^2 \left(\langle \varepsilon^2 \rangle - \langle \varepsilon \rangle^2 \right) = \frac{\sigma^2}{\tau^2}$$

where we have used $\ln(1+x) \approx x - x^2/2$. Note that, in general, the parameter βE_n is not small (since it is proportional to the number of particles), but, subsequently, we obtain another parameter $\beta \varepsilon \ll 1$.

b) For a classical paramagnetic solid:

$$\varepsilon(\theta) = -\mu H \cos \theta = -\mu H z \qquad (-1 \le z \le 1)$$

so

$$\langle \varepsilon \rangle = 0$$

and we have

$$\sigma^2 = \int\limits_{-1}^{1} (\mu H z)^2 \frac{dz}{2} = \frac{1}{3}\mu^2 H^2 \qquad (S.4.82.3)$$

where $dz/2$ is the probability density. Therefore,

$$c_c \approx \frac{1}{3}\frac{\mu^2 H^2}{\tau^2} \qquad (S.4.82.4)$$

For the quantum mechanical case, $\langle \varepsilon \rangle = 0$; there is an equidistant energy spectrum: $E_m = -g\mu_B H m$ (see Problem 4.81) and

$$\sigma^2 = \frac{1}{2J+1}\sum_{m=-J}^{J}(-g\mu_B H m)^2 = \frac{g^2 \mu_B^2 H^2}{2J+1}\sum_{m=-J}^{J} m^2 \qquad (S.4.82.5)$$

To calculate $\sum\limits_{m=-J}^{J} m^2$, we can use the following trick (assuming J integer):

$$(N+1)^3 = \sum_{n=0}^{N}\left[(n+1)^3 - n^3\right] \qquad (S.4.82.6)$$

$$= \sum_{n=0}^{N}(3n^2 + 3n + 1) = 3\sum_{n=0}^{N}n^2 + 3\sum_{n=0}^{N}n + (N+1)$$

From (S.4.82.6) we have

$$\sum_{n=0}^{N} n^2 = \frac{1}{3}\left[(N+1)^3 - 3\sum_{n=0}^{N}n - (N+1)\right] \qquad (S.4.82.7)$$

With the familiar sum

$$\sum_{n=0}^{N} n = \sum_{n=1}^{N} = \frac{N(N+1)}{2}$$

we arrive at

$$\sum_{n=0}^{N} n^2 = \frac{1}{3}\left[(N+1)^3 - \frac{3N(N+1)}{2} - (N+1)\right] \qquad (S.4.82.8)$$

$$= \frac{1}{6}(N+1)\left[2N^2 + 4N + 2 - 3N - 2\right] = \frac{N(N+1)(2N+1)}{6}$$

We wish to perform the sum from $-J$ to J, so

$$\sum_{m=-J}^{J} m^2 = 2 \sum_{m=1}^{J} m^2 = \frac{J(J+1)(2J+1)}{3}$$

and (S.4.82.5) gives

$$\sigma^2 = \frac{J(J+1)g^2\mu_B^2 H^2}{3} \qquad (S.4.82.9)$$

$$c_q = \frac{J(J+1)g^2\mu_B^2 H^2}{3\tau^2} \qquad (S.4.82.10)$$

c) For $J = 1/2$,

$$\sum_{m=-1/2}^{1/2} m^2 = \frac{1}{2}$$

and

$$c_q = \frac{g^2\mu_B H^2}{4\tau^2} \qquad (S.4.82.11)$$

As in Problem 4.81 for $J = 1/2$:

$$Z(\beta) = 2\cosh\frac{x}{2} = 2\cosh\frac{g\mu_B H}{2\tau} = 2\cosh\frac{\beta y}{2} \qquad (S.4.82.12)$$

where $y = g\mu_B H$. We then find

$$c = \beta^2 \frac{\partial^2}{\partial\beta^2} \ln\left(2\cosh\frac{\beta y}{2}\right) = \beta^2 \frac{\partial}{\partial\beta}\left[\frac{(y/2)\sinh(\beta y/2)}{\cosh(\beta y/2)}\right]$$

$$= \left(\frac{\beta y}{2}\right)^2 \frac{1}{\cosh^2(\beta y/2)} \qquad (S.4.82.13)$$

$$= \frac{\beta^2 g^2 \mu_B^2 H^2}{4} \frac{1}{\cosh^2(\beta g\mu_B H/2)}$$

For $\tau \to \infty$, $\beta \to 0$,

$$c \approx \frac{\beta^2 g^2 \mu_B^2 H^2}{4} = \frac{g^2\mu_B^2 H^2}{4\tau^2}$$

which coincides with (S.4.82.11).

4.83   One-Dimensional Ising Model (Tennessee)

a) The partition function is defined as

$$Z = \prod_{n=1}^{N} \sum_{s_n=\pm1} e^{-\beta H} \qquad (S.4.83.1)$$

where the product is taken over the n sites. Define $K = \beta J$ where $\beta = 1/\tau$. Start by evaluating the sum at one end, say for $n = 1$. The answer is independent of the value of $s_2 = \pm1$:

$$\sum_{s_1=\pm1} e^{-K s_1 s_2} = e^K + e^{-K} \qquad (S.4.83.2)$$

Next we evaluate the sum over s_2, which is also independent of the value of s_3:

$$\sum_{s_2=\pm1} e^{-K s_2 s_3} = e^K + e^{-K} \qquad (S.4.83.3)$$

$$Z = (e^K + e^{-K})^N \qquad (S.4.83.4)$$

So each summation over s_n gives the identical factor $e^K + e^{-K}$, and Z is the product of N such factors.

b) The heat capacity per spin is obtained using thermodynamic identities. The partition function is related to the free energy F:

$$F = -\tau \ln Z = -N\tau \ln \left(e^K + e^{-K}\right) \qquad (S.4.83.5)$$

The entropy is given by

$$S = -\frac{\partial F}{\partial \tau} = N \ln \left(e^K + e^{-K}\right) - \frac{NJ}{\tau^2} \tanh K \qquad (S.4.83.6)$$

Now, the heat capacity C is given by

$$C = \tau \left(\frac{\partial S}{\partial \tau}\right) = \frac{NJ^2}{\tau^2} \frac{1}{\cosh^2 K} = \frac{NK^2}{\cosh^2 K} \qquad (S.4.83.7)$$

The heat capacity per spin c is

$$c = \frac{K^2}{\cosh^2 K} \qquad (S.4.83.8)$$

4.84   Three Ising Spins (Tennessee)

a) Define $K = \beta J$ and $f = \beta F$ where $\beta = 1/\tau$. The definition of the partition function is

$$Z = \prod_{n=1}^{3} \sum_{s_n = \pm 1} e^{-\beta H} \qquad (S.4.84.1)$$

A direct calculation gives

$$Z = 2e^{3K} \cosh 3f + 6e^{-K} \cosh f \qquad (S.4.84.2)$$

b) The average value of spin is

$$\langle s \rangle = -\frac{1}{3} \frac{d}{df} \ln Z = -\frac{2}{Z} \left(e^{3K} \sinh 3f + e^{-K} \sinh f \right) \qquad (S.4.84.3)$$

c) The internal energy is

$$\varepsilon = -J \frac{\partial}{\partial K} \ln Z = \frac{6J}{Z} \left(e^{3K} \cosh 3f - e^{-K} \cosh f \right) \qquad (S.4.84.4)$$

4.85 N Independent Spins (Tennessee)

a) The partition function is given by

$$Z = \prod_{n=1}^{N} \sum_{s_n = \pm 1} e^{-s_n \xi} \qquad (S.4.85.1)$$

where $\xi = \beta \mu H$. Each spin is independent, so one has the same result as for one spin, but raised to the Nth power

$$Z = (e^{\xi} + e^{-\xi})^N \qquad (S.4.85.2)$$

b) The internal energy is the derivative of $\ln Z$ with respect to β:

$$\varepsilon = -\frac{\partial}{\partial \beta} \ln Z = -N\mu H \tanh \xi \qquad (S.4.85.3)$$

c) The entropy is derivative of $\tau \ln Z$ with respect to τ:

$$S = \frac{\partial}{\partial \tau} \tau \ln Z = N[\ln(e^{\xi} + e^{-\xi}) - \xi \tanh \xi] \qquad (S.4.85.4)$$

4.86 N Independent Spins, Revisited (Tennessee)

We use the expression $S = \ln w$ where w is the probability of the arrangement of spins. For N spins we assume that N_u are up and N_d are down, where $N = N_u + N_d$. The different arrangements are

$$w = \frac{N!}{N_u! N_d!} \tag{S.4.86.1}$$

$$N_u = \frac{N(1+f)}{2} \tag{S.4.86.2}$$

$$N_d = N\frac{N(1-f)}{2} \tag{S.4.86.3}$$

Use Stirling's approximation for the factorial to obtain

$$S = \ln w = N \ln N - N_u \ln N_u - N_d \ln N_d \tag{S.4.86.4}$$

$$= -N\left\{ \frac{1+f}{2} \ln \frac{1+f}{2} + \frac{1-f}{2} \ln \frac{1-f}{2} \right\}$$

4.87 Ferromagnetism (Maryland, MIT)

Using the mean field approximation, we may write the magnetization M of the lattice as (see Problem 4.81)

$$M = n\mu \tanh\left(\frac{\mu B_{\text{eff}}}{\tau} \right) \tag{S.4.87.1}$$

where n is the density of the spins and B_{eff} is the sum of the imposed field and the field at spin σ_i produced by the neighboring spins:

$$B_{\text{eff}} \equiv B + \lambda M \tag{S.4.87.2}$$

where λ is a constant. We may rewrite (S.4.87.1) as

$$M = n\mu \tanh \frac{\mu(B + \lambda M)}{\tau} \tag{S.4.87.3}$$

The susceptibility χ is given by

$$\chi \equiv \frac{\partial M}{\partial B} = \frac{n\mu^2}{\cosh^2[\mu(B + \lambda M)/\tau]} \left[\frac{1 + \lambda \frac{\partial M}{\partial B}}{\tau} \right] \tag{S.4.87.4}$$

For B and M small we may rearrange (S.4.87.4), yielding

$$\chi = \frac{n\mu^2/\tau}{1 - \lambda n\mu^2/\tau} = \frac{n\mu^2}{\tau - \tau_c} \qquad \text{(S.4.87.5)}$$

where $\tau_c \equiv \lambda n\mu^2$. The divergence of χ at $\tau = \tau_c$ indicates the onset of ferromagnetism. The spins will align spontaneously in the absence of an applied magnetic field at this temperature.

4.88   Spin Waves in Ferromagnets (Princeton, Colorado)

Quantum spins have the commutation relations

$$[s_x, s_y] = is_z \qquad \text{(S.4.88.1)}$$

$$[s_y, s_z] = is_x \qquad \text{(S.4.88.2)}$$

$$[s_z, s_y] = is_y \qquad \text{(S.4.88.3)}$$

a) The time dependences of the spins are given by the equations of motion:

$$\frac{\partial}{\partial t} \mathbf{s}_i = i[H, \mathbf{s}_i] \qquad \text{(S.4.88.4)}$$

$$\frac{\partial}{\partial t} s_{xi} = -iJ \sum_j [s_{xi}s_{xj} + s_{yi}s_{yj} + s_{zi}s_{zj}, s_{xi}] \qquad \text{(S.4.88.5)}$$

$$= J \sum_j \{s_{yi}s_{zj} - s_{zi}s_{yj}\}$$

$$= \frac{\partial}{\partial t} \mathbf{s}_i = J\mathbf{s}_i \times \sum_j \mathbf{s}_j \qquad \text{(S.4.88.6)}$$

b) The classical spin field at point $\mathbf{r}_i$ is $\mathbf{s}(\mathbf{r}_i, t)$. In the simple cubic lattice the six neighboring lattice sites are at the points $\mathbf{r}_j = \mathbf{r}_i \pm a\hat{\mathbf{j}}$, where $\hat{\mathbf{j}}$ is $\hat{\mathbf{x}}, \hat{\mathbf{y}},$ or $\hat{\mathbf{z}}$. We expand the sum in a Taylor series, assuming that a is a small number, and find

$$\sum_j \mathbf{s}(\mathbf{r}_j, t) = 6\mathbf{s}(\mathbf{r}_i, t) + a^2 \nabla^2 \mathbf{s}(\mathbf{r}_i, t) + O(a^4) \qquad \text{(S.4.88.7)}$$

$$\frac{\partial}{\partial t} \mathbf{s}(\mathbf{r}, t) = Ja^2 \mathbf{s}(\mathbf{r}, t) \times \nabla^2 \mathbf{s}(\mathbf{r}_j, t) \qquad \text{(S.4.88.8)}$$

c) Given the form of the spin operator in part (c), one immediately derives
the equation by neglecting terms of order $O(m_0^2)$:

$$\frac{\partial}{\partial t}\mathbf{m}_0(t) = -J(ka)^2\mathbf{M} \times \mathbf{m}_0 \qquad (\text{S.4.88.9})$$

$$\frac{\partial m_{0x}}{\partial t} = -\omega_k m_{0y} \qquad (\text{S.4.88.10})$$

$$\frac{\partial m_{0y}}{\partial t} = \omega_k m_{0x} \qquad (\text{S.4.88.11})$$

$$\omega_k = J(ka)^2 M \qquad (\text{S.4.88.12})$$

The equations of motion have an eigenvalue ω_k, which represents the frequencies of the spin waves.

d) The internal energy per unit volume of the spin waves is given by

$$\varepsilon = \int \frac{d^3k}{(2\pi)^3} \frac{\hbar\omega_k}{e^{\beta\hbar\omega_k} - 1} \qquad (\text{S.4.88.13})$$

where the occupation number is suitable for bosons. At low temperature
we can evaluate this expression by defining the dimensionless variable $s = \beta\hbar\omega_k$, which gives for the integral

$$\varepsilon = \frac{1}{4\pi^2}\frac{(\tau)^{5/2}}{(\hbar Ja^2 M)^{3/2}} \int_0^{s_0} \frac{s^{3/2}\,ds}{e^s - 1} \qquad (\text{S.4.88.14})$$

At low temperature the upper limit of the integral s_0 becomes large, and the
internal energy is proportional to $\tau^{5/2}$. The heat capacity is the derivative
of ε with respect to temperature, so it goes as $C \sim \tau^{3/2}$.

Fluctuations

4.89 Magnetization Fluctuation (Stony Brook)

The energy of a dipole in a magnetic field ε_m may be written

$$\varepsilon_\mathrm{m} = \pm\mu H$$

The partition function Z is simply

$$Z = e^{\mu H/\tau} + e^{-\mu H/\tau} = 2\cosh\left(\frac{\mu H}{\tau}\right) \qquad (\text{S.4.89.1})$$

Since the moments are all independent, we may express the average magnetization $\langle M \rangle$ as

$$\langle M \rangle = N \langle \mu \rangle = N\tau \frac{\partial \ln Z}{\partial H} = N\mu \tanh\left(\frac{\mu H}{\tau}\right) \qquad \text{(S.4.89.2)}$$

On the other hand,

$$\langle M^2 \rangle - \langle M \rangle^2 = \left\langle (M - \langle M \rangle)^2 \right\rangle$$

$$= \left\langle \left[\sum_{i=1}^{N} (\mu_i - \langle \mu_i \rangle) \right] \left[\sum_{j=1}^{N} (\mu_j - \langle \mu_j \rangle) \right] \right\rangle \qquad \text{(S.4.89.3)}$$

$$= \sum_{i,j=1}^{N} \langle (\mu_i - \langle \mu_i \rangle)(\mu_j - \langle \mu_j \rangle) \rangle$$

For $i \neq j$, the ensemble averages are independent, and

$$\sum_{i\neq j=1}^{N} \langle (\mu_i - \langle \mu_i \rangle)(\mu_j - \langle \mu_j \rangle) \rangle = \sum_{i\neq j=1}^{N} \langle (\mu_i - \langle \mu_i \rangle) \rangle \langle (\mu_j - \langle \mu_j \rangle) \rangle$$

$$= \sum_{i\neq j=1}^{N} (\langle \mu_i \rangle - \langle \mu_i \rangle)(\langle \mu_j \rangle - \langle \mu_j \rangle) = 0$$

We are left with $i = j$, so (S.4.89.2) and (S.4.89.3) give

$$\langle M^2 \rangle - \langle M \rangle^2 = \sum_{i=1}^{N} \left(\langle \mu_i^2 \rangle - \langle \mu_i \rangle^2 \right) \qquad \text{(S.4.89.4)}$$

$$= N \left[\mu^2 - \mu^2 \tanh^2\left(\frac{\mu H}{\tau}\right) \right] = N\mu^2 \frac{1}{\cosh^2(\mu H/\tau)}$$

We then obtain

$$\frac{\sqrt{\langle M^2 \rangle - \langle M \rangle^2}}{\langle M \rangle} = \frac{\sqrt{N}\mu}{\cosh(\mu H/\tau)\, N\mu \tanh(\mu H/\tau)} \qquad \text{(S.4.89.5)}$$

$$= \frac{1}{\sqrt{N} \sinh(\mu H/\tau)}$$

4.90 Gas Fluctuations (Moscow Phys-Tech)

a) We can disregard any particles from the high-vacuum part of the setup and consider the problem of molecular flow from the ballast volume into the vacuum chamber. The number of particles was calculated in Problem 4.14:

$$N = \frac{n \langle v \rangle}{4} A \, \Delta t \qquad (S.4.90.1)$$

where n is the particle concentration and $\langle v \rangle$ is the average velocity. Expressing n via the pressure P and using (see Problem 4.13)

$$\langle v \rangle = \sqrt{\frac{8\tau}{\pi m}} \qquad (S.4.90.2)$$

we obtain

$$N = \frac{P \langle v \rangle A \, \Delta t}{4\tau} = \frac{P A \, \Delta t}{\sqrt{2\pi m \tau}} \qquad (S.4.90.3)$$

$$= \frac{P A N_A \Delta t}{\sqrt{2\pi \mu R T}} = \frac{10^{-1} \cdot 10^{-3} \cdot 10^{-4} \cdot 6.02 \cdot 10^{23} \cdot 10^{-3}}{\sqrt{2\pi \cdot 0.004 \cdot 8.31 \cdot 293}} \cdot 10^{-3} \approx 7.7 \cdot 10^{11}$$

b) At the given pressure the molecules are in the Knudsen regime, the mean free path $\lambda \gg A^{1/2}$. Therefore, we can assume that the molecular distribution will not change and N can be obtained from the Poisson distribution. The mean fluctuation (see Problem 4.94)

$$\langle \Delta N^2 \rangle = \langle N \rangle \approx 8 \cdot 10^{11} \qquad (S.4.90.4)$$

The mean relative fluctuation is given by

$$\frac{\sqrt{\langle \Delta N^2 \rangle}}{\langle N \rangle} = \frac{1}{\sqrt{\langle N \rangle}} \approx 10^{-6} \qquad (S.4.90.5)$$

c) The probability of finding N particles as a result of one of the measurements, according to the Poisson distribution (see Problem 4.35), is

$$w(N) = \frac{\langle N \rangle^N}{N!} e^{-\langle N \rangle} \qquad (S.4.90.6)$$

Therefore, the probability of counting zero particles in 1 ms is

$$w(0) = e^{-\langle N \rangle} \approx \exp\left[-8 \cdot 10^{11}\right] \qquad (S.4.90.7)$$

an exceedingly small number. This problem is published in Kozel, S. M., Rashba, E. I., and Slavatinskii, S. A., *Problems of the Moscow Institute of Physics and Technology*.

4.91   Quivering Mirror (MIT, Rutgers, Stony Brook)

a) When the mirror is in thermal equilibrium with gas in the chamber, one may again invoke the equipartition theorem and state that there is $(1/2)\tau$ of energy in the rotational degree of freedom of the torsional pendulum, where the torque is given by $\tau = -D\theta$. The mean square fluctuation in the angle would then be given by (see Chapter 13, Fluctuations, in Pathria)

$$\left\langle \frac{1}{2}D\theta^2 \right\rangle = \frac{1}{2}\tau \qquad (S.4.91.1)$$

So,

$$\langle \theta^2 \rangle = \frac{\tau}{D} = \frac{k_B T}{D} \qquad (S.4.91.2)$$

Now, Avogadro's number $N_A = R/k_B$, and we obtain

$$N_A = \frac{R}{k_B} = \frac{RT}{D\langle \theta^2 \rangle} = \frac{8.31 \cdot 10^7 \cdot 287}{9.43 \cdot 10^{-9} \cdot 4.20 \cdot 10^{-6}} = 6.02 \cdot 10^{23} \qquad (S.4.91.3)$$

b) Even if the gas density were reduced in the chamber, the mean square fluctuation $\langle \theta^2 \rangle$ would not change. However, in order to determine whether individual fluctuations might have larger amplitudes, we cannot rely on the equipartition theorem. We instead will examine the fluctuations in the frequency domain. $\langle \theta^2 \rangle$ may be written

$$\langle \theta^2 \rangle = \int_0^\infty w(f) \, df \qquad (S.4.91.4)$$

where $w(f)$ is the power spectral density of θ. At high gas density, $w(f)$ is broader and smaller in amplitude, while the integral remains constant. This corresponds to more frequent collisions and smaller amplitudes, whereas, at low density, $w(f)$ is more peaked around the natural frequency of the torsional pendulum $\sqrt{D/I}$, where I is its moment of inertia, still keeping the integral constant. It then appears that by reducing the density of the gas we actually increase the amplitude of fluctuations!

4.92   Isothermal Compressibility and Mean Square Fluctuation (Stony Brook)

a) Let us use the Jacobian transformation for thermodynamic variables:

$$\kappa_T = -\frac{1}{V}\left(\frac{\partial V}{\partial P}\right)_{T,N} = -\frac{1}{V}\frac{\partial(V,N)/\partial(V,\mu)}{\partial(P,N)/\partial(V,\mu)} = -\frac{1}{V}\left(\frac{\partial N}{\partial \mu}\right)_V \frac{\partial(V,\mu)}{\partial(P,N)}$$

$$= -\frac{1}{V}\left(\frac{\partial N}{\partial \mu}\right)_{V,\tau}\left[\left(\frac{\partial V}{\partial P}\right)_{N,\tau}\left(\frac{\partial \mu}{\partial N}\right)_{P,\tau} - \left(\frac{\partial V}{\partial N}\right)_{P,\tau}\left(\frac{\partial \mu}{\partial P}\right)_{N,\tau}\right]$$

Since the chemical potential μ is expressed in P,τ and does not depend on N, we can write

$$\mathrm{d}\mu = -s\ \mathrm{d}\tau + v\ \mathrm{d}P \tag{S.4.92.1}$$

where $s = S/N$ and $v = V/N$ are reduced entropy and volume respectively. Using the equation for the Gibbs free energy of a single-component system, $G = \mu N$, we can write

$$\kappa_\tau = \frac{1}{V}\left(\frac{\partial N}{\partial \mu}\right)_{V,\tau}\frac{\partial^2 G}{\partial N\partial P}\left(\frac{\partial \mu}{\partial P}\right)_{N,\tau} = \frac{1}{V}\left(\frac{\partial N}{\partial \mu}\right)_{V,\tau}\left(\frac{\partial \mu}{\partial P}\right)_{N,\tau}^2$$

where we also used $V = (\partial G/\partial P)_\tau$. But from (S.4.92.1),

$$\left(\frac{\partial \mu}{\partial P}\right)_{N,\tau} = v = \frac{V}{N}$$

So finally

$$\kappa_\tau = \frac{1}{V}\left(\frac{\partial N}{\partial \mu}\right)_{V,\tau}\left(\frac{V}{N}\right)^2 = \frac{V}{N^2}\left(\frac{\partial N}{\partial \mu}\right)_{V,\tau}$$

b) By definition the average number of particles in the grand canonical ensemble is

$$\langle N\rangle = e^{\Omega/\tau}\sum_N Ne^{\mu N/\tau}\sum_n e^{-E_{nN}/\tau}$$

where $\Omega = F - \mu N$. Now, from (a),

$$\kappa_\tau = \frac{V}{N^2}\left(\frac{\partial N}{\partial \mu}\right)_{V,\tau}$$

where N is an average number of particles:

$$\frac{\partial \langle N\rangle}{\partial \mu} = e^{\Omega/\tau}\sum_N \frac{N^2}{\tau}e^{\mu N/\tau}\sum_n e^{-E_{nN}/\tau} + \frac{1}{\tau}\frac{\partial\Omega}{\partial\mu}\langle N\rangle \tag{S.4.92.2}$$

$$= \frac{1}{\tau}\left(\langle N^2\rangle - \langle N\rangle^2\right)$$

where we have used $\langle N\rangle = -\partial\Omega/\partial\mu$. So

$$\kappa_\tau = \frac{V}{\langle N\rangle^2}\frac{\langle N^2\rangle - \langle N\rangle^2}{\tau} = \frac{V}{\tau}\frac{\langle(\Delta N)^2\rangle}{\langle N\rangle^2} \tag{S.4.92.3}$$

From (S.4.92.3)

$$\frac{\langle (\Delta N)^2 \rangle}{\langle N \rangle^2} = \frac{\tau}{V} \kappa_\tau$$

and

$$\langle (\Delta N)^2 \rangle = \frac{\tau \kappa_\tau}{V} \langle N \rangle^2$$

Since V is proportional to $\langle N \rangle$,

$$\langle (\Delta N)^2 \rangle \propto \langle N \rangle$$

The relative fluctuation is given by

$$\frac{\sqrt{\langle (\Delta N)^2 \rangle}}{\langle N \rangle} \sim \frac{1}{\sqrt{N}}$$

4.93   Energy Fluctuation in Canonical Ensemble (Colorado, Stony Brook)

First solution: For a canonical ensemble:

$$\langle E^2 \rangle = \frac{\sum_n \epsilon_n^2 e^{-\epsilon_n/\tau}}{\sum_n e^{-\epsilon_n/\tau}} = \frac{\frac{\partial^2 Z}{\partial \beta^2}}{Z} \tag{S.4.93.1}$$

where $\beta = 1/\tau$. On the other hand,

$$\langle E \rangle = \frac{\sum_n \epsilon_n e^{-\beta \epsilon_n}}{Z} = -\frac{\frac{\partial Z}{\partial \beta}}{Z} \tag{S.4.93.2}$$

Differentiating (S.4.93.2), we obtain

$$-\frac{\partial \langle E \rangle}{\partial \beta} = \frac{1}{Z} \frac{\partial^2 Z}{\partial \beta^2} - \frac{1}{Z^2} \left(\frac{\partial Z}{\partial \beta} \right)^2 \tag{S.4.93.3}$$

By inspecting (S.4.93.1)–(S.4.93.3), we find that

$$-\frac{\partial \langle E \rangle}{\partial \beta} = \langle E^2 \rangle - \langle E \rangle^2 \tag{S.4.93.4}$$

Now, the heat capacity at constant volume, C_v, is given by

$$C_v = \left(\frac{\partial \langle E \rangle}{\partial \tau} \right)_V = \left(\frac{\partial \langle E \rangle}{\partial \beta} \right)_V \frac{\partial \beta}{\partial \tau} = -\frac{1}{\tau^2} \left(\frac{\partial \langle E \rangle}{\partial \beta} \right)_V \tag{S.4.93.5}$$

Therefore, comparing (S.4.93.4) and (S.4.93.5), we deduce that, at constant volume,

$$\langle E^2 \rangle - \langle E \rangle^2 = \tau^2 C_v \qquad (S.4.93.6)$$

or in standard units

$$\langle E^2 \rangle - \langle E \rangle^2 = k_B T^2 C_v \qquad (S.4.93.7)$$

Since

$$\langle E^2 \rangle - \langle E \rangle^2 = \left\langle (E - \langle E \rangle)^2 \right\rangle \geq 0$$

then

$$C_v \geq 0$$

Second solution: A more general approach may be followed which is applicable to other problems. Because the probability w of finding that the value of a certain quantity X deviates from its average value $\langle X \rangle$ is proportional to $e^{S(X-\langle X \rangle)}$ and denoting $x = X - \langle X \rangle$, we can write

$$w(x)\, dx \sim e^{S(x)}\, dx \qquad (S.4.93.8)$$

Note that $\langle x \rangle = 0$. The entropy has a maximum at $x = 0$. Expanding $S(x)$, we obtain

$$S(x) = S(0) - \frac{x^2}{2\lambda}$$

where

$$\left. \frac{\partial^2 S}{\partial x^2} \right|_{x=0} < 0$$

so $\lambda > 0$. The probability distribution

$$w(x)\, dx = \frac{1}{\sqrt{2\pi\lambda}} e^{-x^2/2\lambda}\, dx \qquad (S.4.93.9)$$

$\langle x^2 \rangle = \lambda$, so

$$w(x)\, dx = \frac{1}{\sqrt{2\pi \langle x^2 \rangle}} \exp\left(\frac{-x^2}{2\langle x^2 \rangle} \right)\, dx \qquad (S.4.93.10)$$

If we have several variables,

$$w \propto \exp\left(-\frac{x_i x_k}{2\lambda_{ik}} \right) \qquad (S.4.93.11)$$

If the fluctuations of two variables x_i, x_k are statistically independent,

$$\langle x_i x_k \rangle = \langle x_i \rangle \langle x_k \rangle = 0$$

The converse is also true: If $\langle x_i x_k \rangle = 0$, the variables x_i and x_k are statistically independent. Now for a closed system we can write

$$w \propto e^{S_0} \propto e^{\Delta S_0} \qquad \text{(S.4.93.12)}$$

where S_0 is the total entropy of the system and ΔS_0 is the entropy change due to the fluctuation. On the other hand,

$$\Delta S_0 = -\frac{W_{\min}}{\tau} \qquad \text{(S.4.93.13)}$$

where $W_{\min}$ is the minimum work to change reversibly the thermodynamic variables of a small part of a system (the rest of the system works as a heat bath), and τ is the average temperature of the system (and therefore the temperature of the heat bath). Hence,

$$w \propto e^{-W_{\min}/\tau_0}$$

However,

$$W_{\min} = \Delta E - \tau \Delta S + P \Delta V \qquad \text{(S.4.93.14)}$$

where ΔE, ΔS, and ΔV are changes of a small part of a system due to fluctuations and τ, P are the average temperature and pressure. So,

$$w \propto \exp\left(-\frac{\Delta E - \tau \Delta S + P \Delta V}{\tau}\right) \qquad \text{(S.4.93.15)}$$

Expanding ΔE (for small fluctuations) gives

$$\Delta E = \frac{\partial E}{\partial S}\Delta S + \frac{\partial E}{\partial V}\Delta V \qquad \text{(S.4.93.16)}$$

$$+\frac{1}{2}\left[\frac{\partial^2 E}{\partial S^2}(\Delta S)^2 + 2\frac{\partial^2 E}{\partial S \partial V}\Delta S\, \Delta V + \frac{\partial^2}{\partial V^2}(\Delta V)^2\right]$$

Substituting (S.4.93.16) into (S.4.93.14), we obtain

$$W_{\min} = \frac{1}{2}\left(\frac{\partial^2 E}{\partial S^2}\right)(\Delta S)^2 + \frac{\partial^2 E}{\partial S \partial V}\Delta S\, \Delta V + \frac{1}{2}\frac{\partial^2}{\partial V^2}(\Delta V)^2$$

$$= \frac{1}{2}\Delta S\, \Delta\left(\frac{\partial E}{\partial S}\right)_V + \frac{1}{2}\Delta V\, \Delta\left(\frac{\partial E}{\partial V}\right)_S \qquad \text{(S.4.93.17)}$$

$$= \frac{1}{2}\left(\Delta S\, \Delta\tau - \Delta P\, \Delta V\right)$$

where we used

$$\frac{\partial E}{\partial S} = \tau \qquad \frac{\partial E}{\partial V} = -P$$

So, finally

$$w \propto \exp\left(\frac{\Delta P\,\Delta V - \Delta\tau\,\Delta S}{2\tau}\right) \qquad \text{(S.4.93.18)}$$

Using V and τ as independent variables we have

$$\Delta S = \left(\frac{\partial S}{\partial\tau}\right)_V \Delta\tau + \left(\frac{\partial S}{\partial V}\right)_\tau \Delta V = \frac{C_v}{\tau}\Delta\tau + \left(\frac{\partial P}{\partial\tau}\right)_V \Delta V$$

$$\Delta P = \left(\frac{\partial P}{\partial\tau}\right)_V \Delta\tau + \left(\frac{\partial P}{\partial V}\right)_\tau \Delta V \qquad \text{(S.4.93.19)}$$

Substituting (S.4.93.19) into (S.4.93.18), we see that the cross terms with $\Delta V\,\Delta\tau$ cancel (which means that the fluctuations of volume and temperature are statistically independent, $\langle \Delta V \cdot \Delta\tau\rangle = 0$):

$$w \propto \exp\left\{\left(\frac{\partial P}{\partial V}\right)_\tau \frac{(\Delta V)^2}{2\tau} - C_v\frac{(\Delta\tau)^2}{2\tau^2}\right\} \qquad \text{(S.4.93.20)}$$

Comparing (S.4.93.20) with (S.4.93.10), we find that the fluctuations of volume and temperature are given by

$$\left\langle (\Delta V)^2\right\rangle = -\tau\left(\frac{\partial V}{\partial P}\right)_\tau \qquad \text{(S.4.93.21)}$$

$$\left\langle (\Delta\tau)^2\right\rangle = \frac{\tau^2}{C_v}$$

To find the energy fluctuation, we can expand ΔE:

$$\Delta E = \left(\frac{\partial E}{\partial V}\right)_\tau \Delta V + \left(\frac{\partial E}{\partial\tau}\right)_V \Delta\tau$$

$$= \left[\tau\left(\frac{\partial P}{\partial\tau}\right)_V - P\right]\Delta V + C_v\,\Delta\tau \qquad \text{(S.4.93.22)}$$

$$\left\langle (\Delta E)^2\right\rangle = \langle E^2\rangle - \langle E\rangle^2$$

$$= \left[\tau\left(\frac{\partial P}{\partial\tau}\right)_V - P\right]^2 \left\langle (\Delta V)^2\right\rangle + C_v^2\left\langle (\Delta\tau)^2\right\rangle$$

where we used $\langle \Delta V \cdot \Delta \tau \rangle = 0$. Substituting $\left\langle (\Delta V)^2 \right\rangle$ and $\left\langle (\Delta \tau)^2 \right\rangle$ from (S.4.93.21), we obtain a more general formula for $\left\langle (\Delta E)^2 \right\rangle$:

$$\left\langle (\Delta E)^2 \right\rangle = -\tau \left[\tau \left(\frac{\partial P}{\partial \tau} \right)_V - P \right]^2 \left(\frac{\partial V}{\partial P} \right)_\tau + C_v \tau^2 \qquad \text{(S.4.93.23)}$$

At constant volume (S.4.93.23) becomes

$$\left\langle (\Delta E)^2 \right\rangle = C_v \tau^2$$

the same as before.

4.94 Number Fluctuations (Colorado (a,b), Moscow Phys-Tech (c))

a) Using the formula derived in Problem 4.92, we have

$$\left(\frac{\partial \langle N \rangle}{\partial \mu} \right)_\tau = \frac{1}{\tau} \left(\langle N^2 \rangle - \langle N \rangle^2 \right) \qquad \text{(S.4.94.1)}$$

$$(\Delta N)^2 = \tau \left(\frac{\partial \langle N \rangle}{\partial \mu} \right)_\tau$$

Consider an assortment of n_k particles which are in the kth quantum state. They are statistically independent of the other particles in the gas; therefore we can apply (S.4.94.1) in the form

$$\left\langle (\Delta n_k)^2 \right\rangle = \tau \left(\frac{\partial \langle n_k \rangle}{\partial \mu} \right)_\tau \qquad \text{(S.4.94.2)}$$

For a Fermi gas

$$\langle n_k \rangle = \frac{1}{e^{(\varepsilon_k - \mu)/\tau} + 1} \qquad \text{(S.4.94.3)}$$

So, by (S.4.94.2),

$$\left\langle \Delta n_k^2 \right\rangle_F = \frac{e^{(\varepsilon_k - \mu)/\tau}}{\left[e^{(\varepsilon_k - \mu)/\tau} + 1 \right]^2} \qquad \text{(S.4.94.4)}$$

$$= \left[\frac{1}{e^{(\varepsilon_k - \mu)/\tau} + 1} \right] \left[1 - \frac{1}{e^{(\varepsilon_k - \mu)/\tau} + 1} \right] = \langle n_k \rangle \left(1 - \langle n_k \rangle \right)$$

Similarly, for a Bose gas

$$\langle n_k \rangle = \frac{1}{e^{(\varepsilon_k - \mu)/\tau} - 1} \qquad \text{(S.4.94.5)}$$

we have

$$\langle \Delta n_k^2 \rangle_{\text{B}} = \langle n_k \rangle \left(1 + \langle n_k \rangle \right) \qquad \text{(S.4.94.6)}$$

b) First solution: Since a classical ideal gas is a limiting case of both Fermi and Bose gases at $\langle n_k \rangle \ll 1$, we get, from (S.4.94.3) or (S.4.94.6),

$$\langle n_k^2 \rangle \approx \langle n_k \rangle \qquad \text{(S.4.94.7)}$$

Alternatively, we can take the distribution function for an ideal classical gas,

$$\langle n_k \rangle = e^{(\mu - \varepsilon_k)/\tau}$$

and use (S.4.94.2) to get the same result. Since all the numbers n_k of particles in each state are statistically independent, we can write

$$\left\langle (\Delta N)^2 \right\rangle = \sum_k \left\langle (\Delta n_k)^2 \right\rangle = \sum_k \langle n_k \rangle = \langle N \rangle \qquad \text{(S.4.94.8)}$$

Second solution: In Problem 4.93 we derived the volume fluctuation

$$\left\langle (\Delta V)^2 \right\rangle = -\tau \left(\frac{\partial V}{\partial P} \right)_\tau \qquad \text{(S.4.94.9)}$$

This gives the fluctuation of a system containing N particles. If we divide (S.4.94.9) by N^2, we find the fluctuation of the volume per particle:

$$\left\langle \left(\Delta \frac{V}{N} \right)^2 \right\rangle = -\frac{\tau}{N^2} \left(\frac{\partial V}{\partial P} \right)_\tau \qquad \text{(S.4.94.10)}$$

This fluctuation should not depend on which is constant, the volume or the number of particles. If we consider that the volume in (S.4.94.10) is constant, then

$$\Delta \frac{V}{N} = V \Delta \frac{1}{N} = -\frac{V}{N^2} \Delta N \qquad \text{(S.4.94.11)}$$

Substituting (S.4.94.11) into (S.4.94.10) gives

$$\left\langle (\Delta N)^2 \right\rangle = -\frac{\tau \langle N \rangle^2}{V^2} \left(\frac{\partial V}{\partial P} \right)_\tau \qquad \text{(S.4.94.12)}$$

Using the equation for an ideal gas, $V = \langle N \rangle \tau / P$, in (S.4.94.12), we obtain

$$\left\langle (\Delta N)^2 \right\rangle = \frac{\tau \langle N \rangle^2}{V^2} \frac{\langle N \rangle \tau}{P^2} = \langle N \rangle \qquad \text{(S.4.94.13)}$$

Third solution: Use the Poisson distribution, which does not require that the fluctuations be small:

$$w_N = \frac{\langle N \rangle^N e^{-\langle N \rangle}}{N!} \qquad \text{(S.4.94.14)}$$

The average square number of particles is

$$\langle N^2 \rangle = \sum_{N=0}^{\infty} N^2 w_N = e^{-\langle N \rangle} \sum_{N=1}^{\infty} \frac{\langle N \rangle^N N}{(N-1)!}$$

$$= e^{-\langle N \rangle} \sum_{N=2}^{\infty} \frac{\langle N \rangle^N}{(N-2)!} + e^{-\langle N \rangle} \sum_{N=1}^{\infty} \frac{\langle N \rangle^N}{(N-1)!} \qquad \text{(S.4.94.15)}$$

$$= e^{-\langle N \rangle} \sum_{N=2}^{\infty} \frac{\langle N \rangle^2 \langle N \rangle^{N-2}}{(N-2)!} + e^{-\langle N \rangle} \sum_{N=1}^{\infty} \frac{\langle N \rangle \langle N \rangle^{N-1}}{(N-1)!} = \langle N \rangle^2 + \langle N \rangle$$

Thus we recover (S.4.94.8) again:

$$\left\langle (\Delta N)^2 \right\rangle = \langle N \rangle$$

c) Again we will use (S.4.94.1):

$$(\Delta N)^2 = \tau \left(\frac{\partial \langle N \rangle}{\partial \mu} \right)_{\tau}$$

Since the gas is strongly degenerate, $\tau \ll \varepsilon_{\mathrm{F}}$, we can use $\mu = \varepsilon_{\mathrm{F}}$ and $\tau = 0$ (see Problem 4.66):

$$\mu = \varepsilon_{\mathrm{F}} = (3\pi^2)^{2/3} \frac{\hbar^2}{2m} \left(\frac{N}{V} \right)^{2/3} \qquad \text{(S.4.94.16)}$$

Then

$$\left\langle (\Delta N)^2 \right\rangle = \frac{\tau}{\left(\frac{\partial \mu}{\partial \langle N \rangle} \right)_{\tau=0}} = \frac{\tau V^{2/3}}{\frac{2}{3} (3\pi^2)^{2/3} \frac{\hbar^2}{2m} N^{-1/3}} \qquad \text{(S.4.94.17)}$$

$$= \frac{3m\tau V^{2/3} N^{1/3}}{(3\pi^2)^{2/3} \hbar^2} = \frac{3^{1/3} m\tau V}{\pi^{4/3} \hbar^2} \left(\frac{N}{V} \right)^{1/3}$$

4.95 Wiggling Wire (Princeton)

First solution: Consider the midpoint of the wire P fixed at points A and B (see Figure S.4.95). let s be the deviation of the wire from the line segment AB. Then, in equilibrium, the wire will consist of segments AP' and $P'B$. To find $\langle s^2 \rangle$, we will have to find the minimum work $W_{\min}$ to change the shape of the wire from APB to $AP'B$:

$$W_{\min} = f\,\Delta L = 2f\left(\sqrt{\left(\frac{L}{2}\right)^2 + s^2} - \frac{L}{2}\right) \qquad \text{(S.4.95.1)}$$

$$\approx 2f\frac{L}{2}\frac{1}{2}\frac{s^2}{(L/2)^2} = \frac{2fs^2}{L}$$

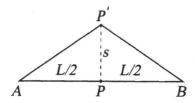

Figure S.4.95

Using a standard formula for the probability of fluctuation (see Problem 4.93),

$$w \propto e^{-W_{\min}/\tau} \qquad \text{(S.4.95.2)}$$

we obtain

$$w \propto \exp\left(-\frac{2fs^2}{L\tau}\right) = \exp\left(-\frac{s^2}{2\langle s^2\rangle}\right) \qquad \text{(S.4.95.3)}$$

So,

$$\langle s^2 \rangle = \frac{\tau L}{4f} \qquad \text{(S.4.95.4)}$$

$$s_{\mathrm{rms}} = \frac{1}{2}\sqrt{\frac{\tau L}{f}}$$

This answer can be easily generalized for an arbitrary point x along the wire (see Landau and Lifshitz, *Statistical Physics*, Sect. 112):

$$\langle s^2 \rangle = \frac{\tau}{fL}x\,(L - x) \qquad \text{(S.4.95.5)}$$

Second solution: We solve the equation of motion for the wire (see derivation in Problem 1.46, Part I). For the boundary conditions $s(x,t)|_{x=0} = s(x,t)|_{x=L} = 0$, we have modes:

$$s_n(x, t) = \sin(\omega_n x/c)(A_n \cos \omega_n t + B_n \sin \omega_n t) \qquad (S.4.95.6)$$

with $\omega_n = ck_n = cn\pi/L$, for $n = 0, 1, \ldots$, where c is the phase velocity, $c = \sqrt{f/\rho}$. Taking for simplicity

$$s_n = A_n \sin(n\pi x/L) \cos \omega_n t \qquad (S.4.95.7)$$

we can find the average kinetic and potential energy in each mode:

$$\langle V_n \rangle = \langle K_n \rangle = \left\langle \frac{1}{2} \int_0^L \rho \dot{s}^2 \, dx \right\rangle$$

$$\left\langle \frac{A_n^2}{2} \int_0^L \rho \omega_n^2 \sin^2\left(\frac{n\pi x}{L}\right) \sin^2(\omega_n t) \, dx \right\rangle = \frac{\rho \omega_n^2 L}{8} A_n^2 \qquad (S.4.95.8)$$

$$= \frac{\rho c^2 n^2 \pi^2}{8L} A_n^2 = \frac{f n^2 \pi^2}{8L} A_n^2$$

The total energy in each mode is

$$E_n = \langle V_n \rangle + \langle K_n \rangle = \frac{f n^2 \pi^2}{4L} A_n^2 = \frac{\tau}{2} + \frac{\tau}{2} = \tau \qquad (S.4.95.9)$$

The fluctuation of the wire is given by

$$\langle s^2 \rangle = \sum_n \langle s_n^2 \rangle = \sum_n A_n^2 \left\langle \sin^2\left(\frac{n\pi x}{L}\right) \cos^2 \omega_n t \right\rangle \qquad (S.4.95.10)$$

$$= \frac{1}{2} \sum_n A_n^2 \sin^2\left(\frac{n\pi x}{L}\right)$$

where we have used $\langle \cos^2 \omega_n t \rangle = 1/2$. For $x = L/2$,

$$\langle s^2 \rangle = \frac{1}{2} \sum_{n=1}^{\infty} A_n^2 \sin^2\left(\frac{n\pi}{2}\right) \qquad (S.4.95.11)$$

$$= \frac{1}{2} \sum_{n=1}^{\infty} \frac{4L\tau}{f\pi^2} \frac{1}{n^2} \sin^2\left(\frac{n\pi}{2}\right)$$

where we substituted A_n from (S.4.95.9). Note that even modes do not contribute to the fluctuation of the midpoint of the wire, as expected from elementary considerations. We may then find the fluctuation:

$$\langle s^2 \rangle = \frac{2L\tau}{f\pi^2} \sum_{n=1}^{\infty} \frac{1}{n^2} \sin^2\left(\frac{n\pi}{2}\right) \qquad (S.4.95.12)$$

$$= \frac{2L\tau}{f\pi^2} \sum_{m=0}^{\infty} \frac{1}{(2m+1)^2} = \frac{2L\tau}{f\pi^2} \frac{\pi^2}{8} = \frac{L\tau}{4f}$$

where we have used the sum given in the problem

$$\sum_{m=0}^{\infty} (2m+1)^{-2} = \frac{\pi^2}{8}$$

So,

$$s_{\rm rms} = \frac{1}{2}\sqrt{\frac{L\tau}{f}}$$

as before.

4.96   *LC* Voltage Noise (MIT, Chicago)

Write the Hamiltonian H for the circuit:

$$H = \frac{1}{2}\frac{Q^2}{C} + \frac{1}{2}L\dot{Q}^2 \qquad (S.4.96.1)$$

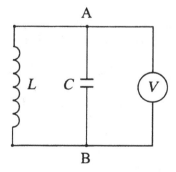

Figure S.4.96

where Q is the charge on the capacitor. This is the Hamiltonian of a harmonic oscillator of frequency $\omega = 1/\sqrt{LC}$, whose energy levels ε_n are

$$\varepsilon_n = \hbar\omega \left(n + \frac{1}{2} \right) \tag{S.4.96.2}$$

The average energy in the circuit is given by (with $\beta \equiv 1/\tau$)

$$\langle E \rangle = \frac{\displaystyle\sum_n \varepsilon_n e^{-\varepsilon_n/\tau}}{\displaystyle\sum_n e^{-\varepsilon_n/\tau}} = \frac{\displaystyle\sum_n \varepsilon_n e^{-\beta\varepsilon_n}}{\displaystyle\sum_n e^{-\beta\varepsilon_n}}$$

$$= -\frac{\partial}{\partial\beta} \ln \left(\sum_n e^{-\beta\varepsilon_n} \right) = -\frac{\partial}{\partial\beta} \ln \left[e^{-\beta\hbar\omega/2} \sum_n \left(e^{-\beta\hbar\omega} \right)^n \right]$$

$$= -\frac{\partial}{\partial\beta} \ln \left(\frac{e^{-\beta\hbar\omega/2}}{1 - e^{-\beta\hbar\omega}} \right) = \frac{\partial}{\partial\beta} \ln \left[2 \sinh \left(\frac{\beta\hbar\omega}{2} \right) \right] \tag{S.4.96.3}$$

$$= \frac{\hbar\omega}{2} \frac{\cosh(\beta\hbar\omega/2)}{\sinh(\beta\hbar\omega/2)} = \frac{\hbar\omega}{2} \coth \left(\frac{\hbar\omega}{2\tau} \right)$$

The average energy is equally distributed between the capacitance and the inductance:

$$\frac{\langle E \rangle}{2} = \frac{1}{2} \langle CV^2 \rangle \tag{S.4.96.4}$$

where V is the voltage across the capacitor (between points A and B; see Figure S.4.96). We then have

$$\langle V^2 \rangle = \frac{\langle E \rangle}{C} = \frac{\hbar\omega}{2C} \coth \left(\frac{\hbar\omega}{2\tau} \right) \tag{S.4.96.5}$$

a) In the classical limit $\hbar\omega \ll \tau$:

$$\langle V^2 \rangle = \left(\frac{\hbar\omega}{2C} \right) \frac{e^{\hbar\omega/2\tau} + e^{-\hbar\omega/2\tau}}{e^{\hbar\omega/2\tau} - e^{-\hbar\omega/2\tau}} \approx \frac{\hbar\omega}{2C} \frac{2\tau}{\hbar\omega} = \frac{\tau}{C} \tag{S.4.96.6}$$

We could equally well have derived the classical result by using the equipartition theorem (see Problem 4.42). For the single degree of freedom, there is an average energy τ, which, as noted, is divided between the capacitor and inductor, so

$$\frac{1}{2} C \langle V^2 \rangle = \frac{\tau}{2} \tag{S.4.96.7}$$

$$\langle V^2 \rangle = \frac{\tau}{C}$$

as found in (S.4.96.6). The mean square noise voltage is

$$\sqrt{\langle V^2 \rangle} \approx \sqrt{\frac{\tau}{C}} \qquad (\text{S.4.96.8})$$

b) If $\hbar\omega \gg \tau$, then (S.4.96.6) becomes

$$\langle V^2 \rangle \approx \frac{\hbar\omega}{2C} \qquad (\text{S.4.96.9})$$

$$\sqrt{\langle V^2 \rangle} \approx \sqrt{\frac{\hbar\omega}{2C}}$$

Applications to Solid State

4.97 Thermal Expansion and Heat Capacity (Princeton)

a) First solution: We can calculate the average displacement of an oscillator:

$$\langle x \rangle = \frac{\int\limits_{-\infty}^{\infty} x e^{-V_0(x)/\tau} \, dx}{\int\limits_{-\infty}^{\infty} e^{-V_0(x)/\tau} \, dx} \qquad (\text{S.4.97.1})$$

Since the anharmonic term is small, $m\lambda x^3/3 \ll \tau$, we can expand the exponent in the integral:

$$\langle x \rangle \approx \frac{\int\limits_{-\infty}^{\infty} e^{-K_0 x^2/2\tau} \left(x + m\lambda x^4/3\tau \right) \, dx}{\int\limits_{-\infty}^{\infty} e^{-K_0 x^2/2\tau} \, dx}$$

$$= \frac{m\lambda}{3\tau} \frac{\int\limits_{-\infty}^{\infty} e^{-K_0 x^2/2\tau} x^4 \, dx}{\int\limits_{-\infty}^{\infty} e^{-K_0 x^2/2\tau} \, dx} \qquad (\text{S.4.97.2})$$

$$= \frac{m\lambda}{3\tau} \frac{\frac{d^2}{d\alpha^2} \int\limits_{-\infty}^{\infty} e^{-\alpha x^2} \, dx}{\int\limits_{-\infty}^{\infty} e^{-\alpha x^2} \, dx}$$

where we set $\alpha \equiv K_0/2\tau$. So,

$$\langle x \rangle = \frac{m\lambda}{3\tau} \frac{\dfrac{d^2}{d\alpha^2}\left(\sqrt{\dfrac{\pi}{\alpha}}\right)}{\sqrt{\dfrac{\pi}{\alpha}}} = \frac{m\lambda}{K_0^2}\tau \qquad (S.4.97.3)$$

Note that, in this approximation, the next term in the potential ($\propto x^4$) would not have introduced any additional shift (only antisymmetric terms do).

Second solution: (see Problem 1.37, Part I) We can solve the equation of motion for the nonlinear harmonic oscillator corresponding to the potential $V_0(x)$:

$$\ddot{x} + \omega_0 x = \lambda x^2 \qquad (S.4.97.4)$$

where $\omega_0 = \sqrt{K_0/m}$ is the principal frequency. The solution (see (S.1.37.10) of Part I) gives

$$x(t) = A' \sin \omega_0 t - \frac{2}{3}\frac{A^2\lambda}{\omega_0^2}\cos\omega_0 t + \frac{1}{2}\frac{A^2\lambda}{\omega_0^2} + \frac{1}{6}\frac{A^2\lambda}{\omega_0^2}\cos 2\omega_0 t \qquad (S.4.97.5)$$

where A' is defined from the initial conditions and A is the amplitude of oscillations of the linear equation. The average $\langle x \rangle$ over a period $T = 2\pi/\omega_0$ is

$$\langle x \rangle = \frac{1}{2}\frac{A^2\lambda}{\omega_0^2} \qquad (S.4.97.6)$$

We need to calculate the thermodynamic average of $\langle x \rangle$:

$$\langle x \rangle = \frac{\displaystyle\int_0^\infty \langle x \rangle e^{-\varepsilon/\tau}\,d\varepsilon}{\displaystyle\int_0^\infty e^{-\varepsilon/\tau}\,d\varepsilon} = \frac{\lambda}{2\omega_0^2\tau}\int_0^\infty A^2(\varepsilon)e^{-\varepsilon/\tau}\,d\varepsilon \qquad (S.4.97.7)$$

Substituting $A^2(\varepsilon) = 2\varepsilon/K_0$, we obtain

$$\langle x \rangle = \frac{\lambda}{K_0\omega_0^2\tau}\int_0^\infty \varepsilon e^{-\varepsilon/\tau}\,d\varepsilon = \frac{\lambda}{K_0\omega_0^2}\tau = \frac{m\lambda}{K_0^2}\tau \qquad (S.4.97.8)$$

the same as before.

b) The partition function of a single oscillator associated with this potential energy is

$$Z = \frac{1}{2\pi\hbar} \int_{-\infty}^{\infty} e^{-p^2/2m\tau} \, dp \int_{-\infty}^{\infty} e^{-V_1(x)/\tau} \, dx$$

$$= \frac{\sqrt{2\pi m\tau}}{2\pi\hbar} \int_{-\infty}^{\infty} \exp\left\{ -\frac{K_0 x^2}{2\tau} + \frac{m\lambda x^3}{3\tau} - \frac{\eta x^4}{4\tau} \right\} \, dx$$

$$\approx \frac{\sqrt{2\pi m\tau}}{2\pi\hbar} \int_{-\infty}^{\infty} e^{-K_0 x^2/2\tau} \left(1 + \frac{m\lambda x^3}{3\tau} - \frac{\eta x^4}{4\tau} \right) \, dx \qquad \text{(S.4.97.9)}$$

$$= \frac{1}{\hbar} \sqrt{\frac{m\tau}{2\pi}} \int_{-\infty}^{\infty} e^{-K_0 x^2/2\tau} \left(1 + \frac{m\lambda x^3}{3\tau} - \frac{\eta x^4}{4\tau} \right) \, dx$$

$$= \frac{1}{\hbar} \sqrt{\frac{m\tau}{2\pi}} \left(\sqrt{\frac{2\tau\pi}{K_0}} - \frac{\eta}{4\tau} \frac{d^2}{d\alpha^2} \sqrt{\frac{\pi}{\alpha}} \right)$$

$$= \frac{\tau}{\hbar} \sqrt{\frac{m}{K_0}} \left(1 - \frac{3\eta\tau}{4K_0^2} \right) = \frac{\tau}{\hbar\omega_0} \left(1 - \frac{3\eta\tau}{4K_0^2} \right)$$

So, the free energy F per oscillator is given by

$$F = -\tau \ln Z = -\tau \ln \frac{\tau}{\hbar\omega_0} \left(1 - \frac{3\eta\tau}{4K_0^2} \right) \qquad \text{(S.4.97.10)}$$

$$\approx -\tau \ln \tau + \tau \ln \hbar\omega_0 + \frac{3\eta\tau^2}{4K_0^2}$$

where we approximated $\ln(1 - x) \approx -x$. The energy per oscillator may be found from

$$E = -\tau^2 \left(\frac{\partial}{\partial\tau} \frac{F}{\tau} \right)_V = \tau - \frac{3\eta\tau^2}{4K_0^2} \qquad \text{(S.4.97.11)}$$

The heat capacity is then

$$C_V = \left(\frac{\partial E}{\partial\tau} \right)_V = 1 - \frac{3\eta\tau}{2K_0^2} \qquad \text{(S.4.97.12)}$$

The anharmonic correction to the heat capacity is negative.

4.98   Schottky Defects (Michigan State, MIT)

When N atoms are displaced to the surface, they leave the same number of vacancies. Now there are N vacancies and $\tilde{N}$ atoms in $\tilde{N} + N$ lattice points. The entropy as a function of N is

$$S(N) = \ln \frac{\left(\tilde{N} + N\right)!}{N! \tilde{N}!} \qquad \text{(S.4.98.1)}$$

$$\approx \left(\tilde{N} + N\right) \ln \left(\tilde{N} + N\right) - \tilde{N} \ln \tilde{N} - N \ln N$$

where we have used Stirling's formula

$$\ln n! \approx n \ln n - n$$

The free energy may be written

$$F(N) = E(N) - \tau S(N) \qquad \text{(S.4.98.2)}$$

$$= N\varepsilon_0 - \tau \left[\left(\tilde{N} + N\right) \ln \left(\tilde{N} + N\right) - \tilde{N} \ln \tilde{N} - N \ln N\right]$$

The minimum of the free energy can be found to be

$$\left(\frac{\partial F}{\partial N}\right)_{\tau, \tilde{N}} = \varepsilon_0 - \tau \ln \left(\tilde{N} + N\right) + \tau \ln N = 0 \qquad \text{(S.4.98.3)}$$

or

$$\frac{N}{\tilde{N} + N} = e^{-\varepsilon_0/\tau} \qquad \text{(S.4.98.4)}$$

Since $N \ll \tilde{N}$, we have

$$N \approx \tilde{N} e^{-\varepsilon_0/\tau} \qquad \text{(S.4.98.5)}$$

which is what one would expect.

4.99   Frenkel Defects (Colorado, MIT)

We assume that the number of defects created around one lattice site does not affect the process of creating new defects. In other words, all configurations of the system are independent (not a very realistic assumption in general, but since $\tau \ll \varepsilon_0$, and the number of defects $n \ll N, \tilde{N}$, it can be used as an approximation). The vacancies and interstices then can be distributed in w_v and w_i ways, respectively:

$$w_v = \frac{N!}{n!\,(N-n)!} \tag{S.4.99.1}$$

$$w_i = \frac{\tilde{N}!}{n!\left(\tilde{N}-n\right)!}$$

The total number of possible configurations of the system, w, is given by

$$w = w_v w_i = \frac{N!\tilde{N}!}{(n!)^2\,(N-n)!\left(\tilde{N}-n\right)!} \tag{S.4.99.2}$$

The entropy, S, may be written

$$S = \ln w = \ln N! + \ln \tilde{N}! - 2\ln n! - \ln(N-n)! - \ln\left(\tilde{N}-n\right)! \tag{S.4.99.3}$$

Using Stirling's formula

$$\ln N! \approx \int_1^N dx\ \ln x = N\ln N - N = N\ln \frac{N}{e} \tag{S.4.99.4}$$

we obtain, from (S.4.99.3),

$$S = N\ln \frac{N}{e} + \tilde{N}\ln \frac{\tilde{N}}{e} - 2n\ln \frac{n}{e} \tag{S.4.99.5}$$

$$- (N-n)\ln \frac{N-n}{e} - \left(\tilde{N}-n\right)\ln \frac{\tilde{N}-n}{e}$$

Using (S.4.99.5) and the fact that the total energy of the system $E = n\varepsilon_0$, we have

$$\frac{1}{\tau} = \left(\frac{\partial S}{\partial E}\right)_V = \frac{1}{\varepsilon_0}\frac{\partial S}{\partial n} \tag{S.4.99.6}$$

$$= \frac{1}{\varepsilon_0}\left[-2\ln n + \ln(N-n) + \ln\left(\tilde{N}-n\right)\right]$$

or

$$-\frac{\varepsilon_0}{2\tau} = \ln \frac{n}{\sqrt{(N-n)(\tilde{N}-n)}} \tag{S.4.99.7}$$

and

$$e^{-\varepsilon_0/2\tau} = \frac{n}{\sqrt{N\tilde{N}}}\left(1 + \frac{1}{2}n\left(\frac{1}{N} + \frac{1}{\tilde{N}}\right) + \cdots\right) \tag{S.4.99.8}$$

The condition $\varepsilon \ll \tau$ implies that $n \ll N, \tilde{N}$, and therefore

$$e^{-\varepsilon_0/2\tau} \approx \frac{n}{\sqrt{N\tilde{N}}} \qquad\qquad n \approx \sqrt{N\tilde{N}}e^{-\varepsilon_0/2\tau} \tag{S.4.99.9}$$

4.100 Two-Dimensional Debye Solid (Columbia, Boston)

a) The number of normal modes in the 2D solid within the interval dk of a wave vector k may be written

$$dN = \frac{2\pi k \, dk}{(2\pi)^2}L^2 \tag{S.4.100.1}$$

In the 2D solid there are only two independent polarizations of the excitations, one longitudinal and one transverse. Therefore,

$$dN = 2\frac{2\pi\omega \, d\omega}{\langle v\rangle^2 (2\pi)^2}L^2 = \frac{\omega \, d\omega}{\pi \langle v\rangle^2}L^2 \tag{S.4.100.2}$$

where $\langle v\rangle$ is the average velocity of sound. To find the Debye frequency ω_D, we use the standard assumption that the integral of (S.4.100.2) from 0 to a certain cut-off frequency ω_D is equal to the total number of vibrational modes; i.e.,

$$\int_0^{\omega_D} dN = \frac{L^2}{\pi \langle v\rangle^2}\int_0^{\omega_D} \omega \, d\omega = \frac{L^2\omega_D^2}{2\pi \langle v\rangle^2} = DN = 2N \tag{S.4.100.3}$$

Therefore,

$$\omega_D = \frac{2\langle v\rangle}{L}\sqrt{\pi N} \tag{S.4.100.4}$$

We can express $\langle v\rangle$ through ω_D:

$$\langle v\rangle = \frac{\omega_D L}{2\sqrt{\pi N}} \tag{S.4.100.5}$$

Then (S.4.100.2) becomes

$$dN = \frac{4N\omega \, d\omega}{\omega_D^2} \tag{S.4.100.6}$$

b) The free energy (see Problem 4.77) then becomes

$$F = N\varepsilon_0 + \tau \frac{4N}{\omega_D^2} \int_0^{\omega_D} \ln\left(1 - e^{-\hbar\omega/\tau}\right) \omega \, d\omega \qquad \text{(S.4.100.7)}$$

Defining $\theta_D = \hbar\omega_D$ and introducing a new variable $x = \hbar\omega/\tau$, we can rewrite (S.4.100.7) in the form

$$F = N\varepsilon_0 + 4N\tau \left(\frac{\tau}{\theta_D}\right)^2 \int_0^{\theta_D/\tau} x \ln\left(1 - e^{-x}\right) \, dx \qquad \text{(S.4.100.8)}$$

Integrating (S.4.100.8) by parts, we obtain

$$F = N\varepsilon_0 + 4N\tau \left(\frac{\tau}{\theta_D}\right)^2 \frac{x^2}{2} \ln\left(1 - e^{-x}\right)\Big|_0^{\theta_D/\tau}$$

$$-2N\tau \left(\frac{\tau}{\theta_D}\right)^2 \int_0^{\theta_D/\tau} \frac{x^2 \, dx}{e^x - 1}$$

$$= N\varepsilon_0 + 2N\tau \left(\frac{\tau}{\theta_D}\right)^2 \left(\frac{\theta_D}{\tau}\right)^2 \ln\left(1 - e^{-\theta_D/\tau}\right) \qquad \text{(S.4.100.9)}$$

$$-2N\tau \left(\frac{\tau}{\theta_D}\right)^2 \int_0^{\theta_D/\tau} \frac{x^2 \, dx}{e^x - 1}$$

$$= N\varepsilon_0 + N\tau \left[2 \ln\left(1 - e^{-\theta_D/\tau}\right) - D_2\left(\theta_D/\tau\right)\right]$$

where the 2D Debye function $D_2(z)$ is

$$D_2(z) \equiv \frac{2}{z^2} \int_0^z \frac{x^2 \, dx}{e^x - 1} \qquad \text{(S.4.100.10)}$$

The energy is given by

$$E = F - \tau \frac{\partial F}{\partial \tau} \qquad \text{(S.4.100.11)}$$

$$= N\varepsilon_0 + \frac{2N\tau^3}{\theta_D^2} \int_0^{\theta_D/\tau} \frac{x^2 \, dx}{e^x - 1} = N\varepsilon_0 + 2N\tau D_2\left(\frac{\theta_D}{\tau}\right)$$

The specific heat C ($C_P \approx C_V$ at low temperatures) is

$$C = \frac{\partial E}{\partial \tau} = 2N \left[D_2 \left(\frac{\theta_D}{\tau} \right) - \frac{\theta_D}{\tau} D_2' \left(\frac{\theta_D}{\tau} \right) \right] \qquad \text{(S.4.100.12)}$$

At low temperatures, $\tau \ll \theta_D$, we can extend the upper limit of integration to infinity:

$$D_2(z) \approx \frac{2}{z^2} \int\limits_0^\infty \frac{x^2 \, dx}{e^x - 1}$$

Therefore, at $\tau \ll \theta_D$,

$$C \approx 2N \left[2 \left(\frac{\tau}{\theta_D} \right)^2 + \frac{4\theta_D}{\tau \, (\theta_D/\tau)^3} \right] \int\limits_0^\infty \frac{x^2 \, dx}{e^x - 1} \qquad \text{(S.4.100.13)}$$

$$= 12N \left(\frac{\tau}{\theta_D} \right)^2 \int\limits_0^\infty \frac{x^2 \, dx}{e^x - 1}$$

where

$$\int\limits_0^\infty \frac{x^2 \, dx}{e^x - 1} = \Gamma(3)\zeta(3)$$

and $\zeta(x)$ is the Riemann ζ function. Note that the specific heat in 2D is

$$C \propto \tau^2$$

(see also Problem 4.75). Note also that you can solve a somewhat different problem: When atoms are confined to the surface but still have three degrees of freedom, the results will, of course, be different.

4.101 Einstein Specific Heat (Maryland, Boston)

a) For a harmonic oscillator with frequency ω the energy

$$\varepsilon = \hbar\omega \, \langle n \rangle \qquad \text{(S.4.101.1)}$$

where

$$\langle n \rangle = \frac{1}{e^{\hbar\omega/\tau} - 1}$$

So,

$$\varepsilon = \frac{\hbar\omega}{e^{\hbar\omega/\tau} - 1} \qquad \text{(S.4.101.2)}$$

b) If we assume that the N atoms of the solid each have three degrees of freedom and the same frequency ω_E, then the total energy

$$E = 3N \frac{\hbar \omega_E}{e^{\hbar \omega_E / \tau} - 1} \qquad \text{(S.4.101.3)}$$

The specific heat

$$C_V = \left(\frac{\partial E}{\partial \tau} \right)_V = 3N \left(\frac{\hbar \omega_E}{\tau} \right)^2 \frac{e^{\hbar \omega_E / \tau}}{\left(e^{\hbar \omega_E / \tau} - 1 \right)^2} \qquad \text{(S.4.101.4)}$$

c) In the high-temperature limit of (S.4.101.4) ($\hbar \omega_E \ll \tau$), we have

$$C_V \approx 3N \left(\frac{\hbar \omega_E}{\tau} \right)^2 \frac{(1 + \hbar \omega_E / \tau)^2}{(\hbar \omega_E / \tau)^2} \approx 3N \qquad \text{(S.4.101.5)}$$

In regular units

$$C_V = 3N k_{\mathrm{B}}$$

which corresponds to the law of Dulong and Petit, does not depend on the composition of the material but only on the total number of atoms and should be a good approximation at high temperatures, especially for one-component elements. From the numbers in the problem,

$$N = nV = 6 \cdot 10^{22} \cdot 5 = 3 \cdot 10^{23} \text{ atoms} \approx \frac{N_A}{2} \qquad \text{(S.4.101.6)}$$

Therefore,

$$C \approx 3 \frac{N_A}{2} k_{\mathrm{B}} = \frac{3}{2} R \qquad \text{(S.4.101.7)}$$

Note that, at high enough temperatures, anharmonic effects calculated in Problem 4.97 may become noticeable. Anharmonic corrections are usually negative and linearly proportional to temperature.

d) At low temperatures (S.4.101.4) becomes

$$C = 3N \left(\frac{\hbar \omega_E}{\tau} \right)^2 e^{-\hbar \omega_E / \tau} \qquad \text{(S.4.101.8)}$$

The heat capacity goes to zero as $\exp(-\hbar \omega_E / \tau)$ at $\tau \to 0$, whereas the experimental results give $C \propto \tau^3$ (see Problem 4.42). The faster falloff of the heat capacity is due to the "freezing out" of the oscillations at $\tau \to 0$ given the single natural frequency ω_E.

4.102   Gas Adsorption (Princeton, MIT, Stanford)

For two systems in equilibrium, the chemical potentials should be equal. Consider one of the systems as an ideal gas (vapor) in a volume, and another as a surface submonolayer film. For an ideal gas the free energy F (see Problem 4.38) is given by

$$F = -N\tau \ln \left[\frac{eV}{N} \left(\frac{m\tau}{2\pi\hbar^2} \right)^{3/2} \sum_k e^{-\varepsilon_k/\tau} \right] \qquad \text{(S.4.102.1)}$$

$$= -N\tau \ln \left[\frac{eV}{N} \left(\frac{m\tau}{2\pi\hbar^2} \right)^{3/2} Z' \right]$$

where ε_k and Z' correspond to the energy states and statistical sum associated with the internal degrees of freedom. If the temperature is reasonably small, $\tau \ll \tau_{\text{ion}}$, where τ_{ion} corresponds to the ionization energy of the atoms, so that the atoms are not ionized and mostly in the ground state, and this state is nondegenerate, we can take $Z' = 1$, and then (S.4.102.1) becomes

$$F = -N\tau \ln \left[\frac{eV}{N} \left(\frac{m\tau}{2\pi\hbar^2} \right)^{3/2} \right] \qquad \text{(S.4.102.2)}$$

The Gibbs free energy G is given by

$$G = F + PV = -N\tau \ln \left[\frac{e\tau}{P} \left(\frac{m\tau}{2\pi\hbar^2} \right)^{3/2} \right] + N\tau \quad \text{(S.4.102.3)}$$

$$= -N\tau \ln \left[\frac{\tau}{P} \left(\frac{m\tau}{2\pi\hbar^2} \right)^{3/2} \right]$$

where we have expressed G as a function of P and τ, using $PV = N\tau$. The chemical potential $\mu = G/N$, so

$$\mu = -\tau \ln \left[\frac{\tau}{P} \left(\frac{m\tau}{2\pi\hbar^2} \right)^{3/2} \right] = \tau \ln \left[\left(\frac{2\pi\hbar^2}{m} \right)^{3/2} \frac{P}{\tau^{5/2}} \right] \quad \text{(S.4.102.4)}$$

Now, consider an adsorption site: we can apply a Gibbs distribution with a variable number of particles to this site:

$$\Omega = -\tau \ln \sum_{n=0,1} e^{\mu n/\tau} \sum_k e^{-E_{kn}/\tau} \qquad \text{(S.4.102.5)}$$

where the possible occupational numbers of the site for a submonolayer $n = 0, 1$ (site is empty, site is occupied), with energy $E_{0n} = -n\varepsilon_0$. Performing the sums, we have

$$\Omega = -\tau \ln \sum_{n=0,1} e^{\mu n/\tau} e^{\varepsilon_0 n/\tau} \qquad (S.4.102.6)$$

$$= -\tau \ln \sum_{n=0,1} \left[e^{(\mu+\varepsilon_0)/\tau} \right]^n = -\tau \ln \left(1 + e^{(\mu+\varepsilon_0)/\tau} \right)$$

The average number of particles per site $\langle n \rangle$ may be written

$$\langle n \rangle = -\left(\frac{\partial \Omega}{\partial \mu} \right)_{T,V} = \frac{e^{(\mu+\varepsilon_0)/\tau}}{1 + e^{(\mu+\varepsilon_0)/\tau}} \qquad (S.4.102.7)$$

The total number of adsorbed particles N is given by

$$N = N_0 \langle n \rangle = \frac{N_0}{e^{-(\mu+\varepsilon_0)/\tau} + 1} \qquad (S.4.102.8)$$

The surface concentration θ is simply

$$\theta = \frac{N}{N_0} = \frac{1}{1 + e^{-(\mu+\varepsilon_0)/\tau}} \qquad (S.4.102.9)$$

Substituting μ for an ideal gas from (S.4.102.4) into (S.4.102.9), we have

$$\frac{1}{\theta} - 1 = e^{-\varepsilon_0/\tau} e^{-\mu/\tau} = e^{-\varepsilon_0/\tau} \left(\frac{m}{2\pi\hbar^2} \right)^{3/2} \frac{\tau^{5/2}}{P} \qquad (S.4.102.10)$$

and

$$P = \frac{\theta}{1-\theta} \left(\frac{m}{2\pi\hbar^2} \right)^{3/2} \tau^{5/2} e^{-\varepsilon_0/\tau} = \frac{\theta}{1-\theta} P_0(\tau) \qquad (S.4.102.11)$$

where

$$P_0(\tau) = \left(\frac{m}{2\pi\hbar^2} \right)^{3/2} \tau^{5/2} e^{-\varepsilon_0/\tau} \qquad (S.4.102.12)$$

(S.4.102.9) can also be derived by considering the canonical ensemble. The number of possible ways of distributing N atoms among N_0 sites is

$$\frac{N_0!}{N!\,(N_0 - N)!}$$

The partition function is then

$$Z = \sum_{N=0}^{N_0} \frac{N_0!}{N!\,(N_0 - N)!} e^{N\varepsilon_0/\tau} e^{N\mu/\tau} \qquad (S.4.102.13)$$

$$= \sum_{N=0}^{N_0} \frac{N_0!}{N!\,(N_0 - N)!} \left[e^{(\varepsilon_0+\mu)/\tau} \right]^N = \left[1 + e^{(\varepsilon_0+\mu)/\tau} \right]^{N_0}$$

and the average number of particles

$$\langle N \rangle = \frac{\partial}{\partial \mu} \tau \ln Z = \frac{N_0}{1 + e^{-(\mu+\varepsilon_0)/\tau}} \qquad (S.4.102.14)$$

the same as (S.4.102.8).

4.103   Thermionic Emission (Boston)

a) We can consider the electron gas outside the metal to be in equilibrium with the electrons inside the metal. Then the number of electrons hitting the surface from the outside should be equal to the number of electrons leaving the metal. Using the formula for chemical potential of a monatomic ideal gas (see Problem 4.39), we can write

$$\mu_{\mathrm{g}} = \tau \ln \left[\frac{P}{g\tau^{5/2}} \left(\frac{2\pi\hbar^2}{m} \right)^{3/2} \right] \qquad (S.4.103.1)$$

where $g = 2s + 1 = 2$ for an electron gas. Rewriting (S.4.103.1), we have

$$P = 2 \left(\frac{m}{2\pi\hbar^2} \right)^{3/2} \tau^{5/2} e^{\mu_{\mathrm{g}}/\tau} \qquad (S.4.103.2)$$

The state of equilibrium requires that this chemical potential be equal to the potential inside the metal, which we can take as $\mu_{\mathrm{s}} = -\phi_0$; i.e., the energy ϕ_0 is required to take an electron from the Fermi level inside the metal into vacuum. So, the pressure of the electron gas is given by

$$P = 2 \left(\frac{m}{2\pi\hbar^2} \right)^{3/2} \tau^{5/2} e^{-\phi_0/\tau} \qquad (S.4.103.3)$$

On the other hand, the number of particles of the ideal gas striking the surface per unit area per unit time is

$$N = \frac{n \langle v \rangle}{4} = \frac{n}{4} \sqrt{\frac{8\tau}{\pi m}} = \frac{P}{\sqrt{2\pi m \tau}} \qquad (S.4.103.4)$$

The current

$$J = eN = \frac{Pe}{\sqrt{2\pi m\tau}} \qquad \text{(S.4.103.5)}$$

where e is the electron charge. Therefore, we can express P from (S.4.103.5):

$$P = \frac{J}{e}\sqrt{2\pi m\tau} \qquad \text{(S.4.103.6)}$$

Equating (S.4.103.6) with (S.4.103.3), we find the current

$$J = e\,(2\pi m\tau)^{-1/2}\,2\left(\frac{m}{2\pi\hbar^2}\right)^{3/2}\tau^{5/2}e^{-\phi_0/\tau} \qquad \text{(S.4.103.7)}$$

$$= \frac{2me}{(2\pi)^2\,\hbar^3}\tau^2 e^{-\phi_0/\tau} = \frac{4\pi me}{h^3}\tau^2 e^{-\phi_0/\tau}$$

Alternatively, we can calculate the current by considering the electrons leaving the metal as if they have a kinetic energy high enough to overcome the potential barrier.

b) For one particle,

$$L = E_g - E_s + P\,(v_g - v_s) \qquad \text{(S.4.103.8)}$$

where E_g and E_s are the energies, and v_g and v_s the volumes per particle, of the gas and solid, respectively. Since $v_g \gg v_s$, we can rewrite (S.4.103.8) in the form

$$L = \frac{3}{2}\tau - \mu_s + Pv_g = \frac{3}{2}\tau - \mu_s + \tau = \frac{5}{2}\tau - \mu_s \qquad \text{(S.4.103.9)}$$

Substituting (S.4.103.9) into the Clausius–Clapeyron equation (see Problem 4.57),

$$\frac{dP}{d\tau} = \frac{L}{\tau\,(v_g - v_s)} \approx \frac{L}{\tau v_g} = \frac{L}{\tau(\tau/P)} \qquad \text{(S.4.103.10)}$$

we obtain

$$\frac{dP}{d\tau} = \frac{(5\tau/2 - \mu_s)\,P}{\tau^2} \qquad \text{(S.4.103.11)}$$

We may rewrite (S.4.103.11) as

$$\frac{dP}{P} = \frac{5}{2}\frac{d\tau}{\tau} - \frac{\mu_s}{\tau^2}d\tau \qquad \text{(S.4.103.12)}$$

Integrating, we recover (S.4.103.2):

$$\ln\frac{P}{\tau^{5/2}} = \frac{\mu_s}{\tau} + \ln A$$

where A is some constant, or

$$P = A\tau^{5/2}e^{\mu_s/\tau} \qquad \text{(S.4.103.13)}$$

4.104   Electrons and Holes (Boston, Moscow Phys-Tech)

a) Let the zero of energy be the bottom of the conduction band, so $\mu \leq 0$ (see Figure S.4.104). The number of electrons may be found from

$$N = 2V \int \frac{d^3p}{(2\pi\hbar)^3} e^{(\mu-\varepsilon)/\tau} \qquad (S.4.104.1)$$

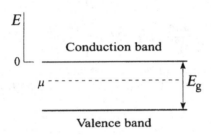

Figure S.4.104

where $2 = 2s+1$ for electrons, and the Fermi distribution formula has been approximated by

$$\langle n \rangle = \frac{1}{e^{(\varepsilon-\mu)/\tau} + 1} \approx e^{(\mu-\varepsilon)/\tau} \qquad (S.4.104.2)$$

The concentration of electrons is then

$$\frac{N}{V} = \frac{8\pi e^{\mu/\tau}}{(2\pi\hbar)^3} \int\limits_{0}^{\infty} e^{-p^2/2m\tau} \, p^2 \, dp$$

$$= \frac{8\pi e^{\mu/\tau}}{(2\pi\hbar)^3} \frac{1}{2} \left(-\frac{d}{d\alpha} \sqrt{\frac{\pi}{\alpha}} \right) = \frac{8\pi e^{\mu/\tau}}{(2\pi\hbar)^3} \frac{1}{2} \frac{\sqrt{\pi}}{2\alpha^{3/2}} \qquad (S.4.104.3)$$

$$= 2 \left(\frac{m_e\tau}{2\pi\hbar^2} \right)^{3/2} e^{\mu/\tau}$$

where $1/\alpha \equiv 2m\tau$.

b) In an intrinsic semiconductor

$$\langle n_h \rangle = 1 - \langle n_e \rangle \qquad (S.4.104.4)$$

since a hole is defined as the absence of an electron. We may then write

$$\langle n_h \rangle = 1 - \frac{1}{e^{(\varepsilon^* - \mu)/\tau} + 1} \tag{S.4.104.5}$$

$$= \frac{1}{e^{(\mu - \varepsilon^*)/\tau} + 1} \approx e^{(\varepsilon^* - \mu)/\tau}$$

where ε^* is the energy of a hole and we have used the nondegeneracy condition for holes $\mu - \varepsilon^* \gg \tau$. The number of holes is

$$N_h = 2V \int_0^\infty \frac{4\pi p^2 \, dp}{(2\pi\hbar)^3} e^{(\varepsilon - \mu)/\tau} \tag{S.4.104.6}$$

The energy of a hole (from the bottom of the conduction band) is

$$\varepsilon = - \left(\frac{p^2}{2m} + E_g \right) \tag{S.4.104.7}$$

Therefore, similar to (a):

$$p = \frac{N_h}{V} = \frac{e^{-(\mu + E_g)/\tau}}{\pi^2 \hbar^3} \int_0^\infty e^{-p^2/2m\tau} \, p^2 \, dp \tag{S.4.104.8}$$

$$= 2 \left(\frac{m_h \tau}{2\pi\hbar^2} \right)^{3/2} e^{-(\mu + E_g)/\tau}$$

The product of the concentrations of electrons and holes does not depend on the chemical potential μ, as we see by multiplying (S.4.104.3) and (S.4.104.8):

$$np = 4 \left(m_e m_h \right)^{3/2} \left(\frac{\tau}{2\pi\hbar^2} \right)^3 e^{-E_g/\tau} \tag{S.4.104.9}$$

We did not use the fact that there are no impurities. The only important assumption is that $\varepsilon - \mu \gg \tau$, which implies that the chemical potential μ is not too close to either the conduction or valence bands.

c) Since, in the case of an intrinsic semiconductor $n_i = p_i$ (every electron in the conduction band leaves behind a hole in the valence band), we can write, using (S.4.104.9),

$$np = n_i^2 = 4 \left(m_e m_h \right)^{3/2} \left(\frac{\tau}{2\pi\hbar^2} \right)^3 e^{-E_g/\tau} \tag{S.4.104.10}$$

Therefore,

$$n_i = 2 \left(m_e m_h\right)^{3/4} \left(\frac{\tau}{2\pi\hbar^2}\right)^{3/2} e^{-E_g/2\tau} \qquad (S.4.104.11)$$

Equating (S.4.104.3) and (S.4.104.11), we can find the chemical potential for an intrinsic semiconductor:

$$\mu = -\frac{E_g}{2} + \frac{1}{2}\ln\left(\frac{m_h}{m_e}\right)^{3/2} = -\frac{E_g}{2} + \frac{3}{4}\ln\frac{m_h}{m_e} \qquad (S.4.104.12)$$

If $m_e = m_h$, then the chemical potential is in the middle of the band gap:

$$\mu = -\frac{E_g}{2}$$

4.105 Adiabatic Demagnetization (Maryland)

a) We start with the usual relation $dE = \tau\, dS - P\, dV$, and substitute $M\, dH$ for $P\, dV$, since the work done in this problem is magnetic rather than mechanical. So

$$dE = \tau\, dS - M\, dH \qquad (S.4.105.1)$$

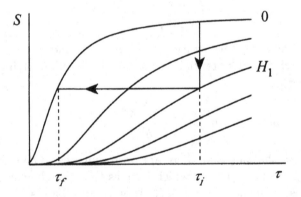

Figure S.4.105

We now want to produce a Maxwell relation whose independent variables are T and H. Write an equation for the free energy F:

$$dF = d(E - \tau S) = -S\, d\tau - M\, dH \qquad (S.4.105.2)$$

We then obtain, from (S.4.105.2),

$$\left(\frac{\partial F}{\partial \tau}\right)_H = -S \qquad \text{(S.4.105.3)}$$

$$\left(\frac{\partial F}{\partial H}\right)_\tau = -M$$

The cross derivatives of (S.4.105.3) are equal so

$$\frac{\partial^2 F}{\partial H\, \partial \tau} = \frac{\partial^2 F}{\partial \tau\, \partial H} = -\left(\frac{\partial S}{\partial H}\right)_\tau = -\left(\frac{\partial M}{\partial \tau}\right)_H \qquad \text{(S.4.105.4)}$$

The heat capacity at constant magnetic field is given by

$$C_H = \tau \left(\frac{\partial S}{\partial \tau}\right)_H \qquad \text{(S.4.105.5)}$$

from which we obtain

$$\left(\frac{\partial C_H}{\partial H}\right)_\tau = \tau \frac{\partial^2 S}{\partial H\, \partial \tau} \qquad \text{(S.4.105.6)}$$

By again exchanging the order of differentiation in (S.4.105.6) and using the result found in (S.4.105.4), we have

$$\left(\frac{\partial C_H}{\partial H}\right)_\tau = \tau \frac{\partial^2 S}{\partial H\, \partial \tau} = \tau \frac{\partial}{\partial \tau}\left(\frac{\partial M}{\partial \tau}\right)_H \qquad \text{(S.4.105.7)}$$

Replacing M by $V\chi H$ in (S.4.105.7) yields the desired

$$\left(\frac{\partial C_H}{\partial H}\right)_\tau = V\tau H \left(\frac{\partial^2 \chi}{\partial \tau^2}\right)_H \qquad \text{(S.4.105.8)}$$

b) For an adiabatic process, the entropy S is constant. Writing $S = S(\tau, H)$, we compose the differential

$$dS = \left(\frac{\partial S}{\partial \tau}\right)_H d\tau + \left(\frac{\partial S}{\partial H}\right)_\tau dH = 0 \qquad \text{(S.4.105.9)}$$

and by (S.4.105.4) and (S.4.105.5),

$$\left(\frac{\partial \tau}{\partial H}\right)_S = \frac{-(\partial S/\partial H)_\tau}{(\partial S/\partial \tau)_H} = \frac{-(\partial M/\partial \tau)_H}{(1/\tau)\, C_H} = -\frac{V\tau H}{C_H}\left(\frac{\partial \chi}{\partial \tau}\right)_H$$

$$\text{(S.4.105.10)}$$

c) The heat capacity $C_H(\tau, H)$ may be written as the integral

$$C_H(\tau, H) = C_H(\tau, 0) + \int_0^H \frac{\partial C_H(\tau, H')}{\partial H'} \, dH' \qquad \text{(S.4.105.11)}$$

Substituting $\chi = a/\tau$ into (S.4.105.8), we have

$$\left(\frac{\partial C_H}{\partial H}\right)_\tau = \frac{2V\tau a H}{\tau^3} = \frac{2V a H}{\tau^2} \qquad \text{(S.4.105.12)}$$

Using the heat capacity at zero magnetic field, $C_H(\tau, 0) = V/\tau^2$, and (S.4.105.12) in (S.4.105.11), we obtain

$$C_H(\tau, H) = \frac{Vb}{\tau^2} + \frac{V a H^2}{\tau^2} = \frac{V}{\tau^2}\left(b + aH^2\right) \qquad \text{(S.4.105.13)}$$

The temperature τ may be written $\tau = \tau(S, H)$, so for our adiabatic process

$$\tau_f(S, H_f) = \tau_i(S, H_i) + \int_{H_i}^{H_f} \left(\frac{\partial \tau}{\partial H}\right)_S \, dH \qquad \text{(S.4.105.14)}$$

The integrand in (S.4.105.14) is found by substituting $\chi = a/\tau$ into (S.4.105.10):

$$\left(\frac{\partial \tau}{\partial H}\right)_S = -\frac{V\tau H}{(V/\tau^2)(b + aH^2)}\left(-\frac{a}{\tau^2}\right) = \frac{\tau a H}{b + aH^2} \qquad \text{(S.4.105.15)}$$

So, for a process at constant entropy, we may write

$$d\tau = \frac{\tau a H}{b + aH^2} \, dH \qquad \text{(S.4.105.16)}$$

Rearranging and integrating give

$$\ln \tau = \frac{1}{2} \ln(b + aH^2) \qquad \text{(S.4.105.17)}$$

and

$$\frac{\tau_f}{\tau_i} = \sqrt{\frac{b + aH_f^2}{b + aH_i^2}} \qquad \text{(S.4.105.18)}$$

d) A possible route to zero temperature is illustrated in Figure S.4.105. During leg 1 the paramagnetic sample is kept in contact with a reservoir at a low temperature, and the magnetic field is raised from 0 to H_1. The contact with the reservoir is then removed, and the field is reduced to zero along leg 2. The sample is thereby cooled.

4.106 Critical Field in Superconductor (Stony Brook, Chicago)

a) If the external field is smaller than the critical field, $H < H_c$, then the B-field inside the superconductor is zero, $B = 0$, and the magnetization M becomes

$$\mathbf{M} = \frac{\mathbf{B} - \mathbf{H}}{4\pi} = -\frac{1}{4\pi}\mathbf{H} \qquad (S.4.106.1)$$

This means that the superconductor displays perfect diamagnetism (with magnetic susceptibility $\chi = -1/4\pi$). The change in free energy of the superconductor due to the increase of the external field H may be written as $\delta F = -V \cdot \mathbf{M} \cdot d\mathbf{H} = (V/4\pi)\mathbf{H} \cdot d\mathbf{H}$. Therefore, the free energy of the superconductor in a field is given by

$$F_S(H) = F_S(0) + \int_0^H dF = F_S(0) + \frac{V}{4\pi}\int_0^H \mathbf{H} \cdot d\mathbf{H} \quad (S.4.106.2)$$

$$= F_S(0) + \frac{V}{8\pi} H^2$$

The transition to a normal state occurs when the free energy of the superconducting state is equal to that of the normal state:

$$F_s(H_c) = F_s(0) + \frac{V}{8\pi} H_c^2 = F_n(H_c) = F_n(0) \qquad (S.4.106.3)$$

Here we used the fact that, because of the negligible magnetic susceptibility, the free energy of the normal state practically does not depend on the applied field. So, we have

$$F_n(0) - F_s(0) = \frac{V}{8\pi} H_c^2 \qquad (S.4.106.4)$$

where $H_c \equiv H_c(\tau)$. Now, it is easy to calculate the entropy discontinuity since

$$S = -\left(\frac{\partial F}{\partial \tau}\right)_{H,V} \qquad (S.4.106.5)$$

so

$$S_n(0) - S_s(0) = -\frac{V}{4\pi} H_c \frac{\partial H_c}{\partial \tau} \qquad (S.4.106.6)$$

If we recall that the dependence of the critical field on the temperature can be approximated by the formula $H_c(\tau) = H(0)[1 - t^2]$, where $t \equiv \tau/\tau_c$ then we can confirm that a superconducting state is a more ordered state, since $\partial H_c/\partial \tau < 0$ and hence $S_n > S_s$.

b) The latent heat q, if the transition occurs at a constant temperature, is given by $q = \tau(S_n - S_s)$. If the transition is from superconducting to normal, then

$$q = -\frac{\tau V}{4\pi} H_c \left(\frac{\partial H_c}{\partial \tau} \right) > 0 \qquad \text{(S.4.106.7)}$$

So, if we have a transition from the superconducting to normal states, then heat is absorbed.

c) The specific heat is defined as $C = \tau(\partial S/\partial \tau)$. Here we disregard any volume and pressure changes due to the transition. Hence, from equation (S.4.106.6), the specific heat per volume discontinuity is

$$c_s - c_n = \frac{\tau}{4\pi} \left[\left(\frac{\partial H_c}{\partial \tau} \right)^2 + H_c \frac{\partial^2 H_c}{\partial \tau^2} \right] \qquad \text{(S.4.106.8)}$$

At zero field the transition is of second order and $H_c(\tau_c) = 0$, so the specific heat per unit volume discontinuity at $\tau = \tau_c$ from (S.4.106.8) is

$$c_s - c_n = \frac{\tau_c}{4\pi} \left(\frac{\partial H_c}{\partial \tau} \right)^2_{\tau=\tau_c} \qquad \text{(S.4.106.9)}$$

Quantum Mechanics

One-Dimensional Potentials

5.1 Shallow Square Well I (Columbia)

The ground state energy E must be less than zero and greater than the bottom of the well, $0 > E > -V_0$. From the expression

$$\frac{\hbar^2}{2m}\psi''(x) = [V(x) - E]\psi(x) \tag{S.5.1.1}$$

one can deduce the form for the eigenfunction. Denote the ground state energy $E = -\hbar^2\alpha^2/2m$, where α is to be determined. The eigenfunction outside the well ($V = 0$) has the form $\exp(-\alpha|x|)$. Inside the well, define $k^2 = k_0^2 - \alpha^2$, where $k_0^2 = 2mV_0/\hbar^2$. One can show that k^2 is positive since $E + V_0 > 0$. Inside the well, the eigenfunction has the form $A\cos kx$, so

$$\psi(x) = \begin{cases} A\cos kx & |x| < a \\ Be^{-\alpha|x|} & |x| > a \end{cases} \tag{S.5.1.2}$$

Matching $\psi(x)$ and its derivative at $x = a$ gives two expressions:

$$A\cos ka = Be^{-\alpha a} \tag{S.5.1.3}$$

$$-Ak\sin ka = -B\alpha e^{-\alpha a} \tag{S.5.1.4}$$

Dividing these two equations produces the eigenvalue equation

$$\alpha = k \tan ka = \sqrt{k_0^2 - k^2} \qquad (S.5.1.5)$$

The equation given by the rightmost equals sign is an equation for the unknown k. Solving it gives the eigenvalue E.

5.2   Shallow Square Well II (Stony Brook)

a) For the bound state we can write the eigenvalue as $E = -\hbar^2\alpha^2/2m$, where α is the decay constant of the eigenfunction outside the square well (see Problem 5.1). Inside the square well we define a wave vector k by

$$k^2 = \frac{2mV_0}{\hbar^2} - \alpha^2 \qquad (S.5.2.1)$$

The infinite potential at the origin requires that all eigenfunctions vanish at $x = 0$. So the lowest eigenfunction must have the form

$$\psi_B(x) = \begin{cases} A\sin kx & x < a \\ Be^{-\alpha x} & x > a \end{cases} \qquad (S.5.2.2)$$

At the point $x = a$, we match the eigenfunctions and their derivatives:

$$A\sin ka = Be^{-\alpha a} \qquad (S.5.2.3)$$

$$Ak\cos ka = -\alpha Be^{-\alpha a} \qquad (S.5.2.4)$$

We eliminate the constants A and B by dividing these two equations:

$$\alpha \tan ka = -k \qquad (S.5.2.5)$$

Earlier we established the relationship between k and α. So the only unknown variable is α, which is determined by this equation.

b) To find the minimum bound state, we take the limit as $\alpha \to 0$ in the eigenvalue equation. From (S.5.2.1) we see that k goes to a nonzero constant, and the eigenvalue equation only makes sense as $\alpha \to 0$ if $\tan ka \to \infty$, which happens at $ka \to \pi/2$. Using (S.5.2.1) gives $\pi^2/4 = k^2a^2 = 2mV_0a^2/\hbar^2$. Thus, we derive the minimum value of V_0 for a bound state:

$$V_0 = \frac{\hbar^2\pi^2}{8ma^2} \qquad (S.5.2.6)$$

c) For a positive energy state set $E = \hbar^2 \tilde{k}^2 / 2m$, where $\tilde{k}$ is the wave vector outside the square well. Inside the square well we again define a wave vector k according to

$$k^2 = \frac{2mV_0}{\hbar^2} + \tilde{k}^2 \qquad (S.5.2.7)$$

$$\psi_{\tilde{k}}(x) = \left\{ \begin{array}{ll} A \sin kx & x < a \\ B \sin(\tilde{k}x + \delta) & x > a \end{array} \right. \qquad (S.5.2.8)$$

Again we have the requirement that the eigenfunction vanish at $x = 0$. For $x > a$ we have an eigenfunction with two unknown parameters B and δ. Alternatively, we may write it as

$$C \sin \tilde{k}x + D \cos \tilde{k}x \qquad (S.5.2.9)$$

in terms of two unknowns C and D. The two forms are equivalent since $C = B \cos \delta$, $D = B \sin \delta$. We prefer to write it with the phase shift δ. Again we match the two wave functions and their derivatives at $x = a$:

$$A \sin ka = B \sin(\tilde{k}a + \delta) \qquad (S.5.2.10)$$

$$kA \cos ka = \tilde{k}B \cos(\tilde{k}a + \delta) \qquad (S.5.2.11)$$

Dividing (S.5.2.10) by (S.5.2.11), we obtain

$$\tilde{k} \tan ka = k \tan(\tilde{k}a + \delta) \qquad (S.5.2.12)$$

Since k is a known function of $\tilde{k}$, the only unknown in this equation is δ, which is determined by this equation.

d) From (S.5.2.12) we derive an expression for the phase shift:

$$\delta = -\tilde{k}a + \tan^{-1} \left(\frac{\tilde{k}}{k} \tan ka \right) \qquad (S.5.2.13)$$

5.3   Attractive Delta Function Potential I (Stony Brook)

a) The bound state is stationary in time: its eigenvalue is E $(E < 0)$, and the time dependence of the wave function is $\Psi(x,t) = \psi(x) \exp(-iEt/\hbar)$. The equation for the bound state is

$$\left\{ -\frac{\hbar^2}{2m} \frac{\partial^2}{\partial x^2} - C\delta(x) - E \right\} \psi(x) = 0 \qquad (S.5.3.1)$$

The bound state for $x \neq 0$ has the form

$$\psi(x) = Ae^{-\alpha|x|} \qquad \text{(S.5.3.2)}$$

We have already imposed the constraint that $\psi(x)$ be continuous at $x = 0$. This form satisfies the requirement that $\psi(x)$ is continuous at the origin and vanishes at infinity. Away from the origin the potential is zero, and the Schrödinger equation just gives $E = -\hbar^2\alpha^2/2m$. A relation between C and E is found by matching the derivatives of the wave functions at $x = 0$. Taking the integral of (S.5.3.1) between 0^+ and 0^- gives

$$C\psi(0) = -\frac{\hbar^2}{2m}\left[\left.\frac{d\psi}{dx}\right|_{0+} - \left.\frac{d\psi}{dx}\right|_{0-}\right] \qquad \text{(S.5.3.3)}$$

Applying (S.5.3.3) to (S.5.3.2) gives the relations

$$CA = \frac{\hbar^2 A}{m}\alpha \qquad \text{(S.5.3.4)}$$

$$\alpha = \frac{mC}{\hbar^2} \qquad \text{(S.5.3.5)}$$

$$E = -\frac{mC^2}{2\hbar^2} \qquad \text{(S.5.3.6)}$$

We have found the eigenvalue for the bound state. Note that the dimensions of C are energy×distance, which makes the eigenvalue have units of energy. Finally, we find the normalization coefficient A:

$$1 = \int\limits_{-\infty}^{\infty} dx\, |\psi(x)|^2 = 2A^2 \int\limits_{0}^{\infty} dx\, e^{-2\alpha x} = \frac{A^2}{\alpha} \qquad \text{(S.5.3.7)}$$

so $A = \sqrt{\alpha} = \sqrt{mC}/\hbar$.

b) When the potential constant changes from $C \rightarrow C'$, the eigenfunction changes from $\psi(x) \rightarrow \psi'(x)$, where the prime denotes the eigenfunction with the potential strength C'. In the sudden approximation the probability P_0 that the particle remains in the bound state is given by

$$P_0 = |I|^2 \qquad \text{(S.5.3.8)}$$

where

$$I = \int\limits_{-\infty}^{\infty} dx\, \psi'(x)\,\psi(x) \qquad \text{(S.5.3.9)}$$

Substituting (S.5.3.2) into (S.5.3.9) and using the result of (S.5.3.7), we obtain

$$I = 2AA' \int_{0}^{\infty} dx \; e^{-(\alpha+\alpha')x} = \frac{2AA'}{\alpha + \alpha'} = \frac{2\sqrt{\alpha\alpha'}}{\alpha + \alpha'} \qquad (S.5.3.10)$$

Finally, using (S.5.3.5) yields

$$I = \frac{2\sqrt{CC'}}{C + C'} \qquad (S.5.3.11)$$

$$P_0 = \frac{4CC'}{(C + C')^2} \qquad (S.5.3.12)$$

It is easy to show that $P_0 \leq 1$ as required by particle conservation. If $C = C'$, then $P_0 = 1$ since there is no change, and the particle must stay in the bound state.

5.4   Attractive Delta Function Potential II (Stony Brook)

a) In order to construct the wave function for the bound state, we first review its properties. It must vanish at the point $x = 0$. At the point $x = a$, it is continuous and its derivative obeys an equation similar to (S.5.3.3):

$$V_0 \, a\psi(a) = -\frac{\hbar^2}{2m} \left[\left. \frac{d\psi(x)}{dx} \right|_{a+} - \left. \frac{d\psi(x)}{dx} \right|_{a-} \right] \qquad (S.5.4.1)$$

Away from the points $x = (0, a)$, it has an energy $E = -\hbar^2\alpha^2/2m$ and wave functions that are combinations of $e^{-\alpha x}$ and $e^{\alpha x}$. These constraints dictate that the eigenfunction has the form

$$\psi(x) = \begin{cases} A \sinh \alpha x & 0 < x < a \\ Be^{-\alpha x} & x > a \end{cases} \qquad (S.5.4.2)$$

At the point $x = a$, we match the two eigenfunctions and their derivatives, using (S.5.4.1). This yields two equations, which are solved to find an equation for α:

$$A \sinh \alpha a = Be^{-\alpha a} \qquad (S.5.4.3)$$

$$V_0 \, aBe^{-\alpha a} = \frac{\hbar^2 \alpha}{2m} \left[Be^{-\alpha a} + A \cosh \alpha a \right] \qquad (S.5.4.4)$$

We use the first equation to eliminate A in the second equation. Then each term has a factor of $Be^{-\alpha a}$ which is canceled:

$$V_0 a = \frac{\hbar^2 \alpha}{2m} [1 + \coth \alpha a] \qquad (S.5.4.5)$$

Multiplying both sides of (S.5.4.5) by $\sinh \alpha a$ gives

$$\sinh \alpha a = \frac{\hbar^2 \alpha}{2mV_0 a} e^{\alpha a} \qquad (S.5.4.6)$$

$$\frac{\hbar^2 \alpha}{mV_0 a} = 1 - e^{-2\alpha a} \qquad (S.5.4.7)$$

This last equation determines α, which determines the bound state energy. There is only one solution for sufficiently large values of V_0.

b) The minimum value of V_0 for creating a bound state is called V_c. It is found by assuming that the binding energy $E \to 0$, which means $\alpha \to 0$. We examine (S.5.4.7) for small values of α and find that

$$2\alpha a = \frac{\hbar^2 \alpha}{mV_c a} \qquad (S.5.4.8)$$

$$V_c = \frac{\hbar^2}{2ma^2} \qquad (S.5.4.9)$$

5.5   Two Delta Function Potentials (Rutgers)

There are two delta function singularities, one at $x = a$ and one at $x = -a$. The potential can be written in an equivalent way as

$$V(x) = -\frac{\hbar^2 P}{2ma} [\delta(x - a) + \delta(x + a)] \qquad (S.5.5.1)$$

At each delta function we match the amplitudes of the eigenfunctions as well as the slopes, using a relation such as (S.5.3.3). A single, isolated, attractive, professional, delta function potential has a single bound state. We expect that a pair of delta function potentials will generally have one or two bound states.

The lowest energy state, for symmetric potentials, is a symmetric eigenfunction. The eigenvalue has the form $E = -\hbar^2 \alpha^2 / 2m$, where α is the decay constant of the eigenfunction. The most general symmetric eigenfunction for a bound state is

$$\psi_s(x) = \begin{cases} A \cosh \alpha x & |x| < a \\ Be^{-\alpha |x|} & |x| > a \end{cases} \qquad (S.5.5.2)$$

Matching at either $x = \pm a$ gives the pair of equations:

$$A \cosh \alpha a = B e^{-\alpha a} \qquad (S.5.5.3)$$

$$\frac{\hbar^2 \alpha}{2m} \left(B e^{-\alpha a} + A \sinh \alpha a \right) = \frac{\hbar^2 P}{2ma} B e^{-\alpha a} \qquad (S.5.5.4)$$

Eliminating the constants A and B gives the final equation for the unknown constant α:

$$P = \alpha a \left(1 + \tanh \alpha a \right) \qquad (S.5.5.5)$$

For large values of αa the hyperbolic tangent is unity, and we have the approximate result that $P \approx 2\alpha a$, which gives for large P the eigenvalue $E \approx -\hbar^2 P^2 / 8ma^2$. For small values of αa we see that $P \approx \alpha a$ and $E \approx -\hbar^2 P^2 / 2ma^2$. This is always the lowest eigenvalue.

The other possible eigenstate is antisymmetric: it has odd parity. When the separate bound states from the two delta functions overlap, they combine into bonding and antibonding states. The bonding state is the symmetric state we calculated above. Now we calculate the antibonding state, which is antisymmetric:

$$\psi_a(x) = \begin{cases} A \sinh \alpha x & |x| < a \\ B e^{-\alpha x} & x > a \\ -B e^{\alpha x} & x < -a \end{cases} \qquad (S.5.5.6)$$

Using the same matching conditions, we find the two equations, which are reduced to the final equation for α:

$$A \sinh \alpha a = B e^{-\alpha a} \qquad (S.5.5.7)$$

$$\frac{\hbar^2 \alpha}{2m} \left(B e^{-\alpha a} + A \cosh \alpha a \right) = \frac{\hbar^2 P}{2ma} B e^{-\alpha a} \qquad (S.5.5.8)$$

$$P = \alpha a \left(1 + \coth \alpha a \right) \qquad (S.5.5.9)$$

For large values of αa the hyperbolic cotangent function (coth) approaches unity, and again we find $P \approx 2\alpha a$ and $E \approx -\hbar^2 P^2 / 8ma^2$. At small values of x the factor of $x \coth x \to 1$. Here we have $x = \alpha a$, so we find at small values of αa that $P \to 1$. The antisymmetric mode only exists for $P > 1$. For $P < 1$ the only bound state is the symmetric one. For $P > 1$ there are two bound states, symmetric and antisymmetric.

5.6   Transmission Through a Delta Function Potential (Michigan State, MIT, Princeton)

On the left the particle has an incident intensity, which we set equal to unity, and a reflected amplitude R. On the right the transmitted amplitude is denoted by T.

$$\psi(x) = \begin{cases} e^{ikx} + Re^{-ikx} & x < 0 \\ Te^{ikx} & x > 0 \end{cases} \qquad \text{(S.5.6.1)}$$

At the point $x = 0$, we match the value of $\psi(x)$ on both sides. We match the derivative according to an expression such as (S.5.3.3) with $C = -C$. This yields two equations for T and R which can be solved for T:

$$1 + R = T \qquad \text{(S.5.6.2)}$$

$$TC = \frac{\hbar^2 ik}{2m} \left[T - (1 - R) \right] \qquad \text{(S.5.6.3)}$$

$$T = \frac{1}{1 + iCm/\hbar^2 k} \qquad \text{(S.5.6.4)}$$

5.7   Delta Function in a Box (MIT)

a) In the absence of the delta function potential, the states with odd parity are

$$\psi_n^{(o)}(x) = \frac{1}{\sqrt{a}} \sin \frac{n\pi x}{a} \qquad \text{(S.5.7.1)}$$

$$E_n^{(o)} = \frac{(\hbar \pi n)^2}{2ma^2} \qquad \text{(S.5.7.2)}$$

These states have zero amplitude at the site of the delta function $x = 0$ and are unaffected by it. So, the states with odd parity have the same eigenfunction and eigenvalues as when the delta function is absent.

b), c) For a delta function potential without a box, the bound states have a wave function of $Ae^{-\alpha|x|}$ (see Problem 5.3). In the box we expect to have similar exponentials, except that the wave function must vanish at the edges of the box ($x = \pm a$). The states which do this are

$$\psi(x) = A \sinh \alpha(a - |x|) \qquad \text{(S.5.7.3)}$$

$$E = -\frac{\hbar^2 \alpha^2}{2m} \tag{S.5.7.4}$$

$$\left.\frac{d\psi}{dx}\right|_{0\pm} = \mp \alpha A \cosh \alpha a \tag{S.5.7.5}$$

Using (S.5.4.1), we match the difference in the derivatives at $x = 0$ with the amplitude of the delta function potential. This leads to the eigenvalue condition

$$\alpha a \coth \alpha a = \frac{C}{C_0} \tag{S.5.7.6}$$

$$C_0 = \frac{\hbar^2}{ma^2} \tag{S.5.7.7}$$

The quantity on the left of (S.5.7.6) has a minimum value of 1, which it obtains at $\alpha \to 0$. This limit produces the eigenvalue $E = 0$. So we must have $C = C_0$ for the zero eigenvalue, which is the answer to part (b). The above eigenfunction, for values of $\alpha > 0$, gives the bound state energy $E < 0$ when $C > C_0$.

5.8 Particle in Expanding Box (Michigan State, MIT, Stony Brook)

a) For a particle confined to a box $-L/2 \le x \le L/2$, the ground state $\psi_0(x)$ and the first excited state $\psi_1(x)$ are

$$\psi_0(x) = \sqrt{\frac{2}{L}} \cos \frac{\pi x}{L} \tag{S.5.8.1}$$

$$\psi_1(x) = \sqrt{\frac{2}{L}} \sin \frac{2\pi x}{L} \tag{S.5.8.2}$$

After the sudden transition $L \to 2L$, the final eigenfunctions are

$$\psi_0'(x) = \frac{1}{\sqrt{L}} \cos \frac{\pi x}{2L} \tag{S.5.8.3}$$

$$\psi_1'(x) = \frac{1}{\sqrt{L}} \sin \frac{\pi x}{L} \tag{S.5.8.4}$$

b) In the sudden approximation let P_{0j} denote the probability that the particle starts in the ground state 0 and ends in the final state j:

$$P_{0j} = |I_{0j}|^2 \tag{S.5.8.5}$$

where the amplitude of the transition I_{0j} is given by

$$I_{0j} = \int_{-L/2}^{L/2} dx\, \psi_j'(x)\, \psi_0(x) \tag{S.5.8.6}$$

The amplitude for the particle to remain in its ground state is then

$$I_{00} = \frac{\sqrt{2}}{L} \int_{-L/2}^{L/2} dx \cos\frac{\pi x}{2L} \cos\frac{\pi x}{L}$$

$$= \frac{1}{\sqrt{2}\,L} \int_{-L/2}^{L/2} dx \left(\cos\frac{\pi x}{2L} + \cos\frac{3\pi x}{2L} \right) \tag{S.5.8.7}$$

$$= \frac{1}{\sqrt{2}\,L} \frac{4L}{\pi} \left(\sin\frac{\pi x}{2L} + \frac{1}{3}\sin\frac{3\pi x}{2L} \right)\Bigg|_{x=L/2} = \frac{8}{3\pi}$$

The probability P_{00} is given by

$$P_{00} = \left(\frac{8}{3\pi} \right)^2 \tag{S.5.8.8}$$

The same calculation for the transition between the initial ground state and final excited state ψ_1' is as follows:

$$P_{0j} = |I_{01}|^2 \tag{S.5.8.9}$$

where

$$I_{01} = \int_{-L/2}^{L/2} dx\, \psi_1'(x)\, \psi_0(x) \tag{S.5.8.10}$$

$$= \frac{\sqrt{2}}{L} \int_{-L/2}^{L/2} dx \, \sin\frac{\pi x}{L} \cos\frac{\pi x}{L} = 0$$

The integral is zero by parity, since the integrand is an odd function of x, so

$$P_{01} = 0 \tag{S.5.8.11}$$

5.9 One-Dimensional Coulomb Potential (Princeton)

a) Since the electron is confined to the right half-space, its wave function must vanish at the origin. So, an eigenfunction such as $\exp(-\alpha x)$ is unsuitable since it does not vanish at $x = 0$. The ground state wave function must be of the form $\psi = Nx\exp(-\alpha x)$, where α needs to be determined. The operator p^2 acting on this form gives

$$p^2(xe^{-\alpha x}) = \hbar^2\alpha(2 - \alpha x)e^{-\alpha x} \tag{S.5.9.1}$$

so that using this wave function in Schrödinger's equation yields

$$\frac{\hbar^2}{2m}\alpha(2 - \alpha x) - \frac{e^2}{4} = Ex \tag{S.5.9.2}$$

For this equation to be satisfied, the first and third terms on the left must be equal, and the second term on the left must equal the term on the right of the equals sign:

$$\frac{\hbar^2\alpha}{m} = \frac{e^2}{4} \tag{S.5.9.3}$$

$$E = -\frac{\hbar^2\alpha^2}{2m} = -\frac{e^4 m}{32\hbar^2} = -\frac{E_R}{16} \tag{S.5.9.4}$$

The answer is one sixteenth of the Rydberg, where E_R is the ground state energy of the hydrogen atom. The parameter $\alpha = 1/4a_0$, where a_0 is the Bohr radius.

b) Next we find the expectation value $\langle x \rangle$. The first integral is done to find the normalization coefficient:

$$1 = N^2 \int_0^\infty dx\, x^2 e^{-2\alpha x} = \frac{N^2}{4\alpha^3} \tag{S.5.9.5}$$

$$\langle x \rangle = N^2 \int_0^\infty dx\, x^3 e^{-2\alpha x} = \frac{4\alpha^3 3!}{(2\alpha)^4} = \frac{3}{2\alpha} = 6a_0 \tag{S.5.9.6}$$

The average value of x is 6 Bohr radii.

5.10 Two Electrons in a Box (MIT)

a) If the box is in the region $0 < x < a$, then the one-electron orbitals are

$$\psi_n(x) = \sqrt{\frac{2}{a}}\sin\frac{n\pi x}{a} \tag{S.5.10.1}$$

If both electrons are in the spin state α (spin up), then the spin part of the wave function $\alpha_1 \alpha_2$ is symmetric under exchange of coordinates. Therefore, the orbital part has to be antisymmetric, and both particles cannot be in the $n = 1$ state. Instead, the lowest energy occurs when one electron is in the $n = 1$ state and the other is in the $n = 2$ state:

$$\psi(x_1, x_2) = \frac{1}{\sqrt{2}} \left[\psi_1(x_1) \psi_2(x_2) - \psi_1(x_2) \psi_2(x_1) \right] \alpha_1 \alpha_2 \qquad \text{(S.5.10.2)}$$

b) The probability that both are in one half, say the left side, is

$$P_{LL} = \int_0^{a/2} dx_1 \int_0^{a/2} dx_2 \, |\psi(x_1, x_2)|^2 \qquad \text{(S.5.10.3)}$$

$$= \int_0^{a/2} dx_1 \int_0^{a/2} dx_2 \left[\psi_1^2(x_1) \psi_2^2(x_2) - \psi_1(x_1) \psi_2(x_2) \psi_1(x_2) \psi_2(x_1) \right]$$

Three integrals must be evaluated:

$$P_{LL} = I_1 I_2 - I_{12}^2 \qquad \text{(S.5.10.4)}$$

$$I_1 = \frac{2}{a} \int_0^{a/2} dx \, \sin^2 \frac{\pi x}{a} = \frac{1}{2} \qquad \text{(S.5.10.5)}$$

$$I_2 = \frac{2}{a} \int_0^{a/2} dx \, \sin^2 \frac{2\pi x}{a} = \frac{1}{2} \qquad \text{(S.5.10.6)}$$

$$I_{12} = \frac{2}{a} \int_0^{a/2} dx \, \sin \frac{\pi x}{a} \sin \frac{2\pi x}{a} \qquad \text{(S.5.10.7)}$$

$$= \frac{4}{a} \int_0^{a/2} dx \, \sin^2 \frac{\pi x}{a} \cos \frac{\pi x}{a} = \frac{4}{3\pi} \left(\sin^3 \frac{\pi x}{a} \right) \Big|_0^{a/2} = \frac{4}{3\pi}$$

$$P_{LL} = \left(\frac{1}{2} \right)^2 - \left(\frac{4}{3\pi} \right)^2 \approx 0.07 \qquad \text{(S.5.10.8)}$$

The result is rather small. Naturally, it is much more favorable to have one particle on each side of the box.

5.11   Square Well (MIT)

a) The most general solution is

$$\psi_n(x) = \sqrt{\frac{2}{a}} \sin \frac{n\pi x}{a} \qquad\qquad \text{(S.5.11.1)}$$

$$E_n = \frac{(\hbar n\pi)^2}{2ma^2} \qquad\qquad \text{(S.5.11.2)}$$

$$\Psi(x,t) = \sum_{n=1}^{\infty} a_n \psi_n(x) e^{-iE_n t/\hbar} \qquad\qquad \text{(S.5.11.3)}$$

We evaluate the coefficients by using the initial condition at $t = 0$:

$$a_n = \int_0^a dx\, \psi_n(x)\, \Psi(x,0) \qquad\qquad \text{(S.5.11.4)}$$

$$= \frac{2}{a} \int_0^{a/2} dx\, \sin \frac{n\pi x}{a} = \frac{2}{n\pi}\left(1 - \cos \frac{n\pi}{2}\right)$$

The term $\cos(n\pi/2)$ is either 1, 0, or -1, depending on the value of n. The answer to (a) is to use the above expression for a_n in (S.5.11.3). The answer to part (b) is that the probability of being in the nth eigenstate is $|a_n|^2$. The answer to part (c) is that the average value of the energy is

$$\langle \Psi | H | \Psi \rangle = \sum_n a_n^2 E_n \qquad\qquad \text{(S.5.11.5)}$$

$$= \frac{2\hbar^2}{ma^2} \sum_n \left(1 - \cos \frac{n\pi}{2}\right)^2$$

This latter series does not converge. It takes an infinite amount of energy to form the initial wave function.

5.12   Given the Eigenfunction (Boston, MIT)

We evaluate the second derivative of the eigenfunction, which gives the kinetic energy:

$$\frac{\hbar^2}{2m}\psi'' = -\frac{\hbar^2\lambda^2}{2m}\frac{A}{\cosh \lambda x}\left(1 - \frac{2\sinh^2 \lambda x}{\cosh^2 \lambda x}\right) \qquad\qquad \text{(S.5.12.1)}$$

$$= [V(x) - E]\psi(x) = [V(x) - E]\frac{A}{\cosh \lambda x}$$

We take the limit that $x \to \infty$ of the function on the left, and this must equal $-E\psi$ since we assumed that the potential vanishes at infinity. Thus, we find that

$$E = -\frac{\hbar^2 \lambda^2}{2m} \tag{S.5.12.2}$$

The energy is negative, which signifies a bound state. The potential $V(x)$ can be deduced from (S.5.12.1) since everything else in this expression is known:

$$V(x) = -\frac{\hbar^2 \lambda^2}{m} \frac{1}{\cosh^2 \lambda x} \tag{S.5.12.3}$$

This potential energy has a bound state which can be found analytically, and the eigenfunction is the function given at the beginning of the problem.

5.13 Combined Potential (Tennessee)

Let the dimensionless distance be $y = x/b$. The kinetic energy has the scale factor $E_b = \hbar^2/2mb^2$. In terms of these variables we write Schrödinger's equation as

$$\left[-E_b \frac{d^2}{dy^2} + V_0 \left(\frac{1}{y^2} - \frac{1}{y} \right) \right] \psi(y) = E\psi(y) \tag{S.5.13.1}$$

Our experience with the hydrogen atom, in one or three dimensions, is that potentials which are combinations of y^{-1} and y^{-2} are solved by exponentials $e^{-\alpha y}$ times a polynomial in y. The polynomial is required to prevent the particle from getting too close to the origin where there is a large repulsive potential from the y^{-2} term. Since we do not yet know which power of y to use in a polynomial, we try

$$\psi(y) = Ay^s e^{-\alpha y} \tag{S.5.13.2}$$

where s and α need to be found, while A is a normalization constant. This form is inserted into the Hamiltonian. First we present the second derivative from the kinetic energy and then the entire Hamiltonian:

$$\frac{d^2 \psi}{dy^2} = Ae^{-\alpha y} \left[\alpha^2 y^s - 2\alpha s y^{s-1} + s(s-1)y^{s-2} \right] \tag{S.5.13.3}$$

$$y^s e^{-\alpha y} \left[E = -E_b \left(\alpha^2 - \frac{2\alpha s}{y} + \frac{s(s-1)}{y^2} \right) + \frac{V_0}{y^2} - \frac{V_0}{y} \right] \tag{S.5.13.4}$$

We equate terms of like powers of y:

$$-E_b \alpha^2 = E \tag{S.5.13.5}$$

$$2\alpha s E_b = V_0 \tag{S.5.13.6}$$

$$s(s-1)E_b = V_0 \tag{S.5.13.7}$$

The last equation defines s. The middle equation defines α once s is known. The top equation gives the eigenvalue:

$$s = \frac{1}{2}\left(1 + \sqrt{1 + 4\frac{V_0}{E_b}}\right) \tag{S.5.13.8}$$

$$\alpha = \frac{V_0}{E_b\left(1 + \sqrt{1 + 4\frac{V_0}{E_b}}\right)} \tag{S.5.13.9}$$

$$E = -\frac{V_0^2}{E_b\left(1 + \sqrt{1 + 4\frac{V_0}{E_b}}\right)^2} \tag{S.5.13.10}$$

Harmonic Oscillator

5.14 Given a Gaussian (MIT)

Denote the eigenfunctions of the harmonic oscillator as $\psi_n(x)$ with eigenvalue E_n. They are a complete set of states, and we can expand any function in this set. In particular, we expand our function $\Psi(x,0)$ in terms of coefficients a_n:

$$\Psi(x,0) = \sum_n a_n \psi_n(x) \tag{S.5.14.1}$$

$$a_n = \langle \psi_n | \Psi(x,0) \rangle = \int dx\ \psi_n(x)\,\Psi(x,0) \tag{S.5.14.2}$$

The expectation value of the energy is the integral of the Hamiltonian H for the harmonic oscillator:

$$\langle E \rangle = \langle \Psi(x,0) | H | \Psi(x,0) \rangle \tag{S.5.14.3}$$

$$= \int dx\ \Psi(x,0)\,H\Psi(x,0) = \sum_n E_n |a_n|^2$$

where we used the fact that $H\psi_n = E_n \psi_n$. The probability P_n of energy E_n is $|a_n|^2$. So the probability of E_0 is given by

$$P_0 = |a_0|^2 \tag{S.5.14.4}$$

where

$$a_0 = \int dx \ \psi_0(x) \ \Psi(x,0) \tag{S.5.14.5}$$

$$= \frac{1}{\sqrt{\pi x_0 \sigma}} \int_{-\infty}^{\infty} dx \exp\left[-\frac{x^2}{2}\left(\frac{1}{x_0^2} + \frac{1}{\sigma^2}\right)\right] = \frac{\sqrt{2x_0 \sigma}}{\sqrt{x_0^2 + \sigma^2}}$$

and finally

$$P_0 = \frac{2x_0 \sigma}{x_0^2 + \sigma^2} \tag{S.5.14.6}$$

It is easy to show that this quantity is less than unity for any value of $\sigma \neq x_0$ and is unity if $\sigma = x_0$.

5.15 Harmonic Oscillator ABCs (Stony Brook)

a) Here we took $m = 1$, $\hbar = 1$, $\omega = 1$.

$$[a, a^\dagger] = aa^\dagger - a^\dagger a = \frac{1}{2}\{(p - ix)(p + ix) - (p + ix)(p - ix)\}$$

$$= \frac{1}{2}\{2i\,[p, x]\} = \frac{1}{2}\{2i\,(-i)\} = 1 \tag{S.5.15.1}$$

since the commutator

$$[p, x] = -i\frac{\partial}{\partial x}x + ix\frac{\partial}{\partial x} = -i$$

b)

$$a^\dagger a + \frac{1}{2} = \frac{p + ix}{\sqrt{2}}\frac{p - ix}{\sqrt{2}} + \frac{1}{2}$$

$$= \frac{p^2}{2} + \frac{x^2}{2} - \frac{i\,[p, x]}{2} + \frac{1}{2} = \frac{p^2}{2} + \frac{x^2}{2} = H$$

c)

$$[a^\dagger a, H] = \left[a^\dagger a, \frac{1}{2} + a^\dagger a\right] = \left[a^\dagger a, \frac{1}{2}\right] + [a^\dagger a, a^\dagger a] = 0 \tag{S.5.15.2}$$

d) In order to demonstrate that $a^\dagger |n\rangle$ and $a |n\rangle$ are also eigenstates of N, compose the commutator

$$[a, N] = \left[a, a^\dagger a\right] = aa^\dagger a - a^\dagger a^2 = \left(aa^\dagger - a^\dagger a\right) a = a \qquad (\text{S.5.15.3})$$

by (S.5.15.1). Similarly,

$$\left[a^\dagger, N\right] = a^\dagger a^\dagger a - a^\dagger a a^\dagger = a^\dagger \left(a^\dagger a - a a^\dagger\right) = -a^\dagger \left[a, a^\dagger\right] = -a^\dagger \quad (\text{S.5.15.4})$$

Now,

$$a^\dagger N |n\rangle = \left(\left[a^\dagger, N\right] + N a^\dagger\right) |n\rangle \qquad (\text{S.5.15.5})$$

Substituting (S.5.15.4) into (S.5.15.5) and replacing $N |n\rangle$ by $n |n\rangle$, we have

$$a^\dagger N |n\rangle = \left(N a^\dagger - a^\dagger\right) |n\rangle = n a^\dagger |n\rangle \qquad (\text{S.5.15.6})$$

Rearranging (S.5.15.6) yields

$$N \left(a^\dagger |n\rangle\right) = (n + 1) \left(a^\dagger |n\rangle\right) \qquad (\text{S.5.15.7})$$

as required. A similar calculation gives

$$N \left(a |n\rangle\right) = (n - 1) \left(a |n\rangle\right) \qquad (\text{S.5.15.8})$$

We see from the above results that the application of the operator $a^\dagger$ on a state $|n\rangle$ has the effect of "raising" the state by 1, and the operator a lowers the state by 1 (see (f) below).

e)
$$E_0 = \langle 0 |H| 0\rangle = \left\langle 0 \left| a^\dagger a + \frac{1}{2} \right| 0 \right\rangle = \langle 0 |a^\dagger a| 0\rangle + \frac{1}{2} \langle 0 | 0\rangle = \frac{1}{2}$$

since, by assumption, $a |0\rangle = 0$.

f) Since by (c), the number operator $N = a^\dagger a$ and the Hamiltonian $H = a^\dagger a + 1/2$ commute, they have simultaneous eigenstates. Starting with $|0\rangle$, we may generate a number state whose energy eigenvalue is $1 + 1/2$ by applying the raising operator $a^\dagger$. Applying $a^\dagger$ again produces a state of eigenvalue $2 + 1/2$. What remains to be done is to see that these eigenstates (number, energy) are properly normalized. If we assume that the state $|n\rangle$ is normalized, then we may compose the inner product

$$\left(\langle n| a\right) \cdot \left(a^\dagger |n\rangle\right) = \langle n |a a^\dagger| n\rangle = |c|^2 \langle n + 1 | n + 1\rangle \quad (\text{S.5.15.9})$$

$$= (n + 1) \langle n | n\rangle$$

Up to an arbitrary phase, we see that

$$|n+1\rangle = \frac{a^\dagger |n\rangle}{\sqrt{n+1}}$$

Starting with the vacuum ket $|0\rangle$, we can write an energy eigenket $|n\rangle$:

$$|n\rangle = \frac{\left(a^\dagger\right)^n |0\rangle}{\sqrt{n!}}$$

g) The energy spectrum is $n+1/2$, where n takes all positive integer values and zero. From (S.5.15.9) and the fact that the norm of the eigenvectors is positive (actually, 1), we see that n cannot be negative, and so no negative eigenvalues are possible.

5.16 Number States (Stony Brook)

a) In this problem it is important to use only the information given. We may write the Hamiltonian as

$$H = \frac{p^2}{2m} + \frac{m\omega^2 q^2}{2} = \hbar\omega \left\{ \frac{p+im\omega q}{\sqrt{2m\hbar\omega}} \frac{p-im\omega q}{\sqrt{2m\hbar\omega}} + \frac{i}{2\hbar} [p,q] \right\} \qquad \text{(S.5.16.1)}$$

where $[p,q] = -i\hbar$ (see Problem 5.15), so

$$H = \hbar\omega \left(a^\dagger a + \frac{1}{2} \right) = \hbar\omega \left(N + \frac{1}{2} \right) \qquad \text{(S.5.16.2)}$$

We may establish the following:

$$[a, a^\dagger] = aa^\dagger - a^\dagger a$$
$$= \left[\frac{p-im\omega q}{\sqrt{2m\hbar\omega}} \frac{p+im\omega q}{\sqrt{2m\hbar\omega}} - \frac{p+im\omega q}{\sqrt{2m\hbar\omega}} \frac{p-im\omega q}{\sqrt{2m\hbar\omega}} \right] \qquad \text{(S.5.16.3)}$$
$$= \frac{2i}{2\hbar} [p,q] = 1$$

Apply the number operator $a^\dagger a$ to the state $|n\rangle$ directly:

$$a^\dagger a |n\rangle = a^\dagger a \frac{1}{\sqrt{n!}} \left(a^\dagger\right)^n |0\rangle$$
$$= \frac{1}{\sqrt{n!}} \left(aa^\dagger\right)\left(a^\dagger\right)^{n-1} |0\rangle = \frac{1}{\sqrt{n!}} \left(1 + a^\dagger a\right)\left(a^\dagger\right)^{n-1} |0\rangle$$

$$= \frac{1}{\sqrt{n!}} \left\{ a^\dagger \left(a^\dagger\right)^{n-1} + \left(a^\dagger\right)^2 a a^\dagger \left(a^\dagger\right)^{n-2} \right\} |0\rangle$$

$$= \frac{1}{\sqrt{n!}}$$

$$\cdot \left\{ a^\dagger \left(a^\dagger\right)^{n-1} + \left(a^\dagger\right)^2 \left(a^\dagger\right)^{n-2} + \cdots + \left(a^\dagger\right)^n \cdot 1 + \left(a^\dagger\right)^{n+1} a \right\} |0\rangle$$

$$= \frac{1}{\sqrt{n!}} n \left(a^\dagger\right)^n |0\rangle + \frac{1}{\sqrt{n!}} \left(a^\dagger\right)^{n+1} a |0\rangle$$

Since $a|0\rangle = 0$, we have

$$a^\dagger a |n\rangle = n \frac{1}{\sqrt{n!}} \left(a^\dagger\right)^n |0\rangle = n|n\rangle \qquad (S.5.16.4)$$

b) We see from (S.5.16.2) that the Hamiltonian is just

$$H = \hbar \omega a^\dagger a + \frac{\hbar \omega}{2}$$

We demonstrated in (a) that $|n\rangle$ is an eigenstate of the number operator $a^\dagger a$, so $|n\rangle$ is also an eigenstate of the Hamiltonian with eigenvalues E_n given by

$$E_n = \langle n|H|n\rangle = \left\langle n \left| \hbar \omega \left(a^\dagger a + \frac{1}{2} \right) \right| n \right\rangle \qquad (S.5.16.5)$$

$$= \hbar \omega \langle n |a^\dagger a| n\rangle + \frac{1}{2} \hbar \omega \langle n | n\rangle = \hbar \omega n + \frac{\hbar \omega}{2} = \hbar \omega \left(n + \frac{1}{2} \right)$$

c) The expectation value $\langle n |q^2| n\rangle$ may be calculated indirectly. Note that

$$q^2 = V(q) \frac{2}{m \omega^2}$$

where $V(q)$ is the potential energy. The expectation values of the potential and kinetic energies are equal for the quantum oscillator, as for time averages in the classical oscillator. Therefore, they are half of the total energy:

$$\langle n |q^2| n\rangle = \frac{2}{m \omega^2} \langle n |U(q)| n\rangle = \frac{1}{2} \frac{2}{m \omega^2} \langle n |H| n\rangle = \frac{E(n)}{m \omega^2}$$

In this problem, however, you are explicitly asked to use the operators a and $a^\dagger$ to calculate $\langle n |q^2| n\rangle$, so we have

$$q = \sqrt{\frac{2\hbar}{m \omega}} \frac{(a - a^\dagger) i}{2} = \sqrt{\frac{\hbar}{2m \omega}} (a - a^\dagger) i$$

We proceed to find

$$\langle n \,|q^2|\, n\rangle = -\frac{\hbar}{2m\omega}\left\langle n\left|a^2 - aa^\dagger - a^\dagger a + \left(a^\dagger\right)^2\right| n\right\rangle$$

$$= \frac{\hbar}{2m\omega}\left\langle n\,|aa^\dagger + a^\dagger a|\,n\right\rangle$$

$$= \frac{\hbar}{2m\omega}\left\langle n\,|n+1|\,n\right\rangle + \frac{\hbar}{2m\omega}\left\langle n\,|n|\,n\right\rangle$$

$$= \left(n+\frac{1}{2}\right)\frac{\hbar\omega}{m\omega^2}$$

$$= \frac{E(n)}{m\omega^2}$$

Thus, the result is the same by both approaches.

5.17   Coupled Oscillators (MIT)

The Hamiltonian of the system is

$$H = \frac{1}{2m}\left[p_1^2 + p_2^2\right] + \frac{m\omega^2}{2}\left[x_1^2 + x_2^2\right] + Cx_1x_2 \qquad (S.5.17.1)$$

The problem is easily solved in center-of-mass coordinates. So define

$$
\begin{aligned}
x &\equiv x_1 - x_2 & p &\equiv \frac{p_1 - p_2}{2}\\
y &\equiv \frac{x_1 + x_2}{2} & \Pi &\equiv p_1 + p_2\\
x_1 &= \frac{x}{2} + y & p_1 &= p + \frac{\Pi}{2}\\
x_2 &= -\frac{x}{2} + y & p_2 &= -p + \frac{\Pi}{2}
\end{aligned}
\qquad (S.5.17.2)
$$

These new coordinates are used to rewrite the Hamiltonian. It now decouples into separate x- and y-parts:

$$H = H_x + H_y \qquad (S.5.17.3)$$

$$H_x = \frac{p^2}{m} + \frac{mx^2}{4}\left[\omega^2 - \frac{C}{m}\right] \qquad (S.5.17.4)$$

$$H_y = \frac{\Pi^2}{4m} + my^2\left[\omega^2 + \frac{C}{m}\right] \qquad (S.5.17.5)$$

The x-oscillator has a frequency $\Omega_x = \sqrt{\omega^2 - C/m}$ and eigenvalues $\hbar\Omega_x(n_x + 1/2)$, where n_x is an integer. The y-oscillator has a frequency of $\Omega_y = \sqrt{\omega^2 + C/m}$ and eigenvalues $\hbar\Omega_y(n_y + 1/2)$, where n_y is an integer.

5.18 Time-Dependent Harmonic Oscillator I (Wisconsin-Madison)

a) At times $t \geq 0$ the wave function is

$$\Psi(x,t) = \sqrt{\frac{1}{3}}\, \psi_0(x) e^{-i\omega t/2} + \sqrt{\frac{2}{3}}\, \psi_2(x) e^{-5i\omega t/2} \qquad (S.5.18.1)$$

b) The state $\Psi(x,t)$ has even parity: it remains the same if one replaces x by $-x$, since $\psi_{2n}(-x) = \psi_{2n}(x)$. This is true for all times.

c) The average value of the energy is

$$\langle\Psi|H|\Psi\rangle = \frac{1}{3}E_0 + \frac{2}{3}E_2 \qquad (S.5.18.2)$$

$$= \left[\frac{1}{3}\cdot\frac{1}{2} + \frac{2}{3}\cdot\frac{5}{2}\right]\hbar\omega = \frac{11}{6}\hbar\omega$$

which is independent of time.

5.19 Time-Dependent Harmonic Oscillator II (Michigan State)

a) The time dependence of the wave function is

$$\Psi(t) = \frac{1}{\sqrt{5}}|1\rangle\, e^{-3i\omega t/2} + \frac{2}{\sqrt{5}}|2\rangle\, e^{-5i\omega t/2} \qquad (S.5.19.1)$$

b) The expectation value for the energy is

$$\langle E\rangle = \langle\Psi(t)|H|\Psi(t)\rangle = \frac{1}{5}E_1 + \frac{4}{5}E_2 \qquad (S.5.19.2)$$

$$= \left[\frac{1}{5}\cdot\frac{3}{2} + \frac{4}{5}\cdot\frac{5}{2}\right]\hbar\omega = \frac{23}{10}\hbar\omega$$

which is independent of time.

c) To find the average value of the position operator, we first need to show that

$$\langle 1 \left| x \right| 1 \rangle = 0 \tag{S.5.19.3}$$

$$\langle 2 \left| x \right| 2 \rangle = 0 \tag{S.5.19.4}$$

$$\langle 1 \left| x \right| 2 \rangle = \langle 2 \left| x \right| 1 \rangle = X_0 \tag{S.5.19.5}$$

Then

$$\langle x(t) \rangle = \langle \Psi(t) \left| x \right| \Psi(t) \rangle = \frac{2}{5} X_0 \left(e^{-i\omega t} + e^{i\omega t} \right) = \frac{4}{5} X_0 \cos \omega t \tag{S.5.19.6}$$

The expectation value of the position operator oscillates in time.

5.20   Switched-on Field (MIT)

a) Operate on the eigenfunction by the kinetic energy term in the Hamiltonian:

$$\frac{p^2}{2m} \psi_0 = -\frac{\hbar^2}{2m} \frac{d^2}{dx^2} \left(N e^{-\alpha^2 x^2 / 2} \right) \tag{S.5.20.1}$$

$$= N \frac{\hbar^2 \alpha^2}{2m} e^{-\alpha^2 x^2 / 2} \left(1 - \alpha^2 x^2 \right)$$

Consider the factor $1 - \alpha^2 x^2$; the 1 must give the eigenvalue and $\alpha^2 x^2$ must cancel the potential energy. These two constraints give the identities

$$\frac{\hbar^2 \alpha^4}{m} = m\omega^2 \rightarrow \alpha^2 = \frac{m\omega}{\hbar} \tag{S.5.20.2}$$

$$E_0 = \frac{\hbar^2 \alpha^2}{2m} = \frac{\hbar\omega}{2} \tag{S.5.20.3}$$

The normalization constant N is determined by

$$1 = \int_{-\infty}^{\infty} dx \ \psi_0(x)^2 = N^2 \int_{-\infty}^{\infty} dx \ e^{-\alpha^2 x^2} = N^2 \frac{\sqrt{\pi}}{\alpha} \tag{S.5.20.4}$$

$$N = \frac{\sqrt{\alpha}}{\pi^{1/4}} \tag{S.5.20.5}$$

b) The solution is given above: $E_0 = \hbar\omega/2$.

c) After the perturbation $e|\mathbf{E}|x$ is added, the Hamiltonian can be solved exactly by completing the square on the x-variable:

$$H + H' = \frac{p^2}{2m} + \frac{m\omega^2}{2}(x + \xi_0)^2 - \frac{e^2|\mathbf{E}|^2}{2m\omega^2} \qquad (S.5.20.6)$$

where the displacement $\xi_0 = e|\mathbf{E}|/m\omega^2$. The new ground state energy and eigenfunction are

$$E' = \hbar\omega\left(n + \frac{1}{2}\right) - \frac{e^2|\mathbf{E}|^2}{2m\omega^2} \qquad (S.5.20.7)$$

$$\psi_0'(x) = Ne^{-\alpha^2(x+\xi_0)^2} \qquad (S.5.20.8)$$

The harmonic oscillator vibrates about the new equilibrium point $-\xi_0$ with the same frequency ω as before. The constants N and α are unchanged by the addition of the perturbation.

d) To find the probability that a particle, initially in the ground state, remains in the ground state after switching on the potential, we employ the sudden approximation. Here we just evaluate the overlap integral of the two eigenfunctions, and the probability is the square of this overlap:

$$I_{00'} = \int dx \ \psi_0(x) \ \psi_0'(x)$$

$$= \frac{\alpha}{\sqrt{\pi}} \int dx \ e^{-\alpha^2[x^2+(x+\xi_0)^2]/2} \qquad (S.5.20.9)$$

$$= \exp\left(-\frac{e^2|\mathbf{E}|^2}{4m\hbar\omega^3}\right)$$

$$P = I_{00'}^2 = \exp\left(-\frac{e^2|\mathbf{E}|^2}{2m\hbar\omega^3}\right) \qquad (S.5.20.10)$$

5.21   Cut the Spring! (MIT)

a) Below we give the Hamiltonian H_2, the frequency ω_2, and the eigenvalues $E_n^{(2)}$ of the particle while coupled to two springs:

$$H_2 = \frac{p^2}{2m} + Kx^2 \qquad (S.5.21.1)$$

$$\omega_2 = \sqrt{\frac{2K}{m}} \qquad (S.5.21.2)$$

$$E_n^{(2)} = \hbar\omega_2\left(n + \frac{1}{2}\right) \qquad (S.5.21.3)$$

The only change from the harmonic oscillator for a single spring is that, with two identical springs, the effective spring constant is $2K$.

b) The eigenfunction of the ground state $(n = 0)$ is

$$\psi_0^{(2)}(x) = \frac{\sqrt{\alpha}}{\pi^{1/4}} \, e^{-\alpha^2 x^2/2} \qquad \text{(S.5.21.4)}$$

$$\alpha = \left(\frac{2mK}{\hbar^2}\right)^{1/4} \qquad \text{(S.5.21.5)}$$

c) When one spring is cut, the particle is now coupled to only a single spring. So we must replace $2K$ in the above equations by K. The ground state eigenfunction is now

$$\psi_0^{(1)}(x) = \frac{\sqrt{\beta}}{\pi^{1/4}} \, e^{-\beta^2 x^2/2} \qquad \text{(S.5.21.6)}$$

$$\beta = \left(\frac{mK}{\hbar^2}\right)^{1/4} \qquad \text{(S.5.21.7)}$$

Notice that $\beta = \alpha/2^{1/4}$. The amplitude I for remaining in the ground state is found, in the sudden approximation, by taking the overlap integral of the two ground state wave functions. The probability $P_{00'}$ of remaining in the ground state is the square of this overlap integral:

$$I = \int\limits_{-\infty}^{\infty} dx \; \psi_0^{(1)}(x) \, \psi_0^{(2)}(x) \qquad \text{(S.5.21.8)}$$

$$= \sqrt{\frac{\alpha\beta}{\pi}} \int\limits_{-\infty}^{\infty} dx \; e^{-x^2(\alpha^2+\beta^2)/2} = \sqrt{\frac{2\alpha\beta}{\alpha^2 + \beta^2}}$$

$$P_{00'} = I^2 = \frac{2\alpha\beta}{\alpha^2 + \beta^2} = \frac{2^{5/4}}{1 + \sqrt{2}} \approx 0.985 \qquad \text{(S.5.21.9)}$$

where we have used $\beta = \alpha/2^{1/4}$ in deriving the last line. The probability of remaining in the ground state is close to unity.

Angular Momentum and Spin

5.22   Given Another Eigenfunction (Stony Brook)

a) The factor $\cos\theta$ indicates that it is a p_z state which has an angular momentum of 1.

b) In order to determine the energy and potential, we operate on the eigenstate with the kinetic energy operator. For $\ell=1$ this gives for the radial part $(s = \sqrt{3})$

$$\nabla^2 \psi(\mathbf{r}) = \left\{ \frac{1}{r} \frac{d^2}{dr^2} r - \frac{2}{r^2} \right\} A r^s e^{-\alpha r} \cos \theta \qquad (S.5.22.1)$$

$$= \left[\alpha^2 - \frac{2\alpha(1+s)}{r} + \frac{s(s+1)-2}{r^2} \right] \psi(\mathbf{r})$$

The constant in the last term can be simplified to $s(s+1) - 2 = 1 + \sqrt{3}$. In the limit $r \to \infty$, the potential vanishes, and only the constant term in the kinetic energy equals the eigenvalue. Thus, we find

$$E = -\frac{\hbar^2 \alpha^2}{2m} \qquad (S.5.22.2)$$

c) To find the potential we subtract the kinetic energy from the eigenvalue and act on the eigenfunction:

$$V(r)\psi(\mathbf{r}) = \left[E + \frac{\hbar^2 \nabla^2}{2m} \right] \psi(\mathbf{r}) \qquad (S.5.22.3)$$

$$V(r) = (1 + \sqrt{3}) \frac{\hbar^2}{2m} \left[-\frac{2\alpha}{r} + \frac{1}{r^2} \right] \qquad (S.5.22.4)$$

The potential has an attractive Coulomb term $\sim 1/r$ and a repulsive $1/r^2$ term.

5.23 Algebra of Angular Momentum (Stony Brook)

a)

$$[J_3, J^2] = [J_3, J_1^2 + J_2^2 + J_3^2]$$

$$= [J_3, J_1^2] + [J_3, J_2^2]$$

$$= J_1 [J_3, J_1] + [J_3, J_1] J_1 + J_2 [J_3, J_2] + [J_3, J_2] J_2$$

$$= J_1 (iJ_2) + (iJ_2) J_1 + J_2 (-iJ_1) + (-iJ_1) J_2 = 0$$

b) Since J^2 and J_3 commute, we will try to find eigenstates with eigenvalues of J^2 and J_3, denoted by $|jm\rangle$, where j, m are real numbers:

$$J^2 |jm\rangle = \lambda |jm\rangle$$

$$J_3 |jm\rangle = m |jm\rangle$$

Since $J_1^2 + J_2^2 + J_3^2 \geq J_3^2$ we know that $\lambda \geq m^2$. Anticipating the result, let $\lambda \equiv j(j+1)$. Form the raising and lowering operators J_+ and J_-:

$$J_+ = J_1 + iJ_2$$

$$J_- = J_1 - iJ_2$$

$$J_+^* = J_-$$

$$J_-^* = J_+$$

Find the commutators

$$[J_3, J_+] = [J_3, J_1 + iJ_2] = iJ_2 - iiJ_1 = J_+ \qquad (S.5.23.1)$$

$$[J_3, J_-] = [J_3, J_1 - iJ_2] = iJ_2 + iiJ_1 = -J_-$$

From part (a) we know that $[J_\pm, J^2] = 0$. We now ask what is the eigenvalue of J^2 for the states $J_\pm |jm\rangle$:

$$J^2 J_\pm |jm\rangle = J_\pm J^2 |jm\rangle = j(j+1)J_\pm |jm\rangle \qquad (S.5.23.2)$$

So, these states have the same eigenvalue of J^2. Now, examine the eigenvalue of J_3 for these states:

$$J_3 J_\pm |jm\rangle = ([J_3, J_\pm] + J_\pm J_3) |jm\rangle \qquad (S.5.23.3)$$

$$= (\pm J_\pm + J_\pm J_3) |jm\rangle = (m \pm 1) J_\pm |jm\rangle$$

In (S.5.23.3) we see that $J_\pm$ has the effect of raising or lowering the m-value of the states $|jm\rangle$ so that

$$J_\pm |jm\rangle = C_{j\pm}^m |j, m \pm 1\rangle$$

where $C_{j\pm}^m$ are the corresponding coefficients. As determined above, we know that $j(j+1) \geq m^2$, so $J_\pm$ cannot be applied indefinitely to the state $|jm\rangle$; i.e., there must be an $m = m_{\max}$, $m = m_{\min}$ such that

$$J_+ |jm_{\max}\rangle = 0 \qquad (S.5.23.4)$$

$$J_- |jm_{\min}\rangle = 0 \qquad (S.5.23.5)$$

Expand $J_- J_+$ and apply J_- to (S.5.23.4):

$$J_- J_+ = (J_1 - i J_2)(J_1 + i J_2) = J_1^2 + J_2^2 + i [J_1, J_2]$$

$$= J^2 - J_3^2 + ii J_3 = J^2 - J_3^2 - J_3$$

$$J_- J_+ |jm_{max}\rangle = (J^2 - J_3^2 - J_3) |jm_{max}\rangle = 0$$

$$= [j(j+1) - m_{max}^2 - m_{max}] |jm_{max}\rangle = 0$$

Either the state $|jm_{max}\rangle$ is zero or $j(j+1) - m_{max}^2 - m_{max} = 0$. So

$$j(j+1) = m_{max}(m_{max} + 1)$$

$$m_{max}^{(1)} = j \qquad m_{max}^{(2)} = -(j+1)$$

Similarly,

$$J_+ J_- |jm_{min}\rangle = (J^2 - J_3^2 + J_3) |jm_{min}\rangle = 0$$

$$= [j(j+1) - m_{min}^2 + m_{min}] |jm_{min}\rangle = 0$$

$$j(j+1) - m_{min}^2 + m_{min} = 0$$

$$j(j+1) = m_{min}(m_{min} - 1)$$

$$m_{min}^{(1)} = j+1 \qquad m_{min}^{(2)} = -j$$

For $j \geq 0$, and since $m_{max} \geq m_{min}$, the only solution is

$$m_{max} = j \qquad m_{min} = -j$$

We knew that j was real, but now we have $m_{max} - m_{min} = 2j =$ integer, so

$$j = 0, \frac{1}{2}, 1, \frac{3}{2}, 2, \ldots \qquad m = -j, -j+1, \ldots, j-1, j$$

5.24 Triplet Square Well (Stony Brook)

Since the two spins are parallel, they are in a spin triplet state with $S = 1$ and $M = \pm 1$. The spin eigenfunction has even parity. The two-electron wave function is written as an orbital part $\psi(x_1, x_2)$ times the spin part. The total wave function must have odd parity. Since the spin has even parity, the orbital part must have odd parity: $\psi(x_2, x_1) = -\psi(x_1, x_2)$. Since the interaction potential acts only between the electrons, it is natural to write the orbital part in center-of-mass coordinates, where $X = (x_1 + x_2)/2$ and $x = x_1 - x_2$.

$$\psi(x_1, x_2) = e^{ikX} \psi(x) \tag{S.5.24.1}$$

The problem stated that the total momentum was zero, so set $k = 0$. We must now determine the form for the relative eigenfunction $\psi(x)$. It obeys the Schrödinger equation with the reduced mass $\mu = m/2$, where m is the electron mass:

$$\left[-\frac{\hbar^2}{2\mu} \frac{d^2}{dx^2} + V(x) - E \right] \psi(x) = 0 \tag{S.5.24.2}$$

We have reduced the problem to solving the bound state of a "particle" in a box. Here the "particle" is the relative motion of two electrons. However, since the orbital part of the wave function must have odd parity, we need to find the lowest energy state which is antisymmetric, $\psi(-x) = -\psi(x)$.

Bound states have $E = -E_B$ where the binding energy $E_B > 0$. Define two wave vectors: $\alpha^2 = 2\mu E_B/\hbar^2$ for outside the box, $|x| > a$, and $k^2 = 2\mu(V_0 - E_B)/\hbar^2$ when the particle is in the box, $|x| < a$. The lowest antisymmetric wave function is

$$\psi(x) = \begin{cases} Be^{-\alpha x} & x > a \\ A\sin kx & |x| < a \\ -Be^{\alpha x} & x < -a \end{cases} \tag{S.5.24.3}$$

We match the wave function and its derivative at one edge, say $x = a$, which gives two equations:

$$A\sin ka = Be^{-\alpha a} \tag{S.5.24.4}$$

$$kA\cos ka = -\alpha Be^{-\alpha a} \tag{S.5.24.5}$$

We divide these two equations, which eliminates the constants A and B. The remaining equation is the eigenvalue equation for α:

$$\alpha = -k\cot ka \tag{S.5.24.6}$$

$$k = \sqrt{k_0^2 - \alpha^2} \tag{S.5.24.7}$$

$$k_0^2 = \frac{2\mu V_0}{\hbar^2} \tag{S.5.24.8}$$

Since α and k are both positive, the cotangent of ka must be negative, which requires that $ka > \pi/2$. This imposes a constraint for the existence of any antisymmetric bound state:

$$V_0 > \frac{\hbar^2 \pi^2}{8\mu a^2} \qquad (S.5.24.9)$$

Any attractive square well has a bound state which is symmetric, but the above condition is required for the antisymmetric bound state.

5.25 Dipolar Interactions (Stony Brook)

a) We assume the magnetic moment is a vector parallel to the spin with a moment $\mu_j = \mu_0 s_j$ where μ_0 is a constant. Then we write the Hamiltonian as

$$H = \frac{\mu_0^2}{a^3} [\mathbf{s}_1 \cdot \mathbf{s}_2 - 3s_{1z}s_{2z}] \qquad (S.5.25.1)$$

The second term contains only s_{jz} components since the vector $\mathbf{a}$ is along the z-direction.

b) We write

$$\mathbf{S} = \mathbf{s}_1 + \mathbf{s}_2 \qquad (S.5.25.2)$$

$$\mathbf{S} \cdot \mathbf{S} = S(S+1) = (\mathbf{s}_1 + \mathbf{s}_2) \cdot (\mathbf{s}_1 + \mathbf{s}_2) \qquad (S.5.25.3)$$

$$S(S+1) = 2s(s+1) + 2\mathbf{s}_1 \cdot \mathbf{s}_2 \qquad (S.5.25.4)$$

$$S_z = s_{1z} + s_{2z} \qquad (S.5.25.5)$$

$$S_z^2 = s_{1z}^2 + s_{2z}^2 + 2s_{1z}s_{2z} \qquad (S.5.25.6)$$

For $s=1/2$ we have $s(s+1) = 3/4$ and $s_{jz}^2 = 1/4$, so we can write

$$\mathbf{s}_1 \cdot \mathbf{s}_2 = \frac{1}{2}S(S+1) - \frac{3}{4} \qquad (S.5.25.7)$$

$$s_{1z}s_{2z} = \frac{1}{2}S_z^2 - \frac{1}{4} \qquad (S.5.25.8)$$

$$H = E_0[S(S+1) - 3S_z^2] \qquad (S.5.25.9)$$

$$E_0 = \frac{\mu_0^2}{2a^3} \qquad (S.5.25.10)$$

c) The addition of two angular momenta with $s = 1/2$ gives values of S which are 0 or 1:

For $S = 1$ there are three possible eigenvalues of $S_z = (-1, 0, 1)$, which gives an energy of $E_0(-1, 2, -1)$.

For $S = 0$ there is one eigenvalue of $S_z = 0$, and this state has zero energy.

5.26 Spin-Dependent $1/r$ Potential (MIT)

a) The spin operator is $\mathbf{s} = \boldsymbol{\sigma}/2$. For spin $1/2$ the expression $\mathbf{s} \cdot \mathbf{s} = s(s+1)\tilde{I} = 3\tilde{I}/4$ becomes, for Pauli matrices, $\boldsymbol{\sigma} \cdot \boldsymbol{\sigma} = 3\tilde{I}$, where $\tilde{I}$ is the unit matrix. The total spin operator for the two-particle system is

$$\mathbf{S} = \frac{1}{2}(\boldsymbol{\sigma}_1 + \boldsymbol{\sigma}_2) \tag{S.5.26.1}$$

$$\mathbf{S} \cdot \mathbf{S} = S(S+1)\tilde{I} = \frac{1}{2}[3\tilde{I} + \boldsymbol{\sigma}_1 \cdot \boldsymbol{\sigma}_2] \tag{S.5.26.2}$$

$$\langle \boldsymbol{\sigma}_1 \cdot \boldsymbol{\sigma}_2 \rangle = 2S(S+1) - 3 \tag{S.5.26.3}$$

For the spin singlet state $S = 0$, then $V_{S=0} = -3g/r$, while for the spin triplet state $(S = 1)$, then $V_{S=1} = g/r$.

b) The potential is repulsive for the triplet state, and there are no bound states. There are bound states for the singlet state since the potential is attractive. For the hydrogen atom the potential is $-e^2/r$, and the eigenvalues are

$$E_n = -\frac{e^4 m}{2\hbar^2 n^2} \quad \text{(hydrogen atom)} \tag{S.5.26.4}$$

Our two-particle bound state has $3g$ instead of e^2 and the reduced mass $m/2$ instead of the mass m, so we have the eigenvalues

$$E_n = -\frac{9g^2 m}{4\hbar^2 n^2} \tag{S.5.26.5}$$

5.27 Three Spins (Stony Brook)

a) We use the notation that the state with three spins up is $|\uparrow_1\uparrow_2\uparrow_3\rangle$. This is the state with $M = 3/2$. We operate on this with the lowering operator J^-, which shows that the states $|3/2, M\rangle$ with lower values of M are

$$|3/2, 3/2\rangle = |\uparrow_1\uparrow_2\uparrow_3\rangle \tag{S.5.27.1}$$

$$|3/2, 1/2\rangle = \frac{1}{\sqrt{3}}[|\uparrow_1\uparrow_2\downarrow_3\rangle + |\uparrow_1\downarrow_2\uparrow_3\rangle + |\downarrow_1\uparrow_2\uparrow_3\rangle] \tag{S.5.27.2}$$

$$|3/2, -1/2\rangle = \frac{1}{\sqrt{3}}[|\uparrow_1\downarrow_2\downarrow_3\rangle + |\downarrow_1\downarrow_2\uparrow_3\rangle + |\downarrow_1\uparrow_2\downarrow_3\rangle] \quad \text{(S.5.27.3)}$$

$$|3/2, -3/2\rangle = |\downarrow_1\downarrow_2\downarrow_3\rangle \quad \text{(S.5.27.4)}$$

b) From the definition of J^+ we deduce that

$$J^+ |3/2, -3/2\rangle = \sqrt{3} |3/2, -1/2\rangle \quad \text{(S.5.27.5)}$$

$$J^+ |3/2, -1/2\rangle = 2 |3/2, 1/2\rangle \quad \text{(S.5.27.6)}$$

$$J^+ |3/2, 1/2\rangle = \sqrt{3} |3/2, 3/2\rangle \quad \text{(S.5.27.7)}$$

$$J^+ = \begin{pmatrix} 0 & \sqrt{3} & 0 & 0 \\ 0 & 0 & 2 & 0 \\ 0 & 0 & 0 & \sqrt{3} \\ 0 & 0 & 0 & 0 \end{pmatrix} \quad \text{(S.5.27.8)}$$

The matrix J^- is the Hermitian conjugate of J^+:

$$J^- = (J^+)^\dagger \quad \text{(S.5.27.9)}$$

c) Because $J_x = (J^+ + J^-)/2$ and $J_y = -i(J^+ - J^-)/2$, we can construct

$$J_x = \frac{1}{2} \begin{pmatrix} 0 & \sqrt{3} & 0 & 0 \\ \sqrt{3} & 0 & 2 & 0 \\ 0 & 2 & 0 & \sqrt{3} \\ 0 & 0 & \sqrt{3} & 0 \end{pmatrix} \quad \text{(S.5.27.10)}$$

$$J_y = -\frac{i}{2} \begin{pmatrix} 0 & \sqrt{3} & 0 & 0 \\ -\sqrt{3} & 0 & 2 & 0 \\ 0 & -2 & 0 & \sqrt{3} \\ 0 & 0 & -\sqrt{3} & 0 \end{pmatrix} \quad \text{(S.5.27.11)}$$

$$J_z = \begin{pmatrix} 3/2 & 0 & 0 & 0 \\ 0 & 1/2 & 0 & 0 \\ 0 & 0 & -1/2 & 0 \\ 0 & 0 & 0 & -3/2 \end{pmatrix} \quad \text{(S.5.27.12)}$$

d) To find the matrix $J^2 = J_x^2 + J_y^2 + J_z^2$, we square each of the three matrices and add them. This gives $J^2 = 15\tilde{I}/4$, where $\tilde{I}$ is the 4×4 unit matrix. This is what one expects, since the eigenvalue of J^2 is $J(J+1)$, which is $15/4$ when $J = 3/2$.

5.28   Constant Matrix Perturbation (Stony Brook)

a) Define $x = \varepsilon - \lambda - G$ where λ is the eigenvalue. We wish to diagonalize the matrix by finding the determinant of

$$H - \lambda I = \begin{pmatrix} x & -G & -G \\ -G & x & -G \\ -G & -G & x \end{pmatrix} \qquad (S.5.28.1)$$

$$0 = x^3 - 3xG^2 - 2G^3 \qquad (S.5.28.2)$$

When confronted by a cubic eigenvalue equation, it is best first to try to guess an eigenvalue. The obvious guesses are $x = \pm G$. The one that works is $x = -G$, so we factor this out to get

$$0 = x^3 - 3xG^2 - 2G^3 \qquad (S.5.28.3)$$

$$= (x + G)(x^2 - xG - 2G^2) = (x + G)^2 (x - 2G)$$

$$x = -G, -G, 2G \qquad (S.5.28.4)$$

$$\lambda = \varepsilon, \varepsilon, \varepsilon - 3G \qquad (S.5.28.5)$$

We call these eigenvalues $\lambda_1, \lambda_2, \lambda_3$, respectively. When we construct the eigenfunctions, only the one for λ_3 is unique. Since the first two have degenerate eigenvalues, their eigenvectors can be constructed in many different ways. One choice is

$$\psi_1 = \frac{1}{\sqrt{2}} \begin{pmatrix} 1 \\ -1 \\ 0 \end{pmatrix} \qquad (S.5.28.6)$$

$$\psi_2 = \frac{1}{\sqrt{6}} \begin{pmatrix} 1 \\ 1 \\ -2 \end{pmatrix} \qquad (S.5.28.7)$$

$$\psi_3 = \frac{1}{\sqrt{3}} \begin{pmatrix} 1 \\ 1 \\ 1 \end{pmatrix} \qquad (S.5.28.8)$$

b) Since the three states ψ_i form a complete set over this space, we can expand the initial state as

$$|i\rangle = \sum_i \psi_i \langle \psi_i | i \rangle \qquad (S.5.28.9)$$

$$|i\rangle = \frac{1}{\sqrt{2}}\psi_1 + \frac{1}{\sqrt{6}}\psi_2 + \frac{1}{\sqrt{3}}\psi_3 \tag{S.5.28.10}$$

$$|i\rangle(t) = \frac{1}{\sqrt{2}}\psi_1 e^{-i\lambda_1 t} + \frac{1}{\sqrt{6}}\psi_2 e^{-i\lambda_2 t} + \frac{1}{\sqrt{3}}\psi_3 e^{-i\lambda_3 t} \tag{S.5.28.11}$$

$$= e^{-i\varepsilon t}\left(\frac{1}{\sqrt{2}}\psi_1 + \frac{1}{\sqrt{6}}\psi_2 + \frac{1}{\sqrt{3}}\psi_3 e^{3iGt}\right) \tag{S.5.28.12}$$

To find the amplitude in state $|f\rangle$, we operate on the above equation with $\langle f|$. The probability P is found from the absolute-magnitude-squared of this amplitude:

$$\langle f|i\rangle(t) = \frac{e^{-i\varepsilon t}}{3}\left(e^{3iGt} - 1\right) \tag{S.5.28.13}$$

$$P = \frac{4}{9}\sin^2\frac{3Gt}{2} \tag{S.5.28.14}$$

5.29 Rotating Spin (Maryland, MIT)

Let us quantize the spin states along the z-axis so that spin up and spin down are denoted by

$$\alpha \equiv \begin{pmatrix} 1 \\ 0 \end{pmatrix} \tag{S.5.29.1}$$

$$\beta \equiv \begin{pmatrix} 0 \\ 1 \end{pmatrix} \tag{S.5.29.2}$$

$$H\alpha = \hbar\omega\alpha \tag{S.5.29.3}$$

$$H\beta = -\hbar\omega\beta \tag{S.5.29.4}$$

$$\hbar\omega = \mu B_0 \tag{S.5.29.5}$$

The eigenstates of σ_x are ψ_x for pointing along the $+x$-axis, and $\psi_{\bar{x}}$ for the $-x$-axis

$$\psi_x = \frac{1}{\sqrt{2}}(\alpha + \beta) \qquad \sigma_x\psi_x = \psi_x \tag{S.5.29.6}$$

$$\psi_{\bar{x}} = -\frac{1}{\sqrt{2}}(\alpha - \beta) \qquad \sigma_x\psi_{\bar{x}} = -\psi_{\bar{x}} \tag{S.5.29.7}$$

$$\psi_y = \frac{1}{\sqrt{2}}(\alpha + i\beta) \qquad \sigma_y\psi_y = \psi_y \tag{S.5.29.8}$$

$$\psi_{\bar{y}} = \frac{1}{\sqrt{2}}(\alpha - i\beta) \qquad \sigma_y\psi_{\bar{y}} = -\psi_{\bar{y}} \qquad \text{(S.5.29.9)}$$

At time $t = 0$ we start in state ψ_x. Later this state becomes

$$\Psi_x(t) = \frac{1}{\sqrt{2}}\left(\alpha e^{-i\omega t} + \beta e^{i\omega t}\right) \qquad \text{(S.5.29.10)}$$

The amplitude for pointing in the negative y-direction is found by taking the matrix element with $\psi_{\bar{y}}$. The probability is the square of the absolute magnitude of this amplitude:

$$\langle \psi_{\bar{y}} | \Psi_x(t) \rangle = \frac{1}{2}\left(e^{-i\omega t} + ie^{i\omega t}\right) \qquad \text{(S.5.29.11)}$$

$$P_{x\bar{y}}(t) = \cos^2\left(\omega t + \frac{\pi}{4}\right) \qquad \text{(S.5.29.12)}$$

5.30 Nuclear Magnetic Resonance (Princeton, Stony Brook)

a) Let $\alpha(t)$ and $\beta(t)$ denote the probability of spin up and spin down as a function of time. The time-dependent Hamiltonian is

$$\chi(t) = \begin{pmatrix} \alpha \\ \beta \end{pmatrix} \qquad . \qquad \text{(S.5.30.1)}$$

$$i\hbar\frac{d}{dt}\chi = \mu\left(B_1\sigma_x\cos\omega t + B_1\sigma_y\sin\omega t + B_0\sigma_z\right)\chi \qquad \text{(S.5.30.2)}$$

$$i\frac{d}{dt}\chi = \begin{pmatrix} \Omega_\parallel & \Omega_\perp e^{-i\omega t} \\ \Omega_\perp e^{i\omega t} & -\Omega_\parallel \end{pmatrix}\chi \qquad \text{(S.5.30.3)}$$

The equations for the individual components are

$$i\dot{\alpha} = \Omega_\parallel\alpha + \Omega_\perp e^{-i\omega t}\beta \qquad \text{(S.5.30.4)}$$

$$i\dot{\beta} = -\Omega_\parallel\beta + \Omega_\perp e^{i\omega t}\alpha \qquad \text{(S.5.30.5)}$$

where the overdots denote time derivatives. We solve the first equation for β and substitute this into the second equation:

$$\Omega_\perp\beta = e^{i\omega t}[i\dot{\alpha} - \Omega_\parallel\alpha] \qquad \text{(S.5.30.6)}$$

$$i\frac{d}{dt}\left[e^{i\omega t}(i\dot{\alpha} - \Omega_\parallel\alpha)\right] = e^{i\omega t}\left[-\Omega_\parallel(i\dot{\alpha} - \Omega_\parallel\alpha) + \Omega_\perp^2\alpha\right] \qquad \text{(S.5.30.7)}$$

$$0 = i^2\ddot{\alpha} - i\omega\dot{\alpha} + \alpha\left[\omega\Omega_\parallel - \Omega_\parallel^2 - \Omega_\perp^2\right] \qquad \text{(S.5.30.8)}$$

We assume that

$$\alpha(t) = \alpha_0 e^{i\lambda t} \qquad \text{(S.5.30.9)}$$

We determine the eigenvalue frequency λ by inserting the above form for $\alpha(t)$ into (S.5.30.8), which gives a quadratic equation for λ that has two roots:

$$0 = \lambda^2 + \lambda\omega + \omega\Omega_\| - \Omega_\|^2 - \Omega_\perp^2 \qquad \text{(S.5.30.10)}$$

$$\lambda = -\frac{\omega}{2} \pm \bar{\omega} \qquad \text{(S.5.30.11)}$$

$$\bar{\omega} = \sqrt{\left(\Omega_\| - \frac{\omega}{2}\right)^2 + \Omega_\perp^2} \qquad \text{(S.5.30.12)}$$

$$\alpha(t) = e^{-i\omega t/2}\left[a_1 e^{i\bar{\omega}t} + a_2 e^{-i\bar{\omega}t}\right] \qquad \text{(S.5.30.13)}$$

$$\beta(t) = e^{i\omega t/2}\left[b_1 e^{i\bar{\omega}t} + b_2 e^{-i\bar{\omega}t}\right] \qquad \text{(S.5.30.14)}$$

We have introduced the constants a_1, a_2, b_1, b_2. They are not all independent. Inserting these forms into the original differential equations, we obtain two relations which can be used to give

$$\Omega_\perp b_1 = a_1\left[\frac{\omega}{2} - \bar{\omega} - \Omega_\|\right] \qquad \text{(S.5.30.15)}$$

$$\Omega_\perp b_2 = a_2\left[\frac{\omega}{2} + \bar{\omega} - \Omega_\|\right] \qquad \text{(S.5.30.16)}$$

This completes the most general solution. Now we apply the initial conditions that the spin was pointing along the z-axis at $t = 0$. This gives $\alpha(0) = 1$, which makes $a_1 + a_2 = 1$ and $\beta(0) = 0$, which gives $b_1 + b_2 = 0$. These two conditions are sufficient to find

$$a_1 = \frac{\frac{\omega}{2} + \bar{\omega} - \Omega_\|}{2\bar{\omega}} \qquad \text{(S.5.30.17)}$$

$$a_2 = -\frac{\frac{\omega}{2} - \bar{\omega} - \Omega_\|}{2\bar{\omega}} \qquad \text{(S.5.30.18)}$$

$$\beta(t) = -i\frac{\Omega_\perp}{\bar{\omega}}e^{i\omega t/2}\sin\bar{\omega}t \qquad \text{(S.5.30.19)}$$

$$\alpha(t) = e^{-i\omega t/2}\left[\cos\bar{\omega}t + \frac{i}{\bar{\omega}}\left(\frac{\omega}{2} - \Omega_\|\right)\sin\bar{\omega}t\right] \qquad \text{(S.5.30.20)}$$

The probability of spin up is $|\alpha(t)|^2$ and that of spin down is $|\beta(t)|^2$:

$$|\alpha(t)|^2 = 1 - \frac{\Omega_\perp^2}{\bar{\omega}^2} \sin^2 \bar{\omega}t \qquad (S.5.30.21)$$

$$|\beta(t)|^2 = \frac{\Omega_\perp^2}{\bar{\omega}^2} \sin^2 \bar{\omega}t \qquad (S.5.30.22)$$

b) In the usual NMR experiment, one chooses the field B_0 so that $2\Omega_\parallel \approx \omega$, in which case $\bar{\omega} \approx \Omega_\perp$, and $|\alpha(t)|^2 \approx \cos^2 \Omega_\perp t$ and $|\beta(t)|^2 \approx \sin^2 \Omega_\perp t$. The spin oscillates slowly between the up and down states.

Variational Calculations

5.31   Anharmonic Oscillator (Tennessee)

Many possible trial functions can be chosen for the variational calculation. Choices such as $\exp(-\alpha|x|)$ are poor since they have an undesirable cusp at the origin. Instead, the best choice is a Gaussian:

$$\psi(x) = Ne^{-\alpha^2 x^2} \qquad (S.5.31.1)$$

where the potential in the problem is

$$V(x) = Ax^4 \qquad (S.5.31.2)$$

We evaluate the three integrals in (A.3.1)–(A.3.4).

$$\int dx\ \psi^2(x) = N^2 \frac{\sqrt{\pi}}{\alpha\sqrt{2}} \qquad (S.5.31.3)$$

$$\int dx\ Ax^4\psi^2(x) = AN^2 \frac{3\sqrt{\pi}}{16\alpha^5\sqrt{2}} \qquad (S.5.31.4)$$

$$\frac{1}{2m} \int dx\ |p\psi^2(x)| = N^2 \frac{\hbar^2\alpha\sqrt{\pi}}{2m\sqrt{2}} \qquad (S.5.31.5)$$

$$E_0 = \frac{\hbar^2\alpha^2}{2m} + \frac{3A}{16\alpha^4} \qquad (S.5.31.6)$$

We have used (A.3.1) to derive the last expression. Now we find the minimum energy for this choice of trial function by taking the derivative with respect to the variational parameter α. Denote by α_0 the value at this minimum:

$$\frac{dE_0}{d\alpha} = 0 = \frac{\hbar^2 \alpha_0}{m} - \frac{3A}{4\alpha_0^5} \qquad \text{(S.5.31.7)}$$

$$\alpha_0^6 = \frac{3Am}{4\hbar^2} \qquad \text{(S.5.31.8)}$$

$$E_0(\alpha_0) = \left(\frac{9\hbar^2}{16m}\right)^{2/3} A^{1/3} \qquad \text{(S.5.31.9)}$$

This result for E_0 is higher than the exact eigenvalue.

5.32   Linear Potential I (Tennessee)

The potential $V(x)$ is symmetric. The ground state eigenfunction must also be symmetric and have no cusps. A simple choice is a Gaussian:

$$\psi(x) = Ae^{-\alpha^2 x^2/2} \qquad \text{(S.5.32.1)}$$

where the variational parameter is α and A is a normalization constant. Again we must evaluate the three integrals in (A.3.1)–(A.3.4):

$$\int dx \; \psi^2(x) = A^2 \frac{\sqrt{\pi}}{\alpha} \qquad \text{(S.5.32.2)}$$

$$\frac{\hbar^2}{2m} \int dx \left(\frac{d\psi}{dx}\right)^2 = A^2 \frac{\hbar^2 \alpha^4}{2m} \int dx \; x^2 e^{-\alpha^2 x^2} = A^2 \frac{\hbar^2 \alpha \sqrt{\pi}}{4m} \qquad \text{(S.5.32.3)}$$

$$\int dx \; V(x)\psi^2(x) = 2FA^2 \int_0^\infty dx \; xe^{-\alpha^2 x^2} = A^2 \frac{F}{\alpha^2} \qquad \text{(S.5.32.4)}$$

$$E(\alpha) = \frac{\hbar^2 \alpha^2}{4m} + \frac{F}{\sqrt{\pi}\alpha} \qquad \text{(S.5.32.5)}$$

The minimum energy is found at the value α_0, where the energy derivative with respect to α is a minimum:

$$\frac{dE(\alpha)}{d\alpha} = 0 = \frac{\hbar^2 \alpha_0}{2m} - \frac{F}{\sqrt{\pi}\alpha_0^2} \qquad \text{(S.5.32.6)}$$

$$\alpha_0^3 = \frac{2mF}{\sqrt{\pi}\hbar^2} \qquad \text{(S.5.32.7)}$$

$$E(\alpha_0) = \frac{3}{2} \left(\frac{\hbar^2 F^2}{2\pi m} \right)^{1/3} \qquad \text{(S.5.32.8)}$$

5.33   Linear Potential II (MIT, Tennessee)

The wave function must vanish in either limit that $x \to (0, \infty)$. Two acceptable variational trial functions are

$$\psi(x) = Axe^{-\alpha x} \qquad \text{(S.5.33.1)}$$

$$\psi(x) = Bxe^{-\alpha^2 x^2} \qquad \text{(S.5.33.2)}$$

where the prefactor x ensures that the trial function vanish at the origin. In both cases the variational parameter is α. We give the solution for the first one, although either is acceptable. It turns out that (S.5.33.2) gives a higher estimate for the ground state energy, so (S.5.33.1) is better, since the estimate of the ground-state energy is always higher than the exact value. The ground state energy is obtained by evaluating the three integrals in (A.3.1)–(A.3.4):

$$I = \int_0^\infty dx\; \psi^2(x) = A^2 \int_0^\infty dx\; x^2 e^{-2\alpha x} = \frac{A^2}{4\alpha^3} \qquad \text{(S.5.33.3)}$$

$$U = \int_0^\infty dx\; V(x)\psi^2(x) = A^2 F \int_0^\infty dx\; x^3 e^{-2\alpha x} \qquad \text{(S.5.33.4)}$$

$$= \frac{3A^2 F}{8\alpha^4}$$

$$K = \frac{\hbar^2}{2m} \int_0^\infty dx\; \left(\frac{d\psi}{dx} \right)^2 \qquad \text{(S.5.33.5)}$$

$$= \frac{\hbar^2 A^2}{2m} \int_0^\infty dx\; e^{-2\alpha x}(1 - \alpha x)^2 = \frac{\hbar^2 A^2}{8\alpha m}$$

$$E(\alpha) = \frac{\hbar^2 \alpha^2}{2m} + \frac{3F}{2\alpha} \qquad \text{(S.5.33.6)}$$

The optimal value of α, called α_0, is obtained by finding the minimum value of $E(\alpha)$:

$$\frac{dE}{d\alpha} = 0 = \frac{\hbar^2 \alpha_0}{m} - \frac{3F}{2\alpha_0^2} \qquad (S.5.33.7)$$

$$\alpha_0^3 = \frac{3Fm}{2\hbar^2} \qquad (S.5.33.8)$$

$$E(\alpha_0) = \left(\frac{3}{2}\right)^{5/3} \left(\frac{\hbar^2 F^2}{m}\right)^{1/3} \qquad (S.5.33.9)$$

Note that this result is also the first asymmetric state of the potential in Problem 5.32.

5.34 Return of Combined Potential (Tennessee)

a) The potential $V(x)$ contains a term which diverges as $O(x^{-2})$ as $x \to 0$. The only way integrals such as $\int dx\ \psi^2 V(x)$ are well defined at the origin is if this divergence is canceled by factors in ψ. In particular, we must have $\psi \sim x$ at small x. This shows that the wave function must vanish at $x = 0$. This means that a particle on the right of the origin stays there.

b) The bound state must be in the region $x > 0$ since only here is the potential $V(x)$ attractive. The trial wave function is

$$\psi(x) = N\frac{x}{b}e^{-\alpha x/b} \qquad (S.5.34.1)$$

where the variational parameter is α. We evaluate the three integrals in (A.3.1)–(A.3.4), where the variable $s = x/b$:

$$I = N^2 b \int_0^\infty ds\ s^2 e^{-2\alpha s} = N^2 b \frac{1}{4\alpha^3} \qquad (S.5.34.2)$$

$$K = E_b N^2 b \int_0^\infty ds\ e^{-2\alpha s}(1 - \alpha s)^2 = N^2 b \frac{E_b}{4\alpha} \qquad (S.5.34.3)$$

$$U = N^2 b V_0 \int_0^\infty ds\ e^{-2\alpha s}(1 - s) \qquad (S.5.34.4)$$

$$= V_0 N^2 b \left[\frac{1}{2\alpha} - \frac{1}{(2\alpha)^2}\right]$$

$$E(\alpha) = E_b \alpha^2 + V_0 \left(2\alpha^2 - \alpha \right) \qquad \text{(S.5.34.5)}$$

$$E_b = \frac{\hbar^2}{2mb^2} \qquad \text{(S.5.34.6)}$$

The minimum energy is obtained by setting to zero the derivative of $E(\alpha)$ with respect to α. This gives the optimal value α_0 and the minimum energy $E(\alpha_0)$:

$$\frac{\mathrm{d}E}{\mathrm{d}\alpha} = 0 = 2\alpha_0 \left[E_b + 2V_0 \right] - V_0 \qquad \text{(S.5.34.7)}$$

$$\alpha_0 = \frac{1}{2} \frac{V_0}{E_b + 2V_0} \qquad \text{(S.5.34.8)}$$

$$E(\alpha_0) = -\frac{1}{4} \frac{V_0^2}{E_b + 2V_0} \qquad \text{(S.5.34.9)}$$

5.35 Quartic in Three Dimensions (Tennessee)

The potential $V(r) = Ar^4$ is spherically symmetric. In this case we can write the wave function as a radial part $R(r)$ times angular functions. We assume that the ground state is an s-wave, and the angular functions are P_0, which is a constant. So we minimize only the radial part of the wave function and henceforth ignore angular integrals. In three dimensions the integral in spherical coordinates is $\mathrm{d}^3 r = 4\pi r^2 \, \mathrm{d}r$. The factor 4π comes from the angular integrals. It occurs in every integral and drops out when we take the ratio in (A.3.1). So we just evaluate the $r^2 \, \mathrm{d}r$ part. Again we choose the trial function to be a Gaussian:

$$R(r) = Ne^{-\alpha^2 r^2 / 2} \qquad \text{(S.5.35.1)}$$

The three integrals in (A.3.1)–(A.3.4) have a slightly different form in three dimensions:

$$I = N^2 \int\limits_0^\infty \mathrm{d}r \; r^2 e^{-\alpha^2 r^2} = N^2 \frac{\sqrt{\pi}}{4\alpha^3} \qquad \text{(S.5.35.2)}$$

$$K = \frac{\hbar^2}{2m} \int\limits_0^\infty \mathrm{d}r \left[\frac{\mathrm{d}}{\mathrm{d}r}(rR) \right]^2 \qquad \text{(S.5.35.3)}$$

$$= N^2 \frac{\hbar^2}{2m} \int\limits_0^\infty \mathrm{d}r \; e^{-\alpha^2 r^2} (1 - \alpha^2 r^2)^2 = N^2 \frac{3\sqrt{\pi}\hbar^2}{16m\alpha}$$

$$U = AN^2 \int_0^\infty dr \; r^6 e^{-\alpha^2 r^2} = AN^2 \frac{15\sqrt{\pi}}{16\alpha^7} \qquad (S.5.35.4)$$

$$E(\alpha) = \frac{3}{4}\left[\frac{\hbar^2\alpha^2}{m} + \frac{5A}{\alpha^4}\right] \qquad (S.5.35.5)$$

Note the form of the kinetic energy integral K, which again is obtained from Rp^2R by an integration by parts. Again set the derivative of $E(\alpha)$ equal to zero. This determines the value α_0 which minimizes the energy:

$$\alpha_0^6 = 10\frac{mA}{\hbar^2} \qquad (S.5.35.6)$$

$$E(\alpha_0) = \frac{9}{8}\left(\frac{\hbar^2}{m}\right)^{2/3}(10A)^{1/3} \qquad (S.5.35.7)$$

5.36 Halved Harmonic Oscillator (Stony Brook, Chicago (b), Princeton (b))

a) Using the Rayleigh–Ritz variation principle, calculate $\langle E(\lambda)\rangle$, the expectation value of the ground state energy as a function of λ:

$$\langle E(\lambda)\rangle = \frac{\langle \psi(x)|H|\psi(x)\rangle}{\langle \psi(x)|\psi(x)\rangle} \qquad (S.5.36.1)$$

First calculate the denominator of (S.5.36.1):

$$\langle \psi(x)|\psi(x)\rangle = \int_{-\infty}^\infty dx \; |\psi(x)|^2 = \int_0^\infty dx \left(2\lambda^{3/2}xe^{-\lambda x}\right)^2$$

$$= \frac{4\lambda^3}{4}\int_0^\infty dx \; \frac{d^2}{d\lambda^2}\left(e^{-2\lambda x}\right) \qquad (S.5.36.2)$$

$$= \lambda^3\frac{d^2}{d\lambda^2}\int_0^\infty dx \; e^{-2\lambda x} = \lambda^3\frac{d^2}{d\lambda^2}\left(\frac{1}{2\lambda}\right) = 1$$

So our trial function is already normalized. Continuing with the numerator of (S.5.36.1), we have

$$\langle E(\lambda)\rangle = \int_0^\infty dx \; 2\lambda^{3/2}xe^{-\lambda x}\left(-\frac{\hbar^2}{2m}\frac{d^2}{dx^2} + \frac{1}{2}m\omega^2x^2\right)2\lambda^{3/2}xe^{-\lambda x}$$

$$= 4\lambda^3 \int\limits_0^\infty dx \ xe^{-\lambda x} \left(-A\frac{d^2}{dx^2} + Bx^2 \right) xe^{-\lambda x} \qquad (S.5.36.3)$$

where we set $A \equiv \hbar^2/2m$ and $B \equiv m\omega^2/2$. Evaluate the integral $I \equiv \langle E(\lambda)\rangle /4\lambda^3$:

$$I = \int\limits_0^\infty dx \ xe^{-\lambda x} \left(-A\frac{d^2}{dx^2} + Bx^2 \right) xe^{-\lambda x}$$

$$= \int\limits_0^\infty dx \ xe^{-\lambda x} \left(-Ax\lambda^2 e^{-\lambda x} + 2A\lambda e^{-\lambda x} + Bx^3 e^{-\lambda x} \right) \quad (S.5.36.4)$$

$$= -A\lambda^2 \left(-\frac{1}{2} \right)^2 \frac{d^2}{d\lambda^2} \int\limits_0^\infty dx \ e^{-2\lambda x} + 2A\lambda \left(-\frac{1}{2} \right) \frac{d}{d\lambda} \int\limits_0^\infty dx \ e^{-2\lambda x}$$

$$+ B \left(-\frac{1}{2} \right)^4 \frac{d^4}{d\lambda^4} \int\limits_0^\infty dx \ e^{-2\lambda x}$$

Now,

$$\int\limits_0^\infty dx \ e^{-2\lambda x} = \frac{1}{2\lambda} \qquad (S.5.36.5)$$

So, we have

$$I = -\frac{A\lambda^2}{4}\frac{1}{2}\frac{d^2}{d\lambda^2}\left(\frac{1}{\lambda}\right) - A\lambda\frac{1}{2}\frac{d}{d\lambda}\left(\frac{1}{\lambda}\right) + \frac{B}{16}\frac{1}{2}\frac{d^4}{d\lambda^4}\left(\frac{1}{\lambda}\right) (S.5.36.6)$$

$$= -\frac{A}{4\lambda} + \frac{A}{2\lambda} + \frac{B\cdot 4!}{32\lambda^5} = \frac{A}{4\lambda} + \frac{3B}{4\lambda^5}$$

Finally,

$$\langle E(\lambda)\rangle = 4\lambda^3 I = A\lambda^2 + \frac{3B}{\lambda^2} \qquad (S.5.36.7)$$

To minimize this function, find λ_0 corresponding to

$$\frac{\partial \langle E(\lambda)\rangle}{\partial \lambda} = 0 \qquad (S.5.36.8)$$

$$2A\lambda_0 - \frac{6B}{\lambda_0^3} = 0 \qquad (S.5.36.9)$$

$$\lambda_0^4 = \frac{3B}{A} = \frac{3m\omega^2}{\hbar^2/m} = \frac{3m^2\omega^2}{\hbar^2} \qquad (S.5.36.10)$$

$$\lambda_0^2 = \frac{\sqrt{3}m\omega}{\hbar} \qquad (S.5.36.11)$$

Therefore,

$$\langle E(\lambda_0) \rangle = \frac{\hbar^2}{2m}\frac{\sqrt{3}m\omega}{\hbar} + \frac{3}{2}\frac{m\omega^2}{\sqrt{3}m\omega/\hbar} = \sqrt{3}\hbar\omega \approx 1.7\hbar\omega \qquad (S.5.36.12)$$

We should have the inequality (see Problem 5.33)

$$\langle E(\lambda_0) \rangle \geq E_0 \qquad (S.5.36.13)$$

where E_0 is the true ground state energy.

b) To find the exact ground state of the system, notice that *odd* wave functions of a symmetric oscillator problem (from $-\infty$ to ∞) will also be solutions for these boundary conditions since they tend to zero at $x = 0$. Therefore, the ground state wave function of this halved oscillator will correspond to the first excited state wave function of the symmetrical oscillator. The wave function can easily be obtained if you take the ground state $|0\rangle$ and act on it by the creation operator $a^\dagger$ (see Problem 5.16):

$$a^\dagger |0\rangle \propto |1\rangle$$

$$|0\rangle \propto e^{-(m\omega/2\hbar)x^2}$$

$$a^\dagger |0\rangle \propto \left(\sqrt{\frac{m\omega}{2\hbar}}x - i\frac{\hbar}{\sqrt{2m\hbar\omega}}\frac{\mathrm{d}}{\mathrm{d}x} \right) |0\rangle \propto xe^{-(m\omega/2\hbar)x^2}$$

The ground state energy of our halved oscillator will in turn correspond to the first excited state energy $\tilde{E}_1$ of the symmetrical oscillator:

$$E_0 = \tilde{E}_1 = \hbar\omega\left(1 + \frac{1}{2}\right) = \frac{3}{2}\hbar\omega = 1.5\hbar\omega$$

Comparing this result with that of (a), we see that the inequality (S.5.36.13) holds and that our trial function is a fairly good approximation, since it gives the ground state energy to within 15% accuracy.

5.37 Helium Atom (Tennessee)

In the ground state of the two-electron system, both orbitals are in $1s$ states. So the spin state must be a singlet χ_s with $S = 0$. The spin plays no role in the minimization procedure, except for causing the orbital state to have even parity under the interchange of spatial coordinates. The two-electron wave function can be written as the product of the two orbital parts times the spin part:

$$\Psi(\mathbf{r}_1, \mathbf{r}_2) = \psi(r_1)\,\psi(r_2)\,\chi_s \qquad (S.5.37.1)$$

$$\psi(r) = \sqrt{\frac{\alpha^3}{a_0^3 \pi}}\; e^{-\alpha r / a_0} \qquad (S.5.37.2)$$

where a_0 is the Bohr radius and α is the variational parameter. The orbitals $\psi(r)$ are normalized to unity. Each electron has kinetic (K) and potential (U) energy terms which can be evaluated:

$$K = \frac{\hbar^2}{2m} \int d^3 r \, [\nabla \psi(r)]^2 = \alpha^2 E_R \qquad (S.5.37.3)$$

$$U = -2e^2 \int d^3 r \, \frac{\psi^2(r)}{r} = -4\alpha E_R \qquad (S.5.37.4)$$

where $E_R = 13.6$ eV is the Rydberg energy. The difficult integral is that due to the electron–electron interaction, which we call V:

$$V = e^2 \int d^3 r_1 \, \psi^2(r_1) \int d^3 r_2 \, \frac{\psi^2(r_2)}{|\mathbf{r}_1 - \mathbf{r}_2|} \qquad (S.5.37.5)$$

First we must do the angular integral over the denominator. If r_m is the larger of r_1 and r_2, then the integral over a 4π solid angle gives

$$\int \frac{d\Omega}{|\mathbf{r}_1 - \mathbf{r}_2|} = \frac{4\pi}{r_m} \qquad (S.5.37.6)$$

$$V = \frac{e^2 \alpha}{2a_0} \int_0^\infty dx \, x^2 e^{-x} \left[\frac{1}{x} \int_0^x dy \, y^2 e^{-y} + \int_x^\infty dy \, y e^{-y} \right]$$

In the second integral we have set $x = 2\alpha r_1 / a_0$ and $y = 2\alpha r_2 / a_0$, which makes the integrals dimensionless. Then we have split the y-integral into two parts, depending on whether y is smaller or greater than x. The first has a factor $1/x$ from the angular integrals, and the second has a factor

$1/y$. One can exchange the order of integration in one of the integrals and demonstrate that it is identical to the other. We evaluate only one and multiply the result by 2:

$$\int_x^\infty dy \; ye^{-y} = e^{-x}(x+1) \qquad (S.5.37.7)$$

$$\int_0^\infty dx \; x^2(x+1)e^{-2x} = \frac{5}{8} \qquad (S.5.37.8)$$

$$V = \frac{5\alpha}{4}E_R \qquad (S.5.37.9)$$

This completes the integrals. The total ground state energy $E(\alpha)$ in Rydbergs is

$$E(\alpha) = 2\left(\alpha^2 - 4\alpha\right) + \frac{5}{4}\alpha \qquad (S.5.37.10)$$

$$= 2\alpha^2 - \frac{27}{4}\alpha$$

We find the minimum energy by varying α. Denote by α_0 the value of α at which $E(\alpha)$ is a minimum. Setting to zero the derivative of $E(\alpha)$ with respect to α yields the result $\alpha_0 = 27/16$. The ground state energy is

$$E(\alpha_0) = -2\left(\frac{27}{16}\right)^2 \approx -5.7E_R \qquad (S.5.37.11)$$

Perturbation Theory

5.38 Momentum Perturbation (Princeton)

The first step is to rewrite the Hamiltonian by completing the square on the momentum operator:

$$H = \frac{p^2}{2m} + \frac{\lambda}{m}p + V(x) = \frac{(p+\lambda)^2}{2m} - \frac{\lambda^2}{2m} + V(x) \qquad (S.5.38.1)$$

The constant λ just shifts the zero of the momentum operator. The rewritten Hamiltonian in (S.5.38.1) suggests the perturbed eigenstates:

$$\tilde{\psi}_n(x) = e^{-ix\lambda/\hbar}\psi_n(x) \qquad (S.5.38.2)$$

The action of the displaced momentum operator $p+\lambda$ on the new eigenstates is

$$(p + \lambda)\,\tilde{\psi}_n(x) = e^{-ix\lambda/\hbar}\,p\,\psi_n(x) \tag{S.5.38.3}$$

so the Hamiltonian gives

$$H\tilde{\psi}_n = e^{-ix\lambda/\hbar}\left[H_0 - \frac{\lambda^2}{2m}\right]\psi_n(x) = \left[E_n - \frac{\lambda^2}{2m}\right]\psi_n \tag{S.5.38.4}$$

and the eigenvalues are simply

$$\tilde{E}_n = E_n - \frac{\lambda^2}{2m} \tag{S.5.38.5}$$

5.39 Ramp in Square Well (Colorado)

a) For a particle bound in a square well that runs from $-a/2 < x < a/2$, the eigenfunction and eigenvalue for the lowest energy state are

$$\psi_0(x) = \sqrt{\frac{2}{a}}\cos\frac{\pi x}{a} \tag{S.5.39.1}$$

$$E_0 = \frac{(\hbar\pi)^2}{2ma^2} \tag{S.5.39.2}$$

The eigenfunction is symmetric and vanishes at the walls of the well.

b) We use first-order perturbation theory to calculate the change in energy from the perturbation:

$$\delta E = \int_{-a/2}^{a/2} dx\, V(x)\psi_0^2(x) = \frac{8\varepsilon}{a^2}\int_0^{a/2} dx\, x\cos^2\frac{\pi x}{a}$$

$$= \frac{4\varepsilon}{a^2}\int_0^{a/2} dx\, x\left(1 + \cos\frac{2\pi x}{a}\right) \tag{S.5.39.3}$$

$$= \frac{\varepsilon}{\pi^2}\int_0^{\pi} d\theta\,\theta\,(1 + \cos\theta) = \frac{\varepsilon}{\pi^2}\left(\frac{\theta^2}{2} + \theta\sin\theta + \cos\theta\right)\Big|_0^{\pi}$$

$$= \frac{\varepsilon}{2\pi^2}\left(\pi^2 - 4\right)$$

5.40   Circle with Field (Colorado, Michigan State)

The perturbation is $V(\phi) = -e|\mathbf{E}|r\cos\phi$ if we assume the field is in the x-direction. The same result is obtained if we assume the perturbation is in the y-direction ($V = -e|\mathbf{E}|r\sin\phi$). In order to do perturbation theory, we need to find the matrix element of the perturbation between different eigenstates. For first-order perturbation theory we need

$$\langle n\,|V|\,n\rangle = -e|\mathbf{E}|r\int\limits_0^{2\pi}\frac{d\phi}{2\pi}\cos\phi = 0 \qquad (S.5.40.1)$$

The eigenvalues are unchanged to first-order in the field $\mathbf{E}$.

To do second-order perturbation theory, we need off-diagonal matrix elements:

$$\langle n\,|V|\,m\rangle = -e|\mathbf{E}|r\int\limits_0^{2\pi}\frac{d\phi}{2\pi}e^{i\phi(m-n)}\cos\phi \qquad (S.5.40.2)$$

$$= -\frac{1}{2}e|\mathbf{E}|r\delta_{m,n\pm1}$$

If we recall that $\cos\phi = \left(e^{i\phi} + e^{-i\phi}\right)/2$, then we see that $n-m$ can only equal ±1 for the integral to be nonzero. In doing second-order perturbation theory for the state $|n\rangle$, the only permissible intermediate states are $m = n\pm1$:

$$\delta E_n = \left\{\frac{\langle n\,|V|\,n+1\rangle^2}{E_n - E_{n+1}} + \frac{\langle n\,|V|\,n-1\rangle^2}{E_n - E_{n-1}}\right\} \qquad (S.5.40.3)$$

$$= \left(\frac{e|\mathbf{E}|r}{2}\right)^2\left(\frac{2mr^2}{\hbar^2}\right)\left[\frac{1}{n^2-(n+1)^2} + \frac{1}{n^2-(n-1)^2}\right]$$

$$= \frac{me^2r^4|\mathbf{E}|^2}{\hbar^2}\frac{1}{4n^2-1}$$

This solution is valid for states $n > 0$. For the ground state, with $n = 0$, the $n-1$ state does not exist, so the answer is

$$\delta E_0 = -\frac{me^2r^4|\mathbf{E}|^2}{2\hbar^2} \qquad (S.5.40.4)$$

5.41   Rotator in Field (Stony Brook)

a) The eigenfunctions and eigenvalues are

$$\psi_n(\phi) = \frac{1}{\sqrt{2\pi}} e^{in\phi} \tag{S.5.41.1}$$

$$E_n = \frac{\hbar^2 n^2}{2I} \tag{S.5.41.2}$$

b) The electric field interacts with the dipole moment to give an interaction

$$V = \mathbf{p} \cdot \mathbf{E} = p|\mathbf{E}| \cos\phi \tag{S.5.41.3}$$

This problem is almost identical to the previous one. The quantity mr^2 of the previous problem is changed to the moment I in the present problem. The perturbation results are similar. The first-order perturbation vanishes since $\langle n|V|n\rangle = 0$. The second-order perturbation is given by (S.5.40.3) and (S.5.40.4) after changing mr^2 to I and er to p:

$$\delta E_n = \begin{cases} \dfrac{p^2|\mathbf{E}|^2 I}{\hbar^2} \dfrac{1}{4n^2-1} & n > 0 \\[3mm] -\dfrac{p^2|\mathbf{E}|^2 I}{2\hbar^2} & n = 0 \end{cases} \tag{S.5.41.4}$$

5.42   Finite Size of Nucleus (Maryland, Michigan State, Princeton, Stony Brook)

a) To find the potential $V(r)$ near the nucleus, we note Gauss's law, which states that for an electron at a distance r from the center of a spherical charge distribution, the electric field is provided only by those electrons inside a sphere of radius r. For $r < R_0$, this is the charge $Z(r/R_0)^3$, whereas for $r > R_0$ it is just the charge Z. Thus, we find for the derivative of the potential energy:

$$-\frac{dV(r)}{dr} = eE(r) = \begin{cases} -\dfrac{Ze^2}{r^2}\left(\dfrac{r}{R_0}\right)^3 & r < R_0 \\[3mm] -\dfrac{Ze^2}{r^2} & r > R_0 \end{cases} \tag{S.5.42.1}$$

where C is a constant of integration. We chose $C = -3$ to make the potential continuous at $r = R_0$:

$$V(r) = \begin{cases} -\dfrac{Ze^2}{2R_0}\left[3 - \left(\dfrac{r}{R_0}\right)^2\right] & r < R_0 \\ -\dfrac{Ze^2}{r} & r > R_0 \end{cases} \qquad \text{(S.5.42.3)}$$

$$\delta V(r) = V(r) + \frac{Ze^2}{r} \qquad \text{(S.5.42.4)}$$

$$= \begin{cases} -\dfrac{Ze^2}{2R_0}\left[3 - \left(\dfrac{r}{R_0}\right)^2 - 2\dfrac{R_0}{r}\right] & r < R_0 \\ 0 & r > R_0 \end{cases}$$

b) For a single electron bound to a point nucleus, we can use hydrogen wave functions:

$$\Psi(r) = \sqrt{\frac{\alpha^3}{\pi}}\, e^{-\alpha r} \qquad \text{(S.5.42.5)}$$

$$\alpha = \frac{Z}{a_0} \qquad \text{(S.5.42.6)}$$

c) The first-order change in the ground state wave energy is

$$\delta E = \int d^3r\ \Psi(r)^2 \delta V(r) = 4\alpha^3 \int_0^\infty dr\ r^2 e^{-2\alpha r}\delta V(r)$$

$$= -\frac{2Ze^2\alpha^3}{R_0}\int_0^{R_0} dr\ r^2 e^{-2\alpha r}\left[3 - \left(\frac{r}{R_0}\right)^2 - 2\frac{R_0}{r}\right] \qquad \text{(S.5.42.7)}$$

$$= -\frac{Z^2e^2}{2a_0\xi}\int_0^\xi dx\ x^2 e^{-x}\left[3 - \frac{x^2}{\xi^2} - 2\frac{\xi}{x}\right]$$

$$\xi = 2Z\frac{R_0}{a_0} \qquad\qquad x = 2\alpha r \qquad \text{(S.5.42.8)}$$

For any physical value of Z, the parameter ξ is very much smaller than unity. One can evaluate the above integral as an expansion in ξ and show that the first term is $-0.2\xi^3$, so the answer is approximately $0.2Z^2E_R\xi^2$.

5.43 U and U^2 Perturbation (Princeton)

The result from first-order perturbation theory is obtained by taking the integral of the perturbation δV with the ground state wave function ψ_g:

$$\psi_g(\mathbf{r}) = \psi_0(x)\,\psi_0(y)\,\psi_0(z) \qquad (S.5.43.1)$$

$$\delta E^{(1)} = \int d^3r\; \psi_g(r)^2 \delta V(r) \qquad (S.5.43.2)$$

The ground state energy is $E_0 = 3\hbar\omega/2$. The first term in δV has odd parity and integrates to zero in the above expression. The second term in δV has even parity and gives a nonzero contribution. In this problem it is easiest to keep the eigenfunctions in the separate basis of x, y, z rather than to combine them into $\mathbf{r}$. In one dimension the average of $x^2 = \langle 0|x^2|0\rangle$, so we have

$$\langle 0|x^2|0\rangle = \int dx\; x^2 \psi_0^2(x) = \frac{x_0^2}{2} \qquad (S.5.43.3)$$

$$\delta E^{(1)} = \frac{U^2}{\hbar\omega}\left(\frac{x_0^2}{2}\right)^3 \qquad (S.5.43.4)$$

where $x_0^2 = \hbar/m\omega$. This is probably the simplest way to leave the answer. This completes the discussion of first-order perturbation theory.

The other term $Uxyz$ in δV contributes an energy of $O(U^2)$ in second-order perturbation theory. The excited state must have the symmetry of xyz, which means it is the state $\psi_{\text{ex}}(\mathbf{r}) = \psi_1(x)\,\psi_1(y)\,\psi_1(z)$. This has three quanta excited, so it has an energy $E_{\text{ex}} = 9\hbar\omega/2$:

$$\delta E^{(2)} = -U^2 \frac{|\langle 1|x|0\rangle^3|^2}{3\hbar\omega} \qquad (S.5.43.5)$$

$$\langle 1|x|0\rangle = \int dx\; \psi_1(x)\, x\, \psi_0(x) = \frac{x_0}{\sqrt{2}} \qquad (S.5.43.6)$$

$$\delta E^{(2)} = -U^2 \frac{x_0^6}{24\hbar\omega} \qquad (S.5.43.7)$$

Now we combine the results from first- and second-order perturbation theory:

$$\delta E\left[O(U^2)\right] = \frac{U^2 x_0^6}{\hbar\omega}\left[\frac{1}{8} - \frac{1}{24}\right] = \frac{U^2 x_0^6}{12\hbar\omega} \qquad (S.5.43.8)$$

5.44 Relativistic Oscillator (MIT, Moscow Phys-Tech, Stony Brook (a))

a) The classical Hamiltonian is given by $H_0 = p^2/2m + V(x)$, whereas the relativistic Hamiltonian may be expanded as follows:

$$H_{\text{rel}} = E + V(x) = \sqrt{m^2 c^4 + p^2 c^2} + V(x)$$

$$= mc^2 \sqrt{1 + \frac{p^2}{m^2 c^2}} + V(x) \qquad \text{(S.5.44.1)}$$

$$\approx mc^2 \left(1 + \frac{1}{2} \frac{p^2}{m^2 c^2} - \frac{1}{8} \frac{p^4}{m^4 c^4} \right) + V(x) = H_0 - \frac{1}{8} \frac{p^4}{m^3 c^2}$$

The perturbation to the classical Hamiltonian is therefore

$$H' = -\frac{1}{8} \frac{p^4}{m^3 c^2}$$

First solution: For the nonrelativistic quantum harmonic oscillator, we have

$$H_0 = \frac{p^2}{2m} + \frac{1}{2} m \omega^2 x^2 = \hbar \omega \left(\frac{m\omega}{2\hbar} x^2 + \frac{1}{2m\omega\hbar} p^2 \right) \qquad \text{(S.5.44.2)}$$

$$= \hbar \omega \left[\left(\sqrt{\frac{m\omega}{2\hbar}} x \right)^2 + \left(\sqrt{\frac{1}{2m\omega\hbar}} p \right)^2 \right]$$

where x, p are operators. Defining new operators Q, P,

$$Q = \sqrt{\frac{m\omega}{2\hbar}} \, x \qquad P = \sqrt{\frac{1}{2m\omega\hbar}} \, p$$

and noting the commutation relations

$$[P, Q] = \frac{1}{2\hbar} [p, x] = \frac{1}{2\hbar} \left[-i\hbar \frac{\partial}{\partial x}, x \right] = \frac{1}{2\hbar} \hbar i = \frac{i}{2}$$

we may rewrite (S.5.44.2) as

$$H_0 = \hbar \omega \left(Q^2 + P^2 \right) = \hbar \omega \left[(Q + iP)(Q - iP) - i [P, Q] \right]$$

$$= \hbar \omega \left[(Q + iP)(Q - iP) + \frac{1}{2} \right] \qquad \text{(S.5.44.3)}$$

Introducing the standard creation and annihilation operators (see Problems 5.15 and 5.16):

$$a^\dagger = Q - iP$$

$$a = Q + iP$$ (S.5.44.4)

$$a^\dagger |n\rangle = \sqrt{n+1}\,|n+1\rangle$$

$$a |n\rangle = \sqrt{n}\,|n-1\rangle$$

we find that

$$P = \frac{a - a^\dagger}{2i}$$

Using these results, we may express the first-order energy shift Δ_0 as

$$\Delta_0 = \langle 0|H'|0\rangle = \left\langle 0 \left| -\frac{1}{8}\frac{p^4}{m^3 c^2} \right| 0 \right\rangle$$

$$= -\frac{1}{8m^3 c^2} \left(\sqrt{2m\omega\hbar}\right)^4 \langle 0|P^4|0\rangle$$ (S.5.44.5)

$$= -\frac{1}{2}\frac{(\hbar\omega)^2}{mc^2} \langle 0|P^4|0\rangle = -\Delta \langle 0|P^4|0\rangle$$

where

$$\Delta \equiv \frac{(\hbar\omega)^2}{2mc^2}$$

The expansion of $\langle 0|P^4|0\rangle$ is simplified by the fact that $a|0\rangle = 0$, so

$$\langle 0|P^4|0\rangle = \left\langle 0 \left| \left(\frac{a - a^\dagger}{2i}\right)^4 \right| 0 \right\rangle$$

$$= \frac{1}{16} \left\langle 0 \left| \left(a^2 - aa^\dagger - a^\dagger a + a^{\dagger^2}\right)^2 \right| 0 \right\rangle$$

$$= \frac{1}{16} \left\langle 0 \left| a^2 a^{\dagger^2} + aa^\dagger aa^\dagger \right| 0 \right\rangle$$

$$= \frac{1}{16}\left(\sqrt{2}\sqrt{2} + 1\right) = \frac{3}{16}$$

Finally, we obtain

$$\Delta_0 = -\frac{3}{16}\Delta = -\frac{3}{32}\frac{(\hbar\omega)^2}{mc^2}$$

Second solution: Instead of using operator algebra, we can find a wave function $a(p)$ in the momentum representation, where

$$x = i\hbar \frac{d}{dp} \qquad (S.5.44.6)$$

The Hamiltonian then is

$$H = \frac{p^2}{2m} + \frac{m\omega^2}{2}x^2 = \frac{p^2}{2m} - \frac{m(\hbar\omega)^2}{2}\frac{d^2}{dp^2} \qquad (S.5.44.7)$$

The Schrödinger equation for $a(p)$ becomes

$$\frac{d^2a(p)}{dp^2} + \frac{2}{m(\hbar\omega)^2}\left(E - \frac{p^2}{2m}\right)a(p) = 0 \qquad (S.5.44.8)$$

This equation has exactly the same form as the standard oscillator Schrödinger equation:

$$\frac{d^2\psi}{dx^2} + \frac{2m}{\hbar^2}\left(E - \frac{m\omega^2 x^2}{2}\right)\psi = 0 \qquad (S.5.44.9)$$

We then obtain for the momentum probability distribution for the ground state:

$$|a_0(p)|^2 \frac{dp}{2\pi\hbar} = \frac{1}{\sqrt{\pi m\omega\hbar}}e^{-p^2/m\omega\hbar}\,dp \qquad (S.5.44.10)$$

Therefore

$$\Delta_0 = \langle 0|H'|0\rangle_p = A \int_{-\infty}^{\infty} \frac{dp}{2\pi\hbar}\,|a_0(p)|^2\,p^4 \qquad (S.5.44.11)$$

where $A \equiv -1/8m^3c^2$. Using the old "differentiate with respect to an innocent parameter method" of simplifying an integral, we may rewrite Δ_0 as

$$\Delta_0 = \frac{A}{\sqrt{\pi m\omega\hbar}}\frac{d^2}{d\xi^2}\int_{-\infty}^{\infty}dp\,e^{-\xi p^2} = \frac{A}{\sqrt{\pi m\omega\hbar}}\frac{d^2}{d\xi^2}\sqrt{\frac{\pi}{\xi}} \qquad (S.5.44.12)$$

where we substituted (S.5.44.10) into (S.5.44.11) and let $\xi \equiv 1/m\omega\hbar$. Finally,

$$\Delta_0 = \frac{A}{\sqrt{m\omega\hbar}}\frac{3}{4\xi^{5/2}} = \frac{3A}{4}m^2(\hbar\omega)^2 = -\frac{3}{32}\frac{(\hbar\omega)^2}{mc^2} \qquad (S.5.44.13)$$

as found in the first solution.

b) The first-order energy shift from αx^3 would be zero (no diagonal elements in the Q^3 matrix). The leading correction would be the second-order shift as defined by the formula

$$\Delta_1 = \sum_m{}' \frac{|V_{m0}|^2}{E_0^{(0)} - E_m^{(0)}}$$

where $\sum'$ means sum over $m \neq n = 0$. From (S.5.44.3) and (S.5.44.4), we have

$$Q = \frac{a^\dagger + a}{2}$$

$$x = \sqrt{\frac{2\hbar}{m\omega}} Q = \sqrt{\frac{\hbar}{2m\omega}} (a^\dagger + a)$$

So,

$$\Delta_1 = \alpha^2 \left(\frac{\hbar}{2m\omega}\right)^3 \sum_m{}' \frac{\left|\left\langle m\left|(a^\dagger + a)^3\right|0\right\rangle\right|^2}{E_0^{(0)} - E_m^{(0)}}$$

$$= \alpha^2 \left(\frac{\hbar}{2m\omega}\right)^3 \sum_m{}' \frac{\left|\left\langle m\left|a^{\dagger 3} + a^\dagger a a^\dagger + a a^{\dagger 2}\right|0\right\rangle\right|^2}{E_0^{(0)} - E_m^{(0)}}$$

$$= \alpha^2 \left(\frac{\hbar}{2m\omega}\right)^3 \left[\frac{\left|\langle 3|\sqrt{3}\sqrt{2}\sqrt{1}|3\rangle\right|^2}{E_0^{(0)} - E_3^{(0)}} + \frac{\left|\langle 1|(1+2)|1\rangle\right|^2}{E_0^{(0)} - E_1^{(0)}}\right]$$

$$= -\alpha^2 \left(\frac{\hbar}{2m\omega}\right)^3 \left[\frac{6}{3\hbar\omega} + \frac{9}{\hbar\omega}\right] = -\frac{11}{8}\alpha^2 \frac{\hbar^2}{m^3\omega^4}$$

As for any second-order correction to the ground state, it is negative. To make this expression equal to the one in part (a), we require that

$$-\frac{11}{8}\frac{\hbar^2}{m^3\omega^4}\alpha^2 = -\frac{3}{32}\frac{(\hbar\omega)^2}{mc^2}$$

$$\alpha = \sqrt{\frac{3}{11}}\frac{m\omega^3}{2c}$$

5.45   Spin Interaction (Princeton)

In first-order perturbation theory the change in energy is

$$\delta E^{(1)} = \langle 0 | H' | 0 \rangle = 0 \qquad\qquad (S.5.45.1)$$

where

$$|0\rangle = \psi_0(x)\psi_0(y)\psi_0(z) \qquad\qquad (S.5.45.2)$$

since $H' = \lambda \mathbf{r} \cdot \boldsymbol{\sigma}$ and the matrix element of $\mathbf{r}$ is zero for the ground state $|0\rangle$. The first excited state is three-fold degenerate: denote these states as

$$|x\rangle \equiv \psi_1(x)\,\psi_0(y)\,\psi_0(z) \qquad\qquad (S.5.45.3)$$

$$|y\rangle \equiv \psi_0(x)\,\psi_1(y)\,\psi_0(z) \qquad\qquad (S.5.45.4)$$

$$|z\rangle \equiv \psi_0(x)\,\psi_0(y)\,\psi_1(z) \qquad\qquad (S.5.45.5)$$

In this notation the matrix elements are

$$\langle x | H' | 0 \rangle = \frac{\lambda x_0 \sigma_x}{\sqrt{2}}, \quad \langle y | H' | 0 \rangle = \frac{\lambda x_0 \sigma_y}{\sqrt{2}}, \quad \text{etc.} \qquad (S.5.45.6)$$

In second-order perturbation theory

$$\delta E^{(2)} = -\sum_\ell \frac{\langle \ell | H' | 0 \rangle^2}{E_\ell - E_0} \qquad\qquad (S.5.45.7)$$

$$= -\frac{\lambda^2 x_0^2}{2\hbar\omega}\left[\sigma_x^2 + \sigma_y^2 + \sigma_z^2\right] = -\frac{3\lambda^2 x_0^2}{2\hbar\omega}$$

where $\sigma^2 = 3\tilde{I}$ where the unit matrix is $\tilde{I}$. Each spin state has the same energy, to second order.

5.46   Spin–Orbit Interaction (Princeton)

a) In three dimensions the lowest eigenvalue of the harmonic oscillator is $(3/2)\hbar\omega$, which can be viewed as $\hbar\omega/2$ from each of the three dimensions. The ground state has s-wave symmetry. The lowest excited states have eigenvalue $(5/2)\hbar\omega$. There are three of them. They have p-wave symmetry and are the states $L = 1$ and $M_L = (1, 0, -1)$.

b) In the spin–orbit interaction we take the derivative $\partial V/\partial r$ and find

$$V_{\text{so}} = \Delta \mathbf{L} \cdot \mathbf{S} \qquad\qquad (\text{S}.5.46.1)$$

$$\Delta = \frac{(\hbar\omega)^2}{2mc^2} \qquad\qquad (\text{S}.5.46.2)$$

The matrix element Δ is a constant, which simplifies the calculation. We evaluate the factor $\mathbf{L} \cdot \mathbf{S}$ by defining the total angular momentum $\mathbf{J}$ as

$$\mathbf{J} = \mathbf{L} + \mathbf{S} \qquad\qquad (\text{S}.5.46.3)$$

$$\mathbf{J} \cdot \mathbf{J} = (\mathbf{L} + \mathbf{S}) \cdot (\mathbf{L} + \mathbf{S}) \qquad\qquad (\text{S}.5.46.4)$$

$$J(J+1) = L(L+1) + S(S+1) + 2\mathbf{L} \cdot \mathbf{S} \qquad\qquad (\text{S}.5.46.5)$$

$$\langle \mathbf{L} \cdot \mathbf{S} \rangle = \frac{1}{2}[J(J+1) - L(L+1) - S(S+1)] \qquad\qquad (\text{S}.5.46.6)$$

For the ground state of the harmonic oscillator, $L = 0$ and $J = S = 1/2$. The above expectation value of $\mathbf{L} \cdot \mathbf{S}$ is zero. The ground state is unaffected by the spin–orbit interaction, although it is affected by relativistic corrections (see Problem 5.44) as well as by other states (see Problem 5.45).

The first excited states have $L = 1, S = 1/2$ so that $J = 3/2, 1/2$. For $J = 3/2$ we find that

$$\tilde{E}_{3/2} = (5/2)\hbar\omega + \Delta/2 \qquad\qquad (\text{S}.5.46.7)$$

For $J = 1/2$ we find that

$$\tilde{E}_{1/2} = (5/2)\hbar\omega - \Delta \qquad\qquad (\text{S}.5.46.8)$$

5.47   Interacting Electrons (MIT)

a) The wave function for a single electron bound to a proton is that of the hydrogen atom, which is

$$\psi_0(r) = \frac{1}{\sqrt{\pi a_0^3}}\, e^{-r/a_0} \qquad\qquad (\text{S}.5.47.1)$$

where a_0 is the Bohr radius. When one can neglect the Coulomb repulsion between the two electrons, the ground state energy and eigenfunctions are

$$E_0 = -2E_R \qquad\qquad (\text{S}.5.47.2)$$

$$\Psi_0(r_1, r_2) = \psi(r_1)\psi(r_2)\chi_0 \qquad\qquad (\text{S}.5.47.3)$$

$$\chi_0 = \frac{1}{\sqrt{2}}[\alpha_1\beta_2 - \alpha_2\beta_1] \qquad\qquad (\text{S}.5.47.4)$$

The last factor in (S.5.47.3) is the spin-wave function for the singlet $S = 0$ in terms of up α and down β spin states. Since the spin state has odd parity, the orbital state has even parity, and a simple product function $\psi(r_1)\psi(r_2)$ is correct. The eigenvalue is twice the Rydberg energy E_R.

b) The change in energy in first-order perturbation theory is $\delta E = \langle i\,|V|\,i\rangle$. The orbital part of the matrix element is

$$\langle V\rangle_\circ = \int d^3r_1\, d^3r_2\, \psi_0^2(r_1)V_0\delta^3(\mathbf{r}_1 - \mathbf{r}_2)\psi_0^2(r_2) \qquad \text{(S.5.47.5)}$$

$$= V_0 \int d^3r\, \psi_0^4(r) = \frac{4\pi V_0}{\pi^2 a_0^3} \int_0^\infty dx\, x^2 e^{-4x} = \frac{V_0}{8\pi a_0^3}$$

where the final integration variable is $x = r/a_0$.

Next we evaluate the spin part of the matrix element. The easiest way is to use the definition of the total spin $\mathbf{S} = \mathbf{s}_1 + \mathbf{s}_2$ to derive

$$\mathbf{S} \cdot \mathbf{S} = \mathbf{s}_1 \cdot \mathbf{s}_1 + \mathbf{s}_2 \cdot \mathbf{s}_2 + 2\mathbf{s}_1 \cdot \mathbf{s}_2 \qquad \text{(S.5.47.6)}$$

$$\langle \mathbf{s}_1 \cdot \mathbf{s}_2\rangle = \frac{1}{2}\left[S(S+1) - \frac{3}{2}\right] \qquad \text{(S.5.47.7)}$$

where for spin-$1/2$ particles, such as electrons, $\mathbf{s}_1 \cdot \mathbf{s}_1 = s(s+1) = 3/4$. Since the two spins are in an $S = 0$ state, the expectation value $\langle \mathbf{s}_1 \cdot \mathbf{s}_2\rangle = -3/4$. Combining this with the orbital contribution, we estimate the perturbed ground state energy $\tilde{E}$ to be

$$\tilde{E}_0 = -2E_R - \frac{3V_0}{32\pi a_0^3} \qquad \text{(S.5.47.8)}$$

5.48   Stark Effect in Hydrogen (Tennessee)

We use the notation $|LM\rangle$ to describe the four orbital states: the s-state is $|00\rangle$ and the three p-orbitals are $|1-1\rangle, |10\rangle, |11\rangle$. Spin is not affected by this perturbation and plays no role in the calculation. For degenerate perturbation theory we must evaluate the 10 different matrix elements $\langle LM\,|V(z)|\,L'M'\rangle$ which occur in the symmetric 4×4 matrix. The interaction potential is $V = -e|\mathbf{E}|z$. One can use parity and other group theory arguments to show that only one matrix element is nonzero, and we call it ξ:

$$\xi = -e|\mathbf{E}|\langle 00\,|z|\,10\rangle \qquad \text{(S.5.48.1)}$$

Since the two states $|1 \pm 1\rangle$ have no matrix elements with the other two states, we can omit them from the remaining steps in the calculation. Thus we must find the eigenvalues of a 2×2 matrix for the states $|00\rangle$ and $|10\rangle$.

$$H' = \begin{pmatrix} 0 & \xi \\ \xi & 0 \end{pmatrix} \qquad \text{(S.5.48.2)}$$

This matrix has eigenvalues $\lambda = \pm \xi$. The perturbation splits the fourfold degenerate $n = 2$ state into states with eigenvalues

$$E = E_{2s}, E_{2s}, E_{2s} - \xi, E_{2s} + \xi \qquad \text{(S.5.48.3)}$$

Since ξ is proportional to the electric field, the energies split linearly with $|\mathbf{E}|$.

The matrix element ξ can be evaluated by using the explicit representation for the $n = 2$ eigenstates of the hydrogen atom:

$$|00\rangle = \frac{1}{\sqrt{8\pi a_0^3}} e^{-r/2a} \left(1 - \frac{r}{2a}\right) \qquad \text{(S.5.48.4)}$$

$$|10\rangle = \frac{z}{\sqrt{32\pi a_0^5}} e^{-r/2a_0} \qquad \text{(S.5.48.5)}$$

yielding

$$\xi = -e|\mathbf{E}| \frac{2\pi}{16\pi a_0^4} \int\limits_0^\infty dr \, r^4 e^{-r/a_0} \left(1 - \frac{r}{2a_0}\right) \int\limits_0^\pi d\theta \, \sin\theta \cos^2\theta$$

$$= -\frac{e|\mathbf{E}|a_0}{12} \int\limits_0^\infty ds \, s^4 \left(1 - \frac{s}{2}\right) e^{-s} = 3e|\mathbf{E}|a_0 \qquad \text{(S.5.48.6)}$$

The angular integral gives $2/3$, and $s = r/a_0$.

5.49 $n = 2$ Hydrogen with Electric and Magnetic Fields (MIT)

We use the same notation as in Problem 5.48 to describe the four orbital states: the s-state is $|00\rangle$ and the three p-orbitals are $|1-1\rangle, |10\rangle, |11\rangle$. Here again, spin is not affected by this perturbation. As in Problem 5.48, we must evaluate the 10 different matrix elements $\langle LM|V(z)|L'M'\rangle$ which occur in the symmetric 4×4 matrix.

One interaction potential is $V = -e|\mathbf{E}|x$. One can use parity and other group theory arguments to show that the only nonzero matrix elements are

$$\zeta = -e|\mathbf{E}|\langle 00\,|x|\,11\rangle \qquad (S.5.49.1)$$

$$\zeta' = -e|\mathbf{E}|\langle 00\,|x|\,1-1\rangle \qquad (S.5.49.2)$$

One can show that ζ and ζ' are equal to within a phase factor. We ignore this phase factor and call them equal. The evaluation of this integral was demonstrated in the previous solution. The result here is $\zeta = \xi/\sqrt{2}$, compared to the one in the previous problem.

To first order in the magnetic field, the interaction is given by

$$U = -\frac{e}{mc}\mathbf{p}\cdot\mathbf{A} \qquad (S.5.49.3)$$

In spherical coordinates the three unit vectors for direction are $\left(\hat{\mathbf{r}}, \hat{\boldsymbol{\theta}}, \hat{\boldsymbol{\phi}}\right)$. In these units the vector potential can be written as $\mathbf{A} = Br\hat{\boldsymbol{\phi}}$. Similarly, the momentum operator in this direction is

$$p_\phi = \frac{\hbar}{ir}\frac{\partial}{\partial\phi} \qquad (S.5.49.4)$$

$$U = \hbar\omega_c\frac{\partial}{i\partial\phi} \qquad (S.5.49.5)$$

$$U\,|LM\rangle = M\hbar\omega_c\,|LM\rangle \qquad (S.5.49.6)$$

where the cyclotron frequency is $\omega_c = eB/mc$. The magnetic field is a diagonal perturbation in the basis $|LM\rangle$.

Now the state $|10\rangle$ has no matrix elements for these interactions and is unchanged by these interactions to lowest order. So we must diagonalize the 3×3 interaction matrix for the three states $|00\rangle, |11\rangle, |1-1\rangle$:

$$H' = \begin{pmatrix} 0 & \zeta & \zeta \\ \zeta & \hbar\omega_c & 0 \\ \zeta & 0 & -\hbar\omega_c \end{pmatrix} \qquad (S.5.49.7)$$

$$\lambda = 0, \pm\sqrt{(\hbar\omega_c)^2 + 2\zeta^2} \qquad (S.5.49.8)$$

The $n = 2$ states are initially fourfold degenerate. The double perturbation leaves two states with the same eigenvalue E_{2s} while the other two are shifted by $\pm\sqrt{(\hbar\omega_c)^2 + 2\zeta^2}$, where $\hbar\omega_c \sim B$ and $\zeta \sim |\mathbf{E}|$. Note that $2\zeta^2 = \xi^2$ so that, in the absence of the magnetic field, the result is the same as in Problem 5.48.

5.50 Hydrogen in Capacitor (Maryland, Michigan State)

For time-dependent perturbations a general wave function is

$$\Psi(\mathbf{r}, t) = \sum_j a_j(t)\psi_j(\mathbf{r})e^{-i\omega_j t} \tag{S.5.50.1}$$

where the ψ_j satisfy

$$H_0\psi_j = \hbar\omega_j\psi_j \tag{S.5.50.2}$$

For the time-dependent perturbation $V(t)$,

$$V(t) = -e|\mathbf{E}_0|ze^{-t/\tau} \tag{S.5.50.3}$$

From Schrödinger's equation we can derive an equation for the time development of the amplitudes $a_j(t)$:

$$i\hbar\frac{\partial}{\partial t}\Psi = [H_0 + V(t)]\Psi \tag{S.5.50.4}$$

$$i\hbar\frac{\partial}{\partial t}a_j(t) = \sum_\ell a_\ell \langle j|V(t)|\ell\rangle e^{it(\omega_j - \omega_\ell)} \tag{S.5.50.5}$$

If the system is initially in the ground state, we have $a_{1S}(0) = 1$ and the other values of $a_j(0)$ are zero. For small perturbations it is sufficient to solve the equation for $j \neq 1S$:

$$\frac{\partial}{\partial t}a_j(t) = \frac{ie|\mathbf{E}_0|}{\hbar}\langle j|z|1S\rangle e^{-t[1/\tau - i(\omega_j - \omega_{1S})]} \tag{S.5.50.6}$$

$$a_j(\infty) = \frac{ie|\mathbf{E}_0|\langle z\rangle}{\hbar}\int_0^\infty dt\, e^{-t[1/\tau - i(\omega_j - \omega_{1S})]} \tag{S.5.50.7}$$

$$= \frac{ie|\mathbf{E}_0|\langle z\rangle\tau}{\hbar[1 - i\tau(\omega_j - \omega_{1S})]}$$

The general probability P_j that a transition is made to state j is given by

$$P_j = |a_j(\infty)|^2 = \frac{(e|\mathbf{E}_0|\tau)^2\langle j|z|1S\rangle^2}{\hbar^2\left[1 + \tau^2(\omega_j - \omega_{1S})^2\right]} \tag{S.5.50.8}$$

This probability is dimensionless. It should be less than unity for this theory to be valid.

a) For the state $j = 2S$ the probability is zero. It vanishes because the matrix element of z is zero: $\langle 2S\,|z|\,1S \rangle = 0$ because of parity. Both S-states have even parity, and z has odd parity.

b) For the state $j = 2P$ the transition is allowed to the $L = 1, M = 0$ orbital state, which is called $2P_z$. The matrix element is similar to the earlier problem for the Stark effect. The $2P$ eigenstate for $L = 1$, $S = 0$ is in (S.5.48.5) and that for the $1S$ state is $\exp\left(-r/a_0\right)/\sqrt{\pi a_0^3}$. The integral is

$$\langle 2P_z\,|z|\,1S \rangle = \frac{2\pi}{\pi a_0^4 \sqrt{32}} \int_0^\infty dr\ r^4 e^{-3r/2a_0} \int_0^\pi d\theta\ \sin\theta \cos^2\theta$$

$$= \frac{1}{3\sqrt{2}\,a_0^4} \int_0^\infty dr\ r^4 e^{-3r/2a_0} = a_0 \left(\frac{2^{3/2}}{3}\right)^5 \qquad (S.5.50.9)$$

where a_0 is the Bohr radius of the hydrogen atom.

5.51   Harmonic Oscillator in Field (Maryland, Michigan State)

We adopt (S.5.50.4) and (S.5.50.5) for the time-dependent perturbation theory. Now we label the eigenstates with the index n for the harmonic oscillator state of energy $E_n = \hbar\omega(n+1/2)$ and write the equation satisfied by the time-dependent amplitudes $a_n(t)$,

$$i\hbar\frac{\partial}{\partial t}a_n(t) = \sum_\ell a_\ell\,\langle n\,|\delta V|\,\ell \rangle\, e^{it(\omega_n - \omega_\ell)} \qquad (S.5.51.1)$$

$$\delta V = -e|\mathbf{E}|x \qquad (S.5.51.2)$$

We need to evaluate the matrix element $\langle n\,|x|\,m \rangle$ of x between the states $|n\rangle$ and $|m\rangle$ of the harmonic oscillator. It is only nonzero if $m = n \pm 1$. In terms of raising and lowering operators,

$$x = X_0(a + a^\dagger) \qquad (S.5.51.3)$$

$$X_0 = \sqrt{\frac{\hbar}{2m\omega}} \qquad (S.5.51.4)$$

$$a^\dagger\,|n\rangle = \sqrt{n+1}\,|n+1\rangle \qquad (S.5.51.5)$$

$$a\,|n\rangle = \sqrt{n}\,|n-1\rangle \qquad (S.5.51.6)$$

a) If the initial state is $n = 0$ at $t = 0$, then the amplitude of the $n = 1$ state for $(t < \tau)$ is given by

$$a_1(t) = -i \frac{e|\mathbf{E}|X_0}{\hbar} \int_0^t dt' \, e^{i\omega t'} \qquad \text{(S.5.51.7)}$$

$$= -\frac{e|\mathbf{E}|X_0}{\hbar\omega} \left(e^{i\omega t} - 1 \right)$$

$$P_1(\tau) = |a_1(\tau)|^2 = 4 \left(\frac{e|\mathbf{E}|X_0}{\hbar\omega} \right)^2 \sin^2 \left(\frac{\omega\tau}{2} \right) \qquad \text{(S.5.51.8)}$$

The last equation is the probability of ending in the state $n = 1$ if the initial state is $n = 0$. This expression is valid as long as it is less than 1 or if $2e|\mathbf{E}|X_0 < \hbar\omega$.

b) The $n = 2$ state cannot be reached by a single transition from $n = 0$ since the matrix element $\langle 2 |x| 0 \rangle = 0$. However, $n = 2$ can be reached by a two-step process. It can be reached from $n = 1$, and $n = 1$ is excited from $n = 0$. The matrix element is $\langle 2 |x| 1 \rangle = \sqrt{2}\,X_0$, so we have that

$$a_2(t) = -i \frac{\sqrt{2}e|\mathbf{E}|X_0}{\hbar} \int_0^t dt' \, a_1(t') e^{i\omega t'}$$

$$= i\sqrt{2}\omega \left(\frac{e|\mathbf{E}|X_0}{\hbar\omega} \right)^2 \int_0^t dt' \, e^{i\omega t'} \left(e^{i\omega t'} - 1 \right) \qquad \text{(S.5.51.9)}$$

$$= \frac{1}{\sqrt{2}} \left(\frac{e|\mathbf{E}|X_0}{\hbar\omega} \right)^2 \left(e^{i\omega t} - 1 \right)^2$$

$$P_2(\tau) = |a_2(\tau)|^2 = 8 \left(\frac{e|\mathbf{E}|X_0}{\hbar\omega} \right)^4 \sin^4 \frac{\omega\tau}{2} \qquad \text{(S.5.51.10)}$$

Note that $P_2 = P_1^2/2!$. Similarly, one can show that $P_n = P_1^n/n!$. However, the total probability, when summed over all transitions, cannot exceed 1. Therefore, we define a normalized probability $\mathcal{P}_n$:

$$\mathcal{P}_n = e^{-P_1} \frac{P_1^n}{n!} \qquad \text{(S.5.51.11)}$$

5.52   β-Decay of Tritium (Michigan State)

We use the sudden approximation to calculate the probability that the electron remains in the ground state. One calculates the overlap integral I of the initial and final wave functions, and its square is the probability. The ground states in the initial and final states are called ψ_i and ψ_f, and a_0 is the Bohr radius:

$$\psi_i(r) = \frac{1}{\sqrt{\pi a_0^3}} e^{-r/a_0} \tag{S.5.52.1}$$

$$\psi_f(r) = \sqrt{\frac{8}{\pi a_0^3}} e^{-2r/a_0} \tag{S.5.52.2}$$

$$I = \int d^3r \, \psi_i(r) \, \psi_f(r) = \frac{4\pi\sqrt{8}}{\pi a_0^3} \int_0^\infty dr \, r^2 e^{-3r/a_0} \tag{S.5.52.3}$$

$$= \left(\frac{2^{3/2}}{3}\right)^3$$

$$P = I^2 = \left(\frac{2^{3/2}}{3}\right)^6 \approx 0.702 \tag{S.5.52.4}$$

WKB

5.53   Bouncing Ball (Moscow Phys-Tech, Chicago)

The potential energy here is $U = mgz$. We can apply the quasi-classical (WKB) approximation between points $[0, a]$, where $a = E/mg$, with the quasi-classical function applicable all the way to $z = 0$. The wave function is given by

$$\psi_{I, z<0} \equiv 0 \tag{S.5.53.1}$$

$$\psi_{I, z>0} = \frac{C}{\sqrt{p}} \sin \frac{1}{\hbar} \int_0^z p(z) \, dz = \frac{C}{\sqrt{p}} \cos \left(\frac{1}{\hbar} \int_0^z p(z) \, dz - \frac{\pi}{2}\right)$$

On the other hand, for $z = a$,

$$\psi_{II, z<a} = \frac{C'}{\sqrt{p}} \cos \left(\frac{1}{\hbar} \int_z^a p(z) \, dz - \frac{\pi}{4}\right) \tag{S.5.53.2}$$

Imposing the condition $\psi_I = \psi_{II}$ yields

$$\frac{1}{\hbar} \int_0^z p(z) \, dz - \frac{\pi}{2} + \frac{1}{\hbar} \int_z^a p(z) \, dz - \frac{\pi}{4} = \pi n \qquad \text{(S.5.53.3)}$$

$$\frac{1}{\hbar} \int_0^a p(z) \, dz = \pi \left(n + \frac{3}{4} \right)$$

We know that in this approximation

$$p(z) = \sqrt{2m(E - mgz)} \qquad \text{(S.5.53.4)}$$

so

$$\int_0^a dz \sqrt{E - mgz} = \frac{\pi \hbar}{\sqrt{2m}} \left(n + \frac{3}{4} \right) = \frac{2}{3} \frac{E^{3/2}}{mg} \qquad \text{(S.5.53.5)}$$

$$E_n = \frac{1}{2} \left(9\pi^2 \hbar^2 g^2 m \right)^{1/3} \left(n + \frac{3}{4} \right)^{2/3}$$

for $n = 0, 1, 2, \ldots$.

5.54 Truncated Harmonic Oscillator (Tennessee)

a) If C is the turning point, to be found later, then the WKB formula in one dimension for bound states is

$$\int_{-C}^{C} dx \, k(x) = \pi \left(n + \frac{1}{2} \right) \qquad \text{(S.5.54.1)}$$

$$k(x) = \sqrt{\frac{2m}{\hbar^2} [E - V(x)]} \qquad \text{(S.5.54.2)}$$

$$= \frac{m\omega}{\hbar} \sqrt{\frac{2E}{m\omega^2} + b^2 - x^2}$$

where we have used the truncated harmonic oscillator potential for $V(x)$. The constant C is the value of x where the argument of the square root changes sign, which is

$$C^2 = \frac{2E}{m\omega^2} + b^2 \qquad \text{(S.5.54.3)}$$

$$\int_0^C dx \sqrt{C^2 - x^2} = \frac{\pi \hbar}{2m\omega}\left(n + \frac{1}{2}\right) \qquad \text{(S.5.54.4)}$$

The integral on the left equals $\pi C^2/4$. The easiest way to see this result is to use the change of variables $x = C \sin \theta$, and the integrand becomes $C^2 \, d\theta \cos^2 \theta$ between 0 and $\pi/2$ (Actually, just note that this is the area of a quadrant of a disk of radius C). We get

$$C^2 = \frac{2E}{m\omega^2} + b^2 = \frac{2\hbar}{m\omega}\left(n + \frac{1}{2}\right) \qquad \text{(S.5.54.5)}$$

$$E_n = \hbar\omega\left[n + \frac{1}{2} - \frac{b^2 m\omega}{2\hbar}\right] \qquad \text{(S.5.54.6)}$$

b) The constraint that there be only one bound state is that $E_0 < 0$ and $E_1 > 0$. This gives the following constraints on the last constant in the energy expression:

$$1 < \frac{b^2 m\omega}{\hbar} < 3 \qquad \text{(S.5.54.7)}$$

5.55 Stretched Harmonic Oscillator (Tennessee)

We use (S.5.54.1) and (S.5.54.2) as the basic equations. The turning point C is where the argument of $k(x) = 0$. For the present potential the turning point is

$$C = a + d \qquad \text{(S.5.55.1)}$$

$$d = \sqrt{\frac{2E}{K}} \qquad \text{(S.5.55.2)}$$

The integral in (S.5.54.1) has three regions. In the interval $-a < x < a$ then $V(x) = 0$, $k(x)$ is a constant, and the integral is just $2ak$. The potential $V(x)$ is nonzero in the two intervals $-C < x < -a$ and $a < x < C$. Since the WKB integral is symmetric, we get

$$\frac{\pi}{2}\left(n + \frac{1}{2}\right) = \int_0^a dx \sqrt{\frac{2mE}{\hbar^2}} + \int_a^C dx \sqrt{\frac{2m}{\hbar^2}\left[E - \frac{K}{2}(x-a)^2\right]} \qquad \text{(S.5.55.3)}$$

To evaluate the second integral, change variables to $y = x - a$:

$$\frac{\pi}{2}\left(n + \frac{1}{2}\right) = a\sqrt{\frac{2mE}{\hbar^2}} + \sqrt{\frac{mK}{\hbar^2}} \int_0^d dx \sqrt{d^2 - y^2} \qquad \text{(S.5.55.4)}$$

The last integral equals $d^2\pi/4$. Writing $K = m\omega^2$, we find

$$\frac{\pi}{2}\left(n + \frac{1}{2}\right) = \sqrt{\frac{E}{E_a}} + \frac{\pi E}{2\hbar\omega} \tag{S.5.55.5}$$

$$E_a = \frac{\hbar^2}{2ma^2} \tag{S.5.55.6}$$

We have to determine E. Equation (S.5.55.5) is a quadratic equation for the variable $\sqrt{E}$. Solving the quadratic by the usual formula gives the final result:

$$\sqrt{E_n} = -\frac{\hbar\omega}{\pi\sqrt{E_a}}\left[1 - \sqrt{1 + \pi^2\left(n + \frac{1}{2}\right)\frac{E_a}{\hbar\omega}}\right] \tag{S.5.55.7}$$

$$E_n = \frac{(\hbar\omega)^2}{\pi^2 E_a}\left[1 - \sqrt{1 + \pi^2\left(n + \frac{1}{2}\right)\frac{E_a}{\hbar\omega}}\right]^2 \tag{S.5.55.8}$$

5.56 Ramp Potential (Tennessee)

We use (S.5.54.1) and (S.5.54.2) as the starting point. In the present problem, $V(x) = F|x|$ and $C = E/F$, so

$$k(x) = \sqrt{\frac{2mF}{\hbar^2}\left(\frac{E}{F} - |x|\right)} \tag{S.5.56.1}$$

$$\int_{-C}^{C} dx\,\sqrt{C - |x|} = \frac{\pi\hbar}{\sqrt{2mF}}\left(n + \frac{1}{2}\right) \tag{S.5.56.2}$$

Since the integral is symmetric, we can write it as

$$2\int_{0}^{C} dx\,\sqrt{C - x} = -\frac{4}{3}[C - x]^{3/2}\bigg|_{0}^{C} = \frac{4}{3}C^{3/2} \tag{S.5.56.3}$$

Remembering that $C = E/F$, we obtain the final result:

$$E_n = \left(\frac{3\pi\hbar F(n + 1/2)}{4\sqrt{2m}}\right)^{2/3} \tag{S.5.56.4}$$

5.57 Charge and Plane (Stony Brook)

a) Since $V(x) = \gamma|x|$, we may write

$$p = \sqrt{2m(E - \gamma|x|)} \tag{S.5.57.1}$$

In the WKB approximation

$$\frac{1}{2\pi\hbar} \oint p \, dx = n + \frac{1}{2}$$

or, between turning points,

$$\int_{-x_0}^{x_0} p \, dx = \pi\hbar \left(n + \frac{1}{2} \right) \tag{S.5.57.2}$$

Substituting (S.5.57.1) into (S.5.57.2) and using the symmetry of the motion about $x = 0$, we obtain

$$\sqrt{2mE} \int_0^{x_0} dx \sqrt{1 - \frac{\gamma}{E}x} = \frac{1}{2}\pi\hbar \left(n + \frac{1}{2} \right) \tag{S.5.57.3}$$

$$= \frac{\sqrt{2m}}{\gamma} E^{3/2} \int_0^1 du \sqrt{1 - u} = \frac{2\sqrt{2m}E^{3/2}}{3\gamma}$$

Thus,

$$E_n^{3/2} = \frac{\gamma}{\sqrt{2m}} \frac{3}{4}\pi\hbar \left(n + \frac{1}{2} \right) \tag{S.5.57.4}$$

$$E_n = \left[\frac{3}{4\sqrt{2m}}\gamma\pi\hbar \left(n + \frac{1}{2} \right) \right]^{2/3} \tag{S.5.57.5}$$

b) For the potential where $C \to \infty$,

$$V(x) = \gamma|x| + \text{``}\infty\text{''} \delta(x)$$

the quantization condition gives

$$\int_0^{x_0} p \, dx = \pi\hbar \left(n + \frac{3}{4} \right) \tag{S.5.57.6}$$

$$E_n^{3/2} = \frac{\gamma}{\sqrt{2m}} \frac{3}{4}\pi\hbar \left(2n + 1 + \frac{1}{2}\right) \qquad \text{(S.5.57.7)}$$

$$E_n = \left[\frac{3}{4\sqrt{2m}}\gamma\pi\hbar \left(2n + \frac{3}{2}\right)\right]^{2/3} \qquad \text{(S.5.57.8)}$$

c) Using the boundary conditions at $x = 0$, we obtain

$$\left.\frac{\partial\psi}{\partial x}\right|_{0+} - \left.\frac{\partial\psi}{\partial x}\right|_{0-} = C\psi(0) \qquad \text{(S.5.57.9)}$$

It implies that the odd states, for which $\psi(0) = 0$, are not affected by $C\delta(x)$, while even states should satisfy the condition

$$\left.\frac{\partial\psi}{\partial x}\right|_{0+} = \frac{1}{2}C\psi(0) \qquad \text{(S.5.57.10)}$$

Since

$$\psi(x) = \cos\left(\alpha - \frac{1}{\hbar}\int_0^x p(x)\,dx\right) \qquad \text{(S.5.57.11)}$$

where

$$\alpha = \frac{1}{\hbar}\int_0^{x_0} p\,dx - \frac{\pi}{4} \qquad \text{(S.5.57.12)}$$

this condition takes the following form:

$$\tan\alpha = \frac{C}{2p(0)} = \frac{C}{2\sqrt{2mE}} \qquad \text{(S.5.57.13)}$$

5.58 Ramp Phase Shift (Tennessee)

The following formula is for the phase shift in one dimension where the particle is free on the right ($V(x) \to 0$ as $x \to \infty$) and encounters an impenetrable barrier near the origin:

$$\delta(k) = \frac{\pi}{4} + \lim_{x\to\infty}\left\{\int_C^x dx'\, k'(x') - kx\right\} \qquad \text{(S.5.58.1)}$$

$$k'(x) = \sqrt{\frac{2m}{\hbar^2}[E - V(x)]} \qquad \text{(S.5.58.2)}$$

$$E = \frac{\hbar^2 k^2}{2m} \qquad \text{(S.5.58.3)}$$

The factor $\pi/4$ is the phase change when the particle goes through the turning point where $k'(C) = 0$.

For the present problem we have that $V = 0$ for $x > 0$, and this part of the integral exactly cancels the term $-kx$. For $x < 0$ the potential is $V = -Fx$, assuming that $F > 0$. The turning point is $C = -E/F$, so we have

$$\delta = \frac{\pi}{4} + \sqrt{\frac{2mF}{\hbar^2}} \int_C^0 dx \sqrt{x - C} \qquad (\text{S.5.58.4})$$

$$= \frac{\pi}{4} + \frac{2}{3} \sqrt{\frac{2mF}{\hbar^2}} \left(\frac{E}{F}\right)^{3/2}$$

5.59 Parabolic Phase Shift (Tennessee)

Again we use (S.5.58.1) for the phase shift. The potential $V(x)$ in the present problem is zero for $x > 0$. The integral in this region cancels the term $-kx$. To the left of the origin, the turning point is $C = -\sqrt{2E/K}$.

$$\delta = \frac{\pi}{4} + \sqrt{\frac{mK}{\hbar^2}} \int_C^0 dx \sqrt{C^2 - x^2} \qquad (\text{S.5.59.1})$$

$$= \frac{\pi}{4} \left[1 + C^2 \sqrt{\frac{mK}{\hbar^2}}\right] = \frac{\pi}{4} \left[1 + 2E \sqrt{\frac{m}{\hbar^2 K}}\right]$$

The integral over dx again equals $C^2 \pi/4$. The phase shift is linear with energy and has a constant term.

5.60 Phase Shift for Inverse Quadratic (Tennessee)

Again we use (S.5.58.1) for the phase shift. The turning point is $C = \lambda/k$. The phase integral is

$$\int_C^x dy \sqrt{\frac{2m}{\hbar^2}[E - V(x)]} = \int_C^x dy \sqrt{k^2 - \frac{\lambda^2}{y^2}}$$

$$= k \int_C^x \frac{dy}{y} \sqrt{y^2 - C^2} \qquad (\text{S.5.60.1})$$

$$= k \left(\sqrt{x^2 - C^2} - C \cos^{-1} \frac{C}{x} \right)$$

The last integral is found in standard tables. To evaluate the phase shift, we need to evaluate this expression in the limit $x \to \infty$, which gives $kx - \pi\lambda/2$. So the final expression for the phase shift is

$$\delta = \frac{\pi}{4}(1 - 2\lambda) \tag{S.5.60.2}$$

The phase shift is independent of energy.

Scattering Theory

5.61 Step-Down Potential (Michigan State, MIT)

Denote by p' the momentum of the particle to the right of the origin, and $\pm p$ is momentum on the left. Since energy is conserved, we have

$$E = \frac{p^2}{2m} + V_0 = \frac{p'^2}{2m} \tag{S.5.61.1}$$

$$p' = \sqrt{p^2 + 2mV_0} \tag{S.5.61.2}$$

Now we set up the most general form for the wave function, assuming the incoming wave has unit amplitude:

$$\psi(x) = \begin{cases} e^{ipx/\hbar} + Re^{-ipx/\hbar} & x < 0 \\ Te^{ip'x/\hbar} & x > 0 \end{cases} \tag{S.5.61.3}$$

Matching the wave function and its derivative at the origin gives two equations for the unknowns R and T which are solved to find R:

$$1 + R = T \tag{S.5.61.4}$$

$$ip(1 - R) = ip'T \tag{S.5.61.5}$$

$$R = \frac{p - \sqrt{p^2 + 2mV_0}}{p + \sqrt{p^2 + 2mV_0}} \tag{S.5.61.6}$$

5.62 Step-Up Potential (Wisconsin-Madison)

Write the energy as $E = \hbar^2 k^2/2m$, where k is the wave vector on the left of zero. Since $E > V_0$, define a wave vector k' on the right as $k'^2 = k^2 - 2mV_0/\hbar^2$.

a) The wave functions on the left and right of the origin are

$$\psi(x) = \begin{cases} e^{ikx} + re^{-ikx} & x < 0 \\ te^{ik'x} & x > 0 \end{cases} \qquad \text{(S.5.62.1)}$$

where r and t are the amplitudes of the reflected and transmitted waves. Matching the wave function and its slope at $x = 0$ gives two equations:

$$1 + r = t \qquad \text{(S.5.62.2)}$$

$$ik(1 - r) = ik't \qquad \text{(S.5.62.3)}$$

These two equations are solved to obtain r and t:

$$r = \frac{k - k'}{k + k'} \qquad \text{(S.5.62.4)}$$

$$t = \frac{2k}{k + k'} \qquad \text{(S.5.62.5)}$$

$$R = |r|^2 = \frac{(k - k')^2}{(k + k')^2} \qquad \text{(S.5.62.6)}$$

$$T = |t|^2 = \frac{4k^2}{(k + k')^2} \qquad \text{(S.5.62.7)}$$

b) The particle currents are the velocities times the intensities. The velocities are $\hbar k/m$ on the left and $\hbar k'/m$ on the right:

$$J_R = \frac{\hbar k}{m} R = \frac{\hbar}{m} \frac{k(k - k')^2}{(k + k')^2} \qquad \text{(S.5.62.8)}$$

$$J_T = \frac{\hbar k'}{m} T = \frac{\hbar}{m} \frac{4k^2 k'}{(k + k')^2} \qquad \text{(S.5.62.9)}$$

$$J_R + J_T = \frac{\hbar k}{m(k + k')^2} \left[(k - k')^2 + 4kk' \right] = \frac{\hbar k}{m} \qquad \text{(S.5.62.10)}$$

The last expression equals the current of the incoming particle.

5.63 Repulsive Square Well (Colorado)

a) If the radial part of the wave function is $R(r)$, then define $\chi(r) = rR(r)$. Since R is well behaved at $r \to 0$, $\chi = 0$ in this limit. The function $\chi(r)$ obeys the following equation for s-waves:

$$\left[\frac{d^2}{dr^2} + k^2 - k_0^2 \right] \chi(r) = 0 \qquad \text{(S.5.63.1)}$$

where

$$k_0^2 = \frac{2mV_0}{\hbar^2}\Theta(a - r) \tag{S.5.63.2}$$

and the theta function $\Theta(a - r)$ is 1 if $a > r$ and 0 if $a < r$. For $r > a$ the solutions are in the form of $\sin kr$ or $\cos kr$. Instead, write it as $\sin(kr + \delta)$ where the phase shift is $\delta(k)$. For $r < a$ define a constant α according to $\alpha^2 = k_0^2 - k^2 > 0$. Then the eigenfunction is

$$\chi(r) = \begin{cases} A \sinh \alpha r & r < a \\ B \sin(kr + \delta) & r > a \end{cases} \tag{S.5.63.3}$$

For $r < a$ the constraint that $\chi(0) = 0$ forces the choice of the hyberbolic sine function. Matching the eigenfunction and slope at $r = a$ gives

$$A \sinh \alpha a = B \sin(ka + \delta) \tag{S.5.63.4}$$

$$\alpha A \cosh \alpha a = kB \cos(ka + \delta) \tag{S.5.63.5}$$

Dividing these equations eliminates the constants A and B. The remaining equation defines the phase shift.

$$k \tanh \alpha a = \alpha \tan(ka + \delta) \tag{S.5.63.6}$$

$$\delta(k) = -ka + \tan^{-1}\left[\frac{k}{\sqrt{k_0^2 - k^2}} \tanh\left(a\sqrt{k_0^2 - k^2}\right)\right] \tag{S.5.63.7}$$

b) In the limit that $V_0 \to \infty$, the argument of the arctangent vanishes, since the hyperbolic tangent goes to unity, and $\delta = -ka$.

c) In the limit of zero energy, we can define

$$\lim_{k \to 0} \delta = -kd \tag{S.5.63.8}$$

$$d = a - \frac{1}{k_0} \tanh k_0 a \tag{S.5.63.9}$$

To find the s-wave part of the cross section at low energy, we start with

$$\sigma = \lim_{k \to 0}\left[\frac{4\pi \sin^2 \delta(k)}{k^2}\right] = 4\pi \lim_{k \to 0}\left[\frac{\delta(k)}{k}\right]^2 \tag{S.5.63.10}$$

$$= 4\pi d^2 = 4\pi \left(a - \frac{1}{k_0} \tanh k_0 a\right)^2$$

where the total cross section is σ.

5.64   3D Delta Function (Princeton)

For a particle of wave vector k, Schrödinger's equation for the radial part
of the wave function is

$$\left\{ -\frac{\hbar^2}{2m}\left[\frac{1}{r}\frac{d^2}{dr^2}r - \frac{\ell(\ell+1)}{r^2} \right] + V(r) - \frac{\hbar^2 k^2}{2m} \right\} R(r) = 0 \qquad (S.5.64.1)$$

Only s-wave scattering is important at very low energies, so solve for $\ell = 0$.
Also define $\chi(r) = rR(r)$ and get

$$0 = \left[\frac{d^2}{dr^2} + \gamma\delta(r - a) + k^2 \right]\chi(r) \qquad (S.5.64.2)$$

$$\gamma = \frac{2mC}{\hbar^2} \qquad (S.5.64.3)$$

At $r \to 0$, $R(r)$ is well behaved, so $\chi = rR \to 0$. Thus we choose our wave
functions to be

$$\chi(r) = \begin{cases} A\sin kr & 0 < r < a \\ B\sin(kr + \delta) & r > a \end{cases} \qquad (S.5.64.4)$$

The quantity δ is the phase shift. We match the wave functions at $r = a$.
The formula for matching the slopes is derived from (S.5.64.2):

$$0 = \left(\frac{d\chi}{dr} \right)_{a+} - \left(\frac{d\chi}{dr} \right)_{a-} + \gamma\chi(a) \qquad (S.5.64.5)$$

Matching the function and slope produces the equations

$$A\sin ka = B\sin(ka + \delta) \qquad (S.5.64.6)$$

$$0 = kB\cos(ka + \delta) - Ak\cos ka + \gamma A\sin ka \qquad (S.5.64.7)$$

which are solved to eliminate A and B and get

$$\tan(ka + \delta) = \frac{\tan ka}{1 - \frac{\gamma}{k}\tan ka} \qquad (S.5.64.8)$$

In the limit of low energy, we want $ka \to 0$. We assume there are no bound
states so that $\delta \to kd$, where d is a constant. We find in this limit:

$$d = \frac{\gamma a^2}{1 - \gamma a} \qquad (S.5.64.9)$$

$$\sigma = 4\pi d^2 \qquad (S.5.64.10)$$

We also give the formula for the cross section in terms of the scattering
length d. The assumption of no bound state is that $\gamma a < 1$.

5.65   Two-Delta-Function Scattering (Princeton)

Let us take an unperturbed wave function of the particle of the form

$$\psi = e^{ikz} \qquad \mathbf{k} = \frac{\mathbf{p}}{m} \qquad (S.5.65.1)$$

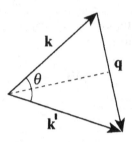

Figure S.5.65

Suppose that, after scattering, the wave vector becomes $\mathbf{k}'$. In the Born approximation, the scattering amplitude $f(\theta)$ is

$$f(\theta) = -\frac{m}{2\pi\hbar^2} \int d^3r \, V(\mathbf{r}) e^{-i\mathbf{q}\cdot\mathbf{r}} \qquad (S.5.65.2)$$

(see, for instance, Landau and Lifshitz, *Quantum Mechanics*, Sect. 126), where $\mathbf{q} = \mathbf{k}' - \mathbf{k}$ and $q = 2k\sin(\theta/2)$ (see Figure S.5.65). Substituting the potential $V(\mathbf{r})$ into (S.5.65.2), we obtain

$$f(\theta) = -\frac{mV_0}{2\pi\hbar^2} \int d^3r \, [\delta(\mathbf{r} - \varepsilon\hat{\mathbf{z}}) - \delta(\mathbf{r} + \varepsilon\hat{\mathbf{z}})] e^{-i\mathbf{q}\cdot\mathbf{r}} \qquad (S.5.65.3)$$

$$= -\frac{mV_0}{2\pi\hbar^2} \left(e^{i\varepsilon\mathbf{q}\cdot\hat{\mathbf{z}}} - e^{-i\varepsilon\mathbf{q}\cdot\hat{\mathbf{z}}} \right) = -\frac{mV_0}{\pi\hbar^2} i \sin q_z\varepsilon = \frac{mV_0}{\pi\hbar^2} i \sin |q_z|\varepsilon$$

where $q_z = -q\sin(\theta/2) = -2k\sin^2(\theta/2)$ is the projection of the vector $\mathbf{q}$ on the z axis. The scattering cross section

$$\frac{d\sigma}{d\Omega} = |f(\theta)|^2 = \frac{m^2V_0^2}{\pi^2\hbar^4} \sin^2 q_z\varepsilon = \frac{m^2V_0^2}{\pi^2\hbar^4} \sin^2\left(2\varepsilon k \sin^2\frac{\theta}{2} \right) \qquad (S.5.65.4)$$

In order to apply the Born approximation, i.e., to use perturbation theory, we must satisfy at least one of two conditions:

$$|V| \ll \frac{\hbar^2}{ma^2} = \frac{\hbar^2}{m\varepsilon^2} \qquad (S.5.65.5)$$

$$|V| \ll \frac{\hbar p}{ma} = \left(\frac{\hbar^2}{m\varepsilon^2} \right) \frac{p\varepsilon}{\hbar} \qquad (S.5.65.6)$$

where $a = \varepsilon$ is the range of the potential. The first condition derives from the requirement that the perturbed wave function be very close to the unperturbed wave function. Inequality (S.5.65.5) may also be considered the requirement that the potential be small compared to the kinetic energy of the particle localized at the source of the perturbation. Even if the first condition is not satisfied, particles with large enough p will also justify the Born approximation.

5.66   Scattering of Two Electrons (Princeton)

We evaluate the scattering in the Born approximation, which is valid when the kinetic energies are much larger than the binding energy. The Fourier transform of the potential is $\bar{V}(q)$:

$$\bar{V}(q) = \frac{4\pi e^2}{\lambda^2 + q^2} \tag{S.5.66.1}$$

and the formula for the total cross section σ of electrons with initial wave vector k is

$$\sigma = \frac{2\pi}{\hbar v} \int \frac{d^3 k'}{(2\pi)^3} \bar{V}^2(\mathbf{k}' - \mathbf{k}) \delta \left(\frac{\hbar^2 k^2}{m} - \frac{\hbar^2 k'^2}{m} \right) \tag{S.5.66.2}$$

This cross section is suitable for classical particles, without regard to spin. The specification to the spin states $S = 0, 1$ is made below. Write $d^3 k' = k'^2 \, dk' \, d\Omega$ where Ω is the solid angle of the scattering. The differential cross section is found by taking the functional derivative of the cross section with respect to this solid angle:

$$\frac{d\sigma}{d\Omega} = \frac{4e^4}{\hbar v} \int \frac{k'^2 \, dk'}{[\lambda^2 + (\mathbf{k}' - \mathbf{k})^2]^2} \delta \left(\frac{\hbar^2 k^2}{m} - \frac{\hbar^2 k'^2}{m} \right) \tag{S.5.66.3}$$

$$= \frac{1}{a_0^2 \left[\lambda^2 + 2k^2 (1 - \cos\theta) \right]^2}$$

where we have used the fact that $v = 2\hbar k/m$, and k is defined by $E = \hbar^2 k^2 / 2m$. The magnitudes of the vectors $\mathbf{k}'$ and $\mathbf{k}$ are the same, so $(\mathbf{k}' - \mathbf{k})^2 = 2k^2(1 - \cos\theta)$ (see Problem 5.65 and Figure S.5.65). All of the dimensional factors are combined into the Bohr radius a_0. Now we consider how this formula is altered by the spin of the electrons. Spin is conserved in the scattering, so the pair of electrons has the same spin state before and after the collision.

a) For $S = 0$ the two electrons are in a spin singlet which has odd parity. Hence, the orbital state must have even parity. The initial and final orbital wave functions are given below, along with the form of the matrix element. The relative coordinate is $\mathbf{r}$:

$$\psi_i = \frac{1}{\sqrt{2}} \left[e^{i\mathbf{k}\cdot\mathbf{r}} + e^{-i\mathbf{k}\cdot\mathbf{r}} \right] \tag{S.5.66.4}$$

$$\psi_f = \frac{1}{\sqrt{2}} \left[e^{i\mathbf{k}'\cdot\mathbf{r}} + e^{-i\mathbf{k}'\cdot\mathbf{r}} \right] \tag{S.5.66.5}$$

$$\langle f \,|V|\, i \rangle = \bar{V}(\mathbf{k}' - \mathbf{k}) + \bar{V}(\mathbf{k}' + \mathbf{k}) \tag{S.5.66.6}$$

$$\frac{d\sigma}{d\Omega} = \frac{1}{a_0^2} \tag{S.5.66.7}$$

$$\cdot \left[\frac{1}{\lambda^2 + 2k^2(1 - \cos\theta)} + \frac{1}{\lambda^2 + 2k^2(1 + \cos\theta)} \right]^2$$

The matrix element has two factors.

b) For $S = 1$ the spins are in a triplet state which has even parity. The orbital part of the wave function has odd parity. There is a minus sign between the two terms in (S.5.66.4) instead of a plus sign, and ditto for the final wave function. Now the differential cross section is

$$\frac{d\sigma}{d\Omega} = \frac{1}{a_0^2} \left[\frac{1}{\lambda^2 + 2k^2(1 - \cos\theta)} - \frac{1}{\lambda^2 + 2k^2(1 + \cos\theta)} \right]^2 \tag{S.5.66.8}$$

There is a relative minus sign between the two term in the matrix element.

5.67 Spin-Dependent Potentials (Princeton)

In the first Born approximation the scattering is proportional to the square of the matrix element between initial and final states. If the initial wave vector is $\mathbf{k}$ and the final one is $\mathbf{k}'$, set $\mathbf{q} = \mathbf{k}' - \mathbf{k}$ and evaluate

$$\bar{V}(\mathbf{q}) = \int d^3r\, e^{i\mathbf{q}\cdot\mathbf{r}} V(r)$$

$$= \frac{\pi^{3/2}}{\mu^{3/2}} e^{-q^2/4\mu} \left[A + \frac{i}{2\mu} B \mathbf{q} \cdot \boldsymbol{\sigma} \right] \tag{S.5.67.1}$$

$$= \frac{\pi^{3/2}}{\mu^{3/2}} e^{-q^2/4\mu} \left[A + \frac{i}{2\mu} B \left(q_z \sigma_z + q^+ \sigma^- + q^- \sigma^+ \right) \right]$$

$$q^{\pm} = q_x \pm iq_y \tag{S.5.67.2}$$

$$\sigma^{\pm} = \frac{1}{2}(\sigma_x \pm i\sigma_y) \tag{S.5.67.3}$$

where we have written the transverse components of momentum in terms of spin raising and lowering operators. The initial spin is pointing along the direction of the initial wave vector $\mathbf{k}$, which we define as the z-direction. Let us quantize the final spins along the same axis. Now consider how the three factors scatter the spins:

a) The term A is spin independent. It puts the final spin in the same state as the initial spin.

b) $Bq_z\sigma_z$ is a diagonal operator, so the final spin is also along the initial direction, and this term has a value of Bq_z.

c) $Bq^+\sigma^-$ flips the spin from $+z$ to $-z$ and contributes a matrix element of Bq^+ to the final state with the spin reversed.

d) $Bq^-\sigma^+$ gives a matrix element of zero since the initial spin cannot be raised.

When we take the magnitude squared of each transition and sum over final states, we get the factors for spins of

$$\left| A + \frac{i}{2\mu}Bq_z \right|^2 + \left| \frac{i}{2\mu}B\left(q_x + iq_y\right) \right|^2 = A^2 + \frac{B^2 q^2}{4\mu^2} \tag{S.5.67.4}$$

The differential cross section is written as

$$\frac{d\sigma}{d\Omega} = \frac{2\pi^4}{\hbar v\mu^3(2\pi)^3} \int k'^2 \, dk' \, e^{-(\mathbf{k}-\mathbf{k}')^2/2\mu} \delta\left(\frac{\hbar^2 k^2}{2m} - \frac{\hbar^2 k'^2}{2m} \right)$$

$$\cdot \left[A^2 + \frac{B^2}{4\mu^2}(\mathbf{k}-\mathbf{k}')^2 \right] \tag{S.5.67.5}$$

$$= \frac{\pi m^2}{4\hbar^4\mu^3} e^{-k^2(1-\cos\theta)/\mu} \left[A^2 + \frac{B^2}{2\mu^2}k^2(1-\cos\theta) \right]$$

We have used the fact that energy is conserved, so $|\mathbf{k}'| = |\mathbf{k}|$, to set $(\mathbf{k}' - \mathbf{k})^2 = 2k^2(1 - \cos\theta)$ (see Problem 5.65 and Figure S.5.65).

5.68   Rayleigh Scattering (Tennessee)

a) The formula for the total cross section σ is

$$\sigma = \frac{2\pi}{\hbar c} \int \frac{d^3 k'}{(2\pi)^3} |M(\mathbf{k}, \mathbf{k}')|^2 \delta(\hbar ck - \hbar ck') \qquad (S.5.68.1)$$

We write $d^3 k' = k'^2 \, dk' \, d\Omega'$, where Ω' is the solid angle. The differential cross section is obtained by taking a functional derivative with respect to $d\Omega'$. There remains only the dk' integral, which is eliminated by the delta function for energy conservation:

$$\frac{d\sigma}{d\Omega'} = \frac{k^2}{(2\pi\hbar c)^2} |M(\mathbf{k}, \mathbf{k}')|^2 \qquad (S.5.68.2)$$

where the vector $\mathbf{k}'$ differs from $\mathbf{k}$ only in direction.

b) With the assigned choice of the matrix element we write our differential cross section as

$$\frac{d\sigma}{d\Omega'} = k^4 \alpha^2(\omega) S \qquad (S.5.68.3)$$

$$S = \left\langle \left(\xi_{\mathbf{k}} \cdot \xi_{\mathbf{k}'} \right)^2 \right\rangle \qquad (S.5.68.4)$$

where the factor S is the average over initial polarizations and the sum over final polarizations. There are two possible polarizations, and both are perpendicular to the direction of the photon. These averages take the form

$$\sum \xi_i \xi_j = \delta_{ij} - \frac{k_i k_j}{k^2} \qquad (S.5.68.5)$$

$$S = \frac{1}{2} \sum_{ij} \left[\delta_{ij} - \frac{k_i k_j}{k^2} \right] \left[\delta_{ji} - \frac{k'_j k'_i}{k'^2} \right] \qquad (S.5.68.6)$$

$$= \frac{1}{2} \left[1 + \left(\hat{\mathbf{k}} \cdot \hat{\mathbf{k}}' \right)^2 \right] = \frac{1}{2} \left[1 + \cos^2 \theta \right]$$

The factor $1/2$ is from the average over initial polarization. The angle θ is between the directions of $\mathbf{k}$ and $\mathbf{k}'$.

5.69 Scattering from Neutral Charge Distribution (Princeton)

a) The particle scatters from the potential energy $V(r)$ which is related to the charge distribution

$$\nabla^2 V(r) = -4\pi e\rho(r) \qquad (S.5.69.1)$$

$$\tilde{V}(q) = \frac{4\pi e}{q^2}\tilde{\rho}(q) \qquad (S.5.69.2)$$

where $\tilde{V}(q)$ is the Fourier transform of $V(r)$ and $\tilde{\rho}(q)$ is the Fourier transform of $\rho(r)$. The differential cross section in the Born approximation is

$$\frac{d\sigma}{d\Omega} = \frac{1}{(2\pi)^2 \hbar v_k} \int dk' \; k'^2 \delta\left(E_k - E_{k'}\right) \left|V\left(k - k'\right)\right|^2 \quad (S.5.69.3)$$

$$= \frac{4e^2 m^2 \; \tilde{\rho}^2 \left(2k\sin\frac{\theta}{2}\right)}{\hbar^4 \; \left(2k\sin\frac{\theta}{2}\right)^4}$$

b) In forward scattering we take $\theta \to 0$. In order that the cross section have a nondivergent result in this limit, we need to find

$$\lim_{q\to 0} \frac{\tilde{\rho}(q)}{q^2} = -\frac{A}{6} \qquad (S.5.69.4)$$

To obtain this result, we examine the behavior of $\tilde{\rho}(q)$ at small values of q:

$$\tilde{\rho}(q) = \int d^3r \; \rho(r)e^{i\mathbf{q}\cdot\mathbf{r}} \qquad (S.5.69.5)$$

$$= \int d^3r \; \rho(r)\left[1 - i\mathbf{q}\cdot\mathbf{r} - \frac{(\mathbf{q}\cdot\mathbf{r})^2}{2} + \cdots\right] = -\frac{q^2 A}{6}$$

Consider the three terms in brackets: (i) the 1 vanishes since the distribution is neutral; (ii) the second term vanishes since the distribution is spherically symmetric; (iii) the last term gives an angular average $(\mathbf{q}\cdot\mathbf{r})^2 = (qr)^2/3$ and the integral of $r^2\rho$ is A. The cross section in forward scattering is

$$\lim_{\theta\to 0} \frac{d\sigma}{d\Omega} = \frac{e^2 m^2 A^2}{9\hbar^4} \qquad (S.5.69.6)$$

c) The charges in a hydrogen atom are the nucleus, which is taken as a delta function at the origin, and the electron, which is given by the square of the ground state wave function $\psi_{1s}^2(r)$:

$$\rho(r) = -e\left[\delta^3(\mathbf{r}) - \psi_{1s}^2(r)\right] \qquad (S.5.69.7)$$

$$\tilde{\rho}(q) = -e\left[1 - \frac{16}{(4 + q^2 a_0^2)^2}\right] \qquad (S.5.69.8)$$

$$A = 3ea_0^2 \qquad (S.5.69.9)$$

$$\lim_{\theta \to 0} \frac{d\sigma}{d\Omega} = a_0^2 \qquad (S.5.69.10)$$

where a_0 is the Bohr radius.

General

5.70   Spherical Box with Hole (Stony Brook)

In spherical coordinates the eigenfunctions for noninteracting particles of wave vector k are of the form

$$[Aj_\ell(kr) + B\eta_\ell(kr)]P_\ell^m(\theta)e^{im\phi} \qquad (S.5.70.1)$$

where j_ℓ and η_ℓ are spherical Bessel functions. The constants A and B are determined by the boundary conditions. Since we were only asked for the states with $\ell=0$, we only need $j_0(z) = \sin z/z$ and $\eta_0(z) = -\cos z/z$. We can take a linear combination of these functions, which is a particular choice of the ratio B/A, to make the wave function vanish at $r = a$:

$$R_{n0}(kr) = A\frac{\sin k(r - a)}{kr} \qquad (S.5.70.2)$$

This satisfies the boundary condition at $r = a$. Requiring that this function vanish at $r = R$ gives $k(R - a) = n\pi$, so

$$k_n = \frac{n\pi}{R - a} \qquad (S.5.70.3)$$

$$E_n = \frac{\hbar^2 k_n^2}{2m} = \frac{(n\hbar\pi)^2}{2m(R - a)^2} \qquad (S.5.70.4)$$

5.71   Attractive Delta Function in 3D (Princeton)

a) The amplitude of the wave function is continuous at the point $r = a$ of the delta function. For the derivative we first note that the eigenfunctions are written in terms of a radial function $R(r)$ and angular functions:

$$\psi(\mathbf{r}) = AR(r)P_\ell^m(\theta)e^{im\phi} \qquad (S.5.71.1)$$

Since the delta function is only for the radial variable r, only the function $R(r)$ has a discontinuous slope. From the radial part of the kinetic energy operator we integrate from $r = a^-$ to $r = a^+$:

$$0 = \int_{a^-}^{a^+} r^2\, dr \left\{ \frac{\hbar^2}{2mr}\frac{d^2}{dr^2}r + \frac{\hbar^2}{2mD}\delta(r-a) + E \right\} R \qquad (S.5.71.2)$$

$$0 = D\left[R'\left(a^+\right) - R'\left(a^-\right) \right] + R(a) \qquad (S.5.71.3)$$

This formula is used to match the slopes at $r = a$.

b) In order to find bound states, we assume that the particle has an energy given by $E = -\hbar^2\alpha^2/2m$, where α needs to be determined by an eigenvalue equation. The eigenfunctions are combinations of $\exp(\pm\alpha r)/r$. In order to be zero at $r = 0$ and to vanish at infinity, we must choose the form

$$R(r) = \begin{cases} \dfrac{A}{r}\sinh\alpha r & r < a \\[2mm] \dfrac{B}{r}e^{-\alpha r} & r > a \end{cases} \qquad (S.5.71.4)$$

We match the values of $R(r)$ at $r = a$. We match the derivative, using the results of part (a):

$$A\sinh\alpha a = Be^{-\alpha a} \qquad (S.5.71.5)$$

$$0 = -\alpha D\left(Be^{-\alpha a} + A\cosh\alpha a\right) + A\sinh\alpha a \qquad (S.5.71.6)$$

We eliminate the constants A and B and obtain the eigenvalue equation for α, which we proceed to simplify:

$$D\alpha\left(\sinh\alpha a + \cosh\alpha a\right) = \sinh\alpha a \qquad (S.5.71.7)$$

$$D\alpha = e^{-\alpha a}\sinh\alpha a \qquad (S.5.71.8)$$

$$\frac{D}{a} = \frac{1 - e^{-2\alpha a}}{2\alpha a} \qquad (S.5.71.9)$$

This is the eigenvalue equation which determines α as a function of parameters such as a, D, m, etc. In order to find the range of allowed values of D for bound states, we examine $\alpha a \to 0$. The right-hand side of (S.5.71.9) goes to 1, which is its largest value. So, the constraint for the existence of bound states is

$$0 < D < a \qquad \text{(S.5.71.10)}$$

5.72   Ionizing Deuterium (Wisconsin-Madison)

The ionization energy of hydrogen is just the binding energy of the electron which is given in terms of the reduced mass μ_{ep} of the electron–proton system. The same expression for deuterium contains the reduced mass μ_{ed} of the electron–deuteron system:

$$E_H = \frac{e^4 \mu_{ep}}{2\hbar^2} \qquad \text{(S.5.72.1)}$$

$$\frac{1}{\mu_{ep}} = \frac{1}{m_e} + \frac{1}{m_p} \qquad \text{(S.5.72.2)}$$

$$E_D = \frac{e^4 \mu_{ed}}{2\hbar^2} \qquad \text{(S.5.72.3)}$$

$$\frac{1}{\mu_{ed}} = \frac{1}{m_e} + \frac{1}{m_d} \qquad \text{(S.5.72.4)}$$

The difference is easily evaluated. The ratio m_e/m_p is a small number and can be used as an expansion parameter:

$$\delta E = E_D - E_H = E_H \left(\frac{\mu_{ed}}{\mu_{ep}} - 1 \right)$$

$$= E_H \left(\frac{1 + m_e/m_p}{1 + m_e/m_d} - 1 \right) \qquad \text{(S.5.72.5)}$$

$$\approx E_H \left(m_e \frac{m_d - m_p}{m_p m_d} \right)$$

The ratio of masses gives $2.72 \cdot 10^{-4}$ and $\delta E \approx 3.700$ meV.

5.73   Collapsed Star (Stanford)

a) Using the 1D Schrödinger equation

$$-\frac{\hbar^2}{2m} \psi'' = E\psi \qquad \text{(S.5.73.1)}$$

with the boundary conditions $\psi(L/2) = \psi(-L/2) = 0$ gives

$$E_n = \frac{\hbar^2 \pi^2 n^2}{2mL^2} \qquad (S.5.73.2)$$

where $n = 1, 2, 3 \dots$. Protons, neutrons, and electrons are spin-$\frac{1}{2}$ fermions, so 2 may occupy each energy level, and we have

$$n_e = n_p = \frac{Z}{2}$$

$$n_n = \frac{N}{2}$$

The kinetic energy of a particle

$$E_n = (\gamma - 1)mc^2 \qquad (S.5.73.3)$$

where

$$\gamma = \frac{1}{\sqrt{1 - v^2/c^2}} \qquad (S.5.73.4)$$

To determine which species are relativistic, we wish to find whether $\gamma \gg 1$. We may extract γ from S.5.73.2. For neutrons:

$$\gamma_n - 1 = \frac{\hbar^2 \pi^2 n^2}{2m^2 c^2 L^2} = \frac{\hbar^2 \pi^2 c^2}{2m^2 c^4} \left(\frac{\lambda}{2}\right)^2 \cdot \left(\frac{N}{A}\right)^2 \qquad (S.5.73.5)$$

$$= \frac{(197 \text{ MeV fm})^2 \pi^2 \left(0.25 \text{ fm}^{-1}\right)^2}{2 \left(10^3 \text{ MeV}\right)^2} \cdot \left(\frac{N}{A}\right)^2 \approx 0.01 \cdot \left(\frac{N}{A}\right)^2$$

Similarly for protons:

$$\gamma_p \approx 0.01 \cdot \left(\frac{Z}{A}\right)^2 \qquad (S.5.73.6)$$

Since $N, Z < A$, both neutrons and protons are non-relativistic. For electrons:

$$\gamma_e - 1 = \frac{(197 \text{ MeV fm})^2 \pi^2 \left(0.25 \text{ fm}^{-1}\right)^2}{2 (0.5 \text{ MeV})^2} \cdot \left(\frac{Z}{A}\right)^2 \approx 5 \cdot 10^4 \cdot \left(\frac{Z}{A}\right)^2 \cdot$$
$$(S.5.73.7)$$

The equilibrium value of Z/A obtained in (c) for relativistic electrons gives $Z/A \sim 0.07$ which still leaves $\gamma_e \gg 1$ in S.5.73.7. Moreover, if we assume that the electrons are non-relativistic and minimize S.5.73.14 below with electron energy (see S.5.73.9)

$$E_e = \frac{\hbar^2 \pi^2 A^3}{24 L^2 m_e} \left(\frac{Z}{A}\right)^3 \qquad (S.5.73.8)$$

we will get $Z/A \sim 0.02$ and $\gamma_e \approx 20$ which contradicts the assumption. So the electrons are relativistic. Alternatively, we can use the result of Problem 4.64,

$$E_F = \frac{\pi^2 \hbar^2}{8m} \lambda^2 \cdot \left(\frac{Z}{A}\right)^2$$

the same as S.5.73.5.

b) The ground state energy of the system is given by the sum of energies of all levels, which we may approximate by an integral. We calculate the total energies of non-relativistic particles (neutrons and protons) and relativistic ones (electrons) separately:

$$\sum_{i=n,p} E_i = 2 \sum_{i=n,p} \sum_{j=1}^{n_i} \frac{\hbar^2 \pi^2 j^2}{2 m_i L^2} \approx \sum_{i=n,p} \frac{\hbar^2 \pi^2}{m_i L^2} \int^{n_i} j^2 \, dj$$

$$= \frac{\hbar^2 \pi^2}{3 L^2} \sum_{i=n,p} \frac{n_i^3}{m_i} = \frac{\hbar^2 \pi^2}{24 L^2} \left(\frac{Z^3}{m_p} + \frac{N^3}{m_n}\right) \qquad \text{(S.5.73.9)}$$

$$= \frac{\hbar^2 \pi^2 A^3}{24 L^2} \left[\left(\frac{Z}{A}\right)^3 \frac{1}{m_p} + \left(1 - \frac{Z}{A}\right)^3 \frac{1}{m_n}\right]$$

$$\approx \frac{\hbar^2 \pi^2 A^3}{24 L^2 m_n} \left[\left(\frac{Z}{A}\right)^3 + \left(1 - \frac{Z}{A}\right)^3\right]$$

For 1-D electrons (see Problem 4.64)

$$dN_e = 2 \frac{dp \, L}{2 \pi \hbar} \qquad \text{(S.5.73.10)}$$

$$N_e = \int_{-p_0}^{p_0} \frac{dp \, L}{\pi \hbar} = 2 \frac{p_0 L}{\pi \hbar} \qquad \text{(S.5.73.11)}$$

$$p_0 = \frac{\pi \hbar N_e}{2L} = \frac{\pi \hbar Z}{2L} \qquad \text{(S.5.73.12)}$$

The total electron energy is

$$E_e = 2 \int_0^{p_0} \frac{dp \, L}{\pi \hbar} cp = \frac{c p_0^2 L}{\pi \hbar} = \frac{c \pi \hbar Z^2}{4L} = \frac{c \pi \hbar A^2}{4L} \left(\frac{Z}{A}\right)^2 \qquad \text{(S.5.73.13)}$$

where we used for an estimate an electron energy of the form $E = cp$ since we have already established that they are relativistic. We can obtain a correct value of γ_e for them:

$$\gamma_e \approx \frac{p}{m_e c} = \frac{\pi \hbar \lambda}{2 m_e c} \left(\frac{Z}{A} \right)$$

$$\approx \pi \cdot 100 \left(\frac{Z}{A} \right) \approx 20$$

where we have used the result of (c), $Z/A \approx 0.07$. The total energy of the star is

$$E = E_n + E_p + E_e \tag{S.5.73.14}$$

$$= \frac{\hbar^2 \pi^2 A^3}{24 L^2 m_n} \left[\left(\frac{Z}{A} \right)^3 + \left(1 - \frac{Z}{A} \right)^3 \right] + \frac{c \pi \hbar A^2}{4L} \left(\frac{Z}{A} \right)^2$$

c) Let $x \equiv Z/A$. We need to find the minimum of the expression

$$f(x) = \alpha x^3 + \alpha (1 - x)^3 + \beta x^2 \tag{S.5.73.15}$$

where

$$\alpha \equiv \frac{\hbar^2 \pi^2 A^3}{24 L^2 m_n} \tag{S.5.73.16}$$

$$\beta \equiv \frac{c \pi \hbar A^2}{4L} \tag{S.5.73.17}$$

Setting the derivative of S.5.73.15 equal to zero gives

$$3 \alpha x^2 - 3 \alpha (1 - x)^2 + 2 \beta x = 0 \tag{S.5.73.18}$$

$$x = \frac{3 \alpha}{6 \alpha + 2 \beta} = \frac{3 (\alpha / \beta)}{6 (\alpha / \beta) + 2} \tag{S.5.73.19}$$

$$\frac{\alpha}{\beta} = \frac{1}{6} \frac{\pi \hbar A}{L c m_n} = \frac{1}{6} \frac{\pi \hbar \lambda c}{m_n c^2} \tag{S.5.73.20}$$

$$= \frac{1}{6} \frac{\pi \cdot 10^{-27} \cdot 0.5 \cdot 10^{13} \cdot 3 \cdot 10^{10}}{1 \cdot 10^9 \cdot 1.6 \cdot 10^{-12}} \approx 0.05$$

Finally,

$$x = \frac{3 \cdot 0.05}{6 \cdot 0.05 + 2} \approx 0.07 \tag{S.5.73.21}$$

So the minimum energy corresponds to a star consisting mostly ($\sim 93\%$) of neutrons.

5.74 Electron in Magnetic Field (Stony Brook, Moscow Phys-Tech)

a) The relationship between the vector potential and magnetic field is $\mathbf{B} = \nabla \times \mathbf{A}$. Using $\mathbf{A} = B(0, x, 0)$ does give $\mathbf{B} = B\hat{\mathbf{z}}$. So this vector potential produces the right field.

b) The vector potential enters the Hamiltonian in the form

$$H = \frac{1}{2m}\left(\mathbf{p} - \frac{e}{c}\mathbf{A}\right)^2 \tag{S.5.74.1}$$

$$= \frac{1}{2m}\left[p_x^2 + \left(p_y - \frac{eB}{c}x\right)^2 + p_z^2\right]$$

One can show easily that p_y and p_z each commute with the Hamiltonian and are constants of motion. Thus, we can write the eigenfunction as plane waves for these two variables, with only the x-dependence yet to be determined:

$$\psi(x, y, z) = e^{i(k_y y + k_z z)}\chi(x) \tag{S.5.74.2}$$

The Hamiltonian operating on ψ gives

$$H\psi = e^{i(k_y y + k_z z)}\left\{E_z + \frac{1}{2m}\left[p_x^2 + \left(\hbar k_y - \frac{eB}{c}x\right)^2\right]\right\}\chi(x) \tag{S.5.74.3}$$

where $E_z = \hbar^2 k_z^2/2m$. We may write the energy E as

$$E = E_z + E_x \tag{S.5.74.4}$$

and find

$$E_x\chi = \left[\frac{p_x^2}{2m} + \frac{m\omega_c^2}{2}(x - x_0)^2\right]\chi \tag{S.5.74.5}$$

$$\omega_c = \frac{eB}{mc} \tag{S.5.74.6}$$

$$x_0 = \frac{\hbar k_y c}{eB} \tag{S.5.74.7}$$

The energy is given by the component E_z along the magnetic field and the energy E_x for motion in the (x, y) plane. The latter contribution is identical to the simple harmonic oscillator in the x-direction. The frequency is the cyclotron frequency ω_c, and the harmonic motion is centered at the point x_0 which depends upon k_y. The eigenvalues and eigenfunctions are

$$E = \frac{\hbar^2 k_z^2}{2m} + \hbar\omega_c\left(n + \frac{1}{2}\right) \tag{S.5.74.8}$$

$$\psi(\mathbf{r}) = e^{i(k_y y + k_z z)}\psi_n(x - x_0) \tag{S.5.74.9}$$

where $\psi_n(x)$ are the eigenfunctions for the one-dimensional harmonic oscillator.

5.75 Electric and Magnetic Fields (Princeton)

a) Many vector potentials $\mathbf{A}(\mathbf{r})$ can be chosen so that $\nabla \times \mathbf{A} = B\hat{\mathbf{y}}$. For the present problem the most convenient choice is $\mathbf{A} = Bz\hat{\mathbf{x}}$. Thus the Hamiltonian is

$$H = \frac{1}{2m}\left[\left(p_x - \frac{eB}{c}z\right)^2 + p_y^2 + p_z^2\right] - e|\mathbf{E}|z \tag{S.5.75.1}$$

The above choice is convenient since only p_z fails to commute with H, so p_x and p_y are constants of motion. Both potentials have been made to depend on z.

b) Since p_x and p_y are constants of motion, we can write the eigenstates and energies as

$$\psi(\mathbf{r}) = e^{i(k_x x + k_y y)}\psi(z) \tag{S.5.75.2}$$

$$E = E_y + E_z \tag{S.5.75.3}$$

$$E_y = \frac{\hbar^2 k_y^2}{2m} \tag{S.5.75.4}$$

$$E_z \psi(z) = \left\{\frac{1}{2m}\left[p_z^2 + \left(\hbar k_x - \frac{eB}{c}z\right)^2\right] - e|\mathbf{E}|z\right\}\psi(z) \tag{S.5.75.5}$$

The last equation determines the eigenvalue E_z and eigenfunctions $\psi(z)$. The potential is a combination of linear and quadratic terms in z. So the motion behaves as a simple harmonic oscillator, where the terms linear in z determine the center of vibration. After some algebra we can write the above expression as

$$E_z \, \psi(z) = \left\{ \frac{p_z^2}{2m} + \frac{m\omega_c^2}{2}(z - z_0)^2 - \frac{mc^2|\mathbf{E}|^2}{2B^2} - \frac{\hbar k_x c|\mathbf{E}|}{B} \right\}$$

$$\cdot \psi(z) \qquad \qquad \text{(S.5.75.6)}$$

$$z_0 = \frac{\hbar k_x c}{eB} + \frac{mc^2|\mathbf{E}|}{eB^2} \qquad \qquad \text{(S.5.75.7)}$$

So, we obtain

$$E_z = \hbar \omega_c \left(n + \frac{1}{2} \right) - \frac{mc^2|\mathbf{E}|^2}{2B^2} - \frac{\hbar k_x c|\mathbf{E}|}{B} \qquad \text{(S.5.75.8)}$$

The total energy is E_z plus the kinetic energy along the y-direction. The z-part of the eigenfunction is a harmonic oscillator $\psi_n(z - z_0)$.

c) In order to find the average velocity, we take a derivative with respect to the wave vector k_x:

$$v_x = \frac{1}{\hbar} \frac{\partial E}{\partial k_x} = -c \frac{|\mathbf{E}|}{B} \qquad \qquad \text{(S.5.75.9)}$$

This is the drift velocity in the x-direction. It agrees with the classical answer.

5.76   Josephson Junction (Boston)

a) Take the first of equations (P.5.76.1),

$$i\hbar \frac{\partial \Psi_1}{\partial t} = U_1 \Psi_1 + K\Psi_2 + K\frac{\Psi_1 \Psi_2^*}{\Psi_1^*} \qquad \text{(S.5.76.1)}$$

and its complex conjugate and multiply them by Ψ_1^* and Ψ_1, respectively:

$$i\hbar \Psi_1^* \frac{\partial \Psi_1}{\partial t} = U_1 \Psi_1^* \Psi_1 + K\Psi_1^* \Psi_2 + K\Psi_1 \Psi_2^* \qquad \text{(S.5.76.2)}$$

$$-i\hbar \Psi_1 \frac{\partial \Psi_1^*}{\partial t} = U_1 \Psi_1^* \Psi_1 + K\Psi_1 \Psi_2^* + K\Psi_1^* \Psi_2 \qquad \text{(S.5.76.3)}$$

Subtracting (S.5.76.3) from (S.5.76.2) yields

$$i\hbar \frac{\partial |\Psi_1|^2}{\partial t} = i\hbar \frac{\partial \rho_1}{\partial t} = 0 \qquad \qquad \text{(S.5.76.4)}$$

Similarly, from the second of (P.5.76.1),

$$i\hbar\frac{\partial|\Psi_2|^2}{\partial t} = i\hbar\frac{\partial\rho_2}{\partial t} = 0 \qquad\qquad (S.5.76.5)$$

b) Substituting the solutions $\Psi_1 = \sqrt{\rho_0}e^{i\theta_1}$ and $\Psi_2 = \sqrt{\rho_0}e^{i\theta_2}$ into (S.5.76.1), we obtain the expression for θ_1:

$$i\hbar\sqrt{\rho_0}i\dot{\theta}_1 e^{i\theta_1} = U_1\sqrt{\rho_0}e^{i\theta_1} + K\sqrt{\rho_0}e^{i\theta_2} + K\sqrt{\rho_0}e^{i(2\theta_1-\theta_2)} \qquad (S.5.76.6)$$

Taking (S.5.76.6), the analogous expression for θ_2 gives

$$\dot{\theta}_1 = -\frac{U_1}{\hbar} - \frac{2K}{\hbar}\cos(\theta_1 - \theta_2) \qquad\qquad (S.5.76.7)$$

$$\dot{\theta}_2 = -\frac{U_2}{\hbar} - \frac{2K}{\hbar}\cos(\theta_2 - \theta_1) \qquad\qquad (S.5.76.8)$$

Subtracting (S.5.76.7) from (S.5.76.8), we obtain

$$\dot{\theta}_2 - \dot{\theta}_1 = \dot{\delta} = \frac{U_1 - U_2}{\hbar} \qquad\qquad (S.5.76.9)$$

where $\delta = \theta_2 - \theta_1$. So

$$\delta = \frac{U_1 - U_2}{\hbar}t + \delta_0 = \frac{eV}{\hbar}t + \delta_0 \qquad\qquad (S.5.76.10)$$

where $eV = U_1 - U_2$.

c) The battery current

$$J_1 = \frac{K}{i\hbar}\left(\Psi_1\Psi_2^* - \Psi_1^*\Psi_2\right)$$

$$= \frac{K}{i\hbar}\left(\sqrt{\rho_0}\right)^2\left[e^{i(\theta_1-\theta_2)} - e^{-i(\theta_1-\theta_2)}\right] \qquad (S.5.76.11)$$

$$= -\frac{2K\rho_0}{\hbar}\sin(\theta_2 - \theta_1)$$

$$= -\frac{2K\rho_0}{\hbar}\sin\delta = -\frac{2K\rho_0}{\hbar}\sin\left(\frac{eV}{\hbar}t + \delta_0\right)$$

APPENDIXES

Appendix 1:

Approximate Values of Physical Constants

Constant	Symbol	SI	CGS
Speed of light	c	$3.00 \cdot 10^8$ m/s	$3.00 \cdot 10^{10}$ cm/s
Planck's constant	h	$6.63 \cdot 10^{-34}$ J · s	$6.63 \cdot 10^{-27}$ erg · s
Reduced Planck's constant	$\hbar = \dfrac{h}{2\pi}$	$1.05 \cdot 10^{-34}$ J · s	$1.05 \cdot 10^{-27}$ erg · s
Avogadro's number	N_A	$6.02 \cdot 10^{26}$ kmol^{-1}	$6.02 \cdot 10^{23}$ mol^{-1}
Boltzmann's constant	k	$1.38 \cdot 10^{-23}$ J/K	$1.38 \cdot 10^{-16}$ erg/K
Electron charge	e	$1.60 \cdot 10^{-19}$ C	$4.80 \cdot 10^{-10}$ esu
Electron mass	m_e	$9.11 \cdot 10^{-31}$ kg	$9.11 \cdot 10^{-28}$ g
Electron charge to mass ratio	$\dfrac{e}{m_e}$	$1.76 \cdot 10^{11}$ C/kg	$5.27 \cdot 10^{17}$ esu/g
Neutron mass	m_n	$1.675 \cdot 10^{-27}$ kg	$1.675 \cdot 10^{-24}$ g
Proton mass	m_p	$1.673 \cdot 10^{-27}$ kg	$1.673 \cdot 10^{-24}$ g
Gravitational constant	G	$6.67 \cdot 10^{-11}$ N · m^2/kg^2	$6.67 \cdot 10^{-8}$ dyn · cm^2/g^2
Acceleration of gravity	g	9.81 m/s^2	981 cm/s^2
Stefan–Boltzmann constant	σ	5.67 W/(m^{2}K^4)	$5.67 \cdot 10^{-5}$ erg/(s · cm^{2}K^4)
Fine structure constant	α	$1/137$	$1/137$
Bohr radius	a_0	$5.29 \cdot 10^{-11}$ m	$5.29 \cdot 10^{-9}$ cm
Classical electron radius	$r_e = \dfrac{e^2}{m_e c^2}$	$2.82 \cdot 10^{-15}$ m	$2.82 \cdot 10^{-13}$ cm
Electron Compton wavelength	$\dfrac{\hbar}{m_e c}$	$3.86 \cdot 10^{-13}$ m	$3.86 \cdot 10^{-11}$ cm
Bohr magneton	$\mu_B = \dfrac{e\hbar}{2m_e c}$	$9.27 \cdot 10^{-24}$ J/T	$9.27 \cdot 10^{-21}$ erg/G
Rydberg constant	R_∞	$1.10 \cdot 10^7$ m^{-1}	$1.10 \cdot 10^5$ cm^{-1}
Universal gas constant	R	$8.31 \cdot 10^3$ J/(kmol · K)	$8.31 \cdot 10^7$ erg/(mol · K)
Josephson constant	$\dfrac{2e}{h}$	$4.84 \cdot 10^{14}$ Hz/V	$1.45 \cdot 10^{17}$ Hz/statvolt
Permittivity of free space	ε_0	$8.85 \cdot 10^{-12}$ F/m	1
Permeability of free space	μ_0	$4\pi \cdot 10^{-7}$ N/A^2	1

Some Astronomical Data

Mass of the Sun $M_S \approx 2 \cdot 10^{30}$ kg

Radius of the Sun $R_S \approx 6.7 \cdot 10^8$ m

Average Distance between the Earth and the Sun $\approx 1.5 \cdot 10^{11}$ m

Average Radius of the Earth $R_E \approx 6.4 \cdot 10^6$ m

Mass of the Earth $M_E \approx 6 \cdot 10^{24}$ kg

Average Velocity of the Earth in Orbit about the Sun $V_E \approx 3 \cdot 10^4$ m/s

Average Distance between the Earth and the Moon $\approx 3.8 \cdot 10^8$ m

Other Commonly Used Units

Angstrom (Å) $= 10^{-8}$ cm $= 10^{-10}$ m

Fermi (Fm) $= 10^{-13}$ cm $= 10^{-15}$ m

Barn $= 10^{-24}$ cm^2 $= 10^{-28}$ m^2

Year $\approx 3.16 \cdot 10^7$ s

Astronomical Year $\approx 9.5 \cdot 10^{17}$ cm $= 9.5 \cdot 10^{15}$ m

Parsec $= 3.1 \cdot 10^{18}$ cm $= 3.1 \cdot 10^{16}$ m

eV $= 1.6 \cdot 10^{-19}$ J $= 1.6 \cdot 10^{-12}$ erg

Room Temperature (294 K) ≈ 0.025 eV

Horsepower (hp) $= 746$ W

Calorie ≈ 4.2 J

Atmosphere $= 10^6$ dynes/cm^2

Appendix 2:

Conversion Table from Rationalized MKSA to Gaussian Units

Physical Quantities	Rationalized MKSA	Conversion Coefficients	Gaussian
Charge	coulomb	$3 \cdot 10^9$	esu
Charge Density	coulomb/m^3	$3 \cdot 10^3$	esu/cm^3
Current	ampere	$3 \cdot 10^9$	esu/sec
Electric Field	volt/m	$\dfrac{1}{3 \cdot 10^4}$	statvolt/cm
Potential (Voltage)	volt	$1/300$	statvolt
Magnetic Flux	weber	10^8	gauss $\cdot$ cm^2 (maxwell)
Magnetic Induction	tesla	10^4	gauss
Magnetic Field	ampere-turn/m	$4\pi \cdot 10^{-4}$	oersted
Inductance	henry	$\dfrac{1}{9 \cdot 10^{11}}$	sec^2/cm
Capacitance	farad	$9 \cdot 10^{11}$	cm
Resistance	ohm	$\dfrac{1}{9 \cdot 10^{11}}$	sec/cm
Conductivity	mho/m	$9 \cdot 10^9$	sec^{-1}

Appendix 3:

Vector Identities

$$\nabla \left(\Phi \Psi \right) = \Phi \nabla \Psi + \Psi \nabla \Phi$$

$$\nabla \cdot \left(\Phi \mathbf{A} \right) = \mathbf{A} \cdot \nabla \Phi + \Phi \nabla \cdot \mathbf{A}$$

$$\nabla \times \left(\Phi \mathbf{A} \right) = \Phi \nabla \times \mathbf{A} + \nabla \Phi \times \mathbf{A}$$

$$\nabla \cdot \left(\mathbf{A} \times \mathbf{B} \right) = \mathbf{B} \cdot \nabla \times \mathbf{A} - \mathbf{A} \cdot \nabla \times \mathbf{B}$$

$$\nabla \times \left(\mathbf{A} \times \mathbf{B} \right) = \left(\mathbf{B} \cdot \nabla \right) \mathbf{A} - \left(\mathbf{A} \cdot \nabla \right) \mathbf{B} + \mathbf{A} \left(\nabla \cdot \mathbf{B} \right) - \mathbf{B} \left(\nabla \cdot \mathbf{A} \right)$$

$$\left(\mathbf{A} \cdot \nabla \right) = A_x \frac{\partial}{\partial x} + A_y \frac{\partial}{\partial y} + A_z \frac{\partial}{\partial z}$$

$$\nabla\left(\mathbf{A}\cdot\mathbf{B}\right) = \left(\mathbf{B}\cdot\nabla\right)\mathbf{A} + \left(\mathbf{A}\cdot\nabla\right)\mathbf{B} + \mathbf{B}\times\left(\nabla\times\mathbf{A}\right) + \mathbf{A}\times\left(\nabla\times\mathbf{B}\right)$$

$$\nabla\cdot\nabla\Phi = \nabla^2\Phi \equiv \Delta\Phi = \left(\frac{\partial^2}{\partial x^2} + \frac{\partial^2}{\partial y^2} + \frac{\partial^2}{\partial z^2}\right)\Phi$$

$$\nabla\times\nabla\times\mathbf{A} = \nabla\left(\nabla\cdot\mathbf{A}\right) - \left(\nabla^2\right)\mathbf{A}$$

Vector Formulas in Spherical and Cylindrical Coordinates

Spherical Coordinates

- **Transformation of Coordinates**

$$r = \sqrt{x^2 + y^2 + z^2}$$

$$\theta = \cos^{-1}\frac{z}{r}$$

$$\tan\varphi = \frac{y}{x}$$

$$x = r\sin\theta\cos\varphi$$

$$y = r\sin\theta\sin\varphi$$

$$z = r\cos\theta$$

- **Transformation of Differentials**

$$dx = \sin\theta\cos\varphi\ dr + r\cos\theta\cos\varphi\ d\theta - r\sin\theta\sin\varphi\ d\varphi$$

$$dy = \sin\theta\sin\varphi\ dr + r\cos\theta\sin\varphi\ d\theta + r\sin\theta\cos\varphi\ d\varphi$$

$$dz = \cos\theta\ dr - r\sin\theta\ d\theta$$

- **Square of the Element of Length**

$$ds^2 = dr^2 + r^2\ d\theta^2 + r^2\sin^2\theta\ d\varphi^2$$

- **Transformation of the Coordinates of a Vector**

$$F_r = F_x \sin\theta \cos\varphi + F_y \sin\theta \sin\varphi + F_z \cos\theta$$

$$F_\theta = F_x \cos\theta \cos\varphi + F_y \cos\theta \sin\varphi - F_z \sin\theta$$

$$F_\varphi = -F_x \sin\varphi + F_y \cos\varphi$$

$$F_x = F_r \sin\theta \cos\varphi + F_\theta \cos\theta \cos\varphi - F_\varphi \sin\varphi$$

$$F_y = F_r \sin\theta \sin\varphi + F_\theta \cos\theta \sin\varphi + F_\varphi \cos\varphi$$

$$F_z = F_r \cos\theta - F_\theta \sin\theta$$

- **Divergence**

$$\nabla \cdot \mathbf{F} = \frac{\partial F_x}{\partial x} + \frac{\partial F_y}{\partial y} + \frac{\partial F_z}{\partial z}$$

$$= \frac{1}{r^2}\frac{\partial}{\partial r}\left(r^2 F_r\right) + \frac{1}{r\sin\theta}\frac{\partial}{\partial\theta}\left(F_\theta \sin\theta\right) + \frac{1}{r\sin\theta}\frac{\partial F_\varphi}{\partial\varphi}$$

- **Curl**

$$\nabla \times \mathbf{F} = \begin{vmatrix} \hat{\mathbf{x}} & \hat{\mathbf{y}} & \hat{\mathbf{z}} \\ \dfrac{\partial}{\partial x} & \dfrac{\partial}{\partial y} & \dfrac{\partial}{\partial z} \\ F_x & F_y & F_z \end{vmatrix}$$

$$= \frac{1}{r\sin\theta}\left[\frac{\partial}{\partial\theta}\left(F_\varphi \sin\theta\right) - \frac{\partial F_\theta}{\partial\varphi}\right]\mathbf{e}_r + \frac{1}{r}\left[\frac{1}{\sin\theta}\frac{\partial F_r}{\partial\varphi} - \frac{\partial\left(rF_\varphi\right)}{\partial r}\right]\mathbf{e}_\theta$$

$$+ \frac{1}{r}\left[\frac{\partial\left(rF_\theta\right)}{\partial r} - \frac{\partial F_r}{\partial\theta}\right]\mathbf{e}_\varphi$$

- **Gradient**

$$\nabla\Phi = \frac{\partial\Phi}{\partial r}\mathbf{e}_r + \frac{1}{r}\frac{\partial\Phi}{\partial\theta}\mathbf{e}_\theta + \frac{1}{r\sin\theta}\frac{\partial\Phi}{\partial\varphi}\mathbf{e}_\varphi$$

- **Laplacian**

$$\nabla^2 \Phi = \frac{\partial^2 \Phi}{\partial x^2} + \frac{\partial^2 \Phi}{\partial y^2} + \frac{\partial^2 \Phi}{\partial z^2}$$

$$= \frac{1}{r^2} \frac{\partial}{\partial r} \left(r^2 \frac{\partial \Phi}{\partial r} \right) + \frac{1}{r^2 \sin \theta} \frac{\partial}{\partial \theta} \left(\sin \theta \frac{\partial \Phi}{\partial \theta} \right) + \frac{1}{r^2 \sin^2 \theta} \frac{\partial^2 \Phi}{\partial \varphi^2}$$

Cylindrical Coordinates

- **Transformation of Coordinates**

$$\rho = \sqrt{x^2 + y^2}$$

$$\tan \varphi = \frac{y}{x}$$

$$z = z$$

$$x = \rho \cos \varphi$$

$$y = \rho \sin \varphi$$

$$z = z$$

- **Transformation of Differentials**

$$dx = \cos \varphi \, d\rho - \rho \sin \varphi \, d\varphi$$

$$dy = \sin \varphi \, d\rho + \rho \cos \varphi \, d\varphi$$

$$dz = dz$$

- **Square of the Element of Length**

$$ds^2 = d\rho^2 + \rho^2 \, d\varphi^2 + dz^2$$

- **Transformation of the Coordinates of a Vector**

$$F_\rho = F_x \cos \varphi + F_y \sin \varphi$$

$$F_\varphi = -F_x \sin \varphi + F_y \cos \varphi$$

$$F_z = F_z$$

$$F_x = F_\rho \cos\varphi - F_\varphi \sin\varphi$$

$$F_y = F_\rho \sin\varphi + F_\varphi \cos\varphi$$

$$F_z = F_z$$

- **Divergence**

$$\nabla \cdot \mathbf{F} = \frac{1}{\rho}\frac{\partial(\rho F_\rho)}{\partial\rho} + \frac{1}{\rho}\frac{\partial F_\varphi}{\partial\varphi} + \frac{\partial F_z}{\partial z}$$

- **Curl**

$$\nabla \times \mathbf{F} = \left(\frac{1}{\rho}\frac{\partial F_z}{\partial\varphi} - \frac{\partial F_\varphi}{\partial z}\right)\mathbf{e}_\rho$$

$$+ \left(\frac{\partial F_\rho}{\partial z} - \frac{\partial F_z}{\partial\rho}\right)\mathbf{e}_\varphi + \frac{1}{\rho}\left(\frac{\partial(\rho F_\varphi)}{\partial\rho} - \frac{\partial F_\rho}{\partial\varphi}\right)\hat{\mathbf{z}}$$

- **Gradient**

$$\nabla\Phi = \frac{\partial\Phi}{\partial\rho}\mathbf{e}_\rho + \frac{1}{\rho}\frac{\partial\Phi}{\partial\varphi}\mathbf{e}_\varphi + \frac{\partial\Phi}{\partial z}\hat{\mathbf{z}}$$

- **Laplacian**

$$\nabla^2\Phi = \frac{1}{\rho}\frac{\partial}{\partial\rho}\left(\rho\frac{\partial\Phi}{\partial\rho}\right) + \frac{1}{\rho^2}\frac{\partial^2\Phi}{\partial\varphi^2} + \frac{\partial^2\Phi}{\partial z^2}$$

Appendix 4:

Legendre Polynomials

$$P_0(x) = 1$$

$$P_1(x) = x$$

$$P_2(x) = \frac{1}{2}\left(3x^2 - 1\right)$$

$$P_3(x) = \frac{1}{2}\left(5x^3 - 3x\right)$$

$$P_4(x) = \frac{1}{8}\left(35x^4 - 30x^2 + 3\right)$$

Rodrigues' Formula

$$P_l(x) = \frac{1}{2^l l!} \frac{d^l}{dx^l} \left(x^2 - 1\right)^l$$

Spherical Harmonics

$$Y_0^0 (\theta, \varphi) = \frac{1}{\sqrt{4\pi}}$$

$$Y_1^1 (\theta, \varphi) = -\sqrt{\frac{3}{8\pi}} \sin \theta \, e^{i\varphi}$$

$$Y_1^0 (\theta, \varphi) = \sqrt{\frac{3}{4\pi}} \cos \theta$$

$$Y_1^{-1} (\theta, \varphi) = \sqrt{\frac{3}{8\pi}} \sin \theta \, e^{-i\varphi}$$

$$Y_2^2 (\theta, \varphi) = \frac{1}{4}\sqrt{\frac{15}{2\pi}} \sin^2 \theta \, e^{2i\varphi}$$

$$Y_2^1 (\theta, \varphi) = -\sqrt{\frac{15}{8\pi}} \sin \theta \cos \theta \, e^{i\varphi}$$

$$Y_2^0 (\theta, \varphi) = \frac{1}{2}\sqrt{\frac{5}{4\pi}} \left(3\cos^2 \theta - 1\right)$$

$$Y_2^{-1} (\theta, \varphi) = \sqrt{\frac{15}{8\pi}} \sin \theta \cos \theta \, e^{-i\varphi}$$

$$Y_2^{-2} (\theta, \varphi) = \frac{1}{4}\sqrt{\frac{15}{2\pi}} \sin^2 \theta \, e^{-2i\varphi}$$

Appendix 5:

Harmonic Oscillator

The first three eigenfunctions of the harmonic oscillator in one dimension are

$$x_0^2 = \frac{\hbar}{m\omega} \tag{A.5.1}$$

$$\psi_0(x) = \frac{1}{\pi^{1/4}\sqrt{x_0}}e^{-x^2/2x_0^2} \tag{A.5.2}$$

$$\psi_1(x) = \frac{\sqrt{2}x}{\pi^{1/4}x_0^{3/2}}e^{-x^2/2x_0^2} \tag{A.5.3}$$

$$\psi_2(x) = \frac{1}{\pi^{1/4}\sqrt{2x_0}}\left[\frac{2x^2}{x_0^2} - 1\right]e^{-x^2/2x_0^2} \tag{A.5.4}$$

where ω is the oscillator frequency.

Appendix 6:

Angular Momentum and Spin

The spin $\frac{1}{2}$ (Pauli) matrices are

$$\sigma_x = \begin{pmatrix} 0 & 1 \\ 1 & 0 \end{pmatrix} \tag{A.6.1}$$

$$\sigma_y = \begin{pmatrix} 0 & -i \\ i & 0 \end{pmatrix} \tag{A.6.2}$$

$$\sigma_z = \begin{pmatrix} 1 & 0 \\ 0 & -1 \end{pmatrix} \tag{A.6.3}$$

while the vector $\boldsymbol{\sigma} = (\sigma_x, \sigma_y, \sigma_z)$.
   The spin 1 matrices are

$$S_x = \frac{1}{\sqrt{2}}\begin{pmatrix} 0 & 1 & 0 \\ 1 & 0 & 1 \\ 0 & 1 & 0 \end{pmatrix} \tag{A.6.4}$$

$$S_y = \frac{1}{\sqrt{2}}\begin{pmatrix} 0 & -i & 0 \\ i & 0 & -i \\ 0 & i & 0 \end{pmatrix} \tag{A.6.5}$$

$$S_z = \frac{1}{\sqrt{2}}\begin{pmatrix} 1 & 0 & 0 \\ 0 & 0 & 0 \\ 0 & 0 & -1 \end{pmatrix} \tag{A.6.6}$$

Appendix 7:

Variational Calculations

The general procedure for solving variational problems in one dimension is to first evaluate three integrals which are functions of the variational parameter α:

$$E(\alpha) = \frac{K + U}{I} \tag{A.7.1}$$

$$I = \int dx \ \psi^2(x) \tag{A.7.2}$$

$$U = \int dx \ V(x)\psi^2(x) \tag{A.7.3}$$

$$K = -\frac{\hbar^2}{2m} \int dx \ \psi(x)\frac{d^2\psi}{dx^2} = \frac{\hbar^2}{2m} \int dx \ \left(\frac{d\psi}{dx}\right)^2 \tag{A.7.4}$$

The two expressions for the kinetic energy K can be shown to be equal by an integration by parts. The second expression is usually easier to use, since one has to take a single derivative of the trial function $\psi(x)$ and then square it.

Appendix 8:

Normalized Eigenstates of Hydrogen Atom

$$\varphi_{100} = \frac{1}{\sqrt{\pi}a_0{}^{3/2}}e^{-r/a_0} \tag{A.8.1}$$

$$\varphi_{200} = \frac{1}{(2a_0)^{3/2}\sqrt{\pi}}\left(1 - \frac{r}{2a_0}\right)e^{-r/2a_0} \tag{A.8.2}$$

$$\varphi_{210} = \frac{1}{(2a_0)^{3/2}\sqrt{\pi}}\left(\frac{r}{2a_0}\right)e^{-r/2a_0}\cos\theta \tag{A.8.3}$$

$$\varphi_{21\pm1} = \frac{1}{8a_0{}^{3/2}}\left(\frac{r}{a_0}\right)e^{-r/2a_0}\sin\theta e^{\pm\phi} \tag{A.8.4}$$

$$a_0 \equiv \frac{\hbar^2}{\mu e^2}$$

Appendix 9:

Conversion Table for Pressure Units

	N/m^2	dyn/cm^2	bar	atm	mm Hg
1 N/m^2 (pascal)	1	10	10^{-5}	$9.87 \cdot 10^{-6}$	$7.50 \cdot 10^{-3}$
1 dyn/cm^2	0.1	1	10^{-6}	$9.87 \cdot 10^{-7}$	$7.50 \cdot 10^{-4}$
1 bar	10^5	10^6	1	0.987	$7.5 \cdot 10^2$
1 atm	$1.01 \cdot 10^5$	$1.01 \cdot 10^6$	1.01	1	$7.6 \cdot 10^2$
1 mm Hg	$1.33 \cdot 10^2$	$1.33 \cdot 10^3$	$1.33 \cdot 10^{-3}$	$1.31 \cdot 10^{-3}$	1

Appendix 10:

Useful Constants

Resistivity of copper (T = 300 K) $\sim 2 \cdot 10^{-6}$ $\Omega \cdot$ cm

Linear expansion coefficient of copper $\sim 2 \cdot 10^{-5}$ K^{-1}

Surface tension of water (at 293 K) ~ 70 dyn $\cdot$ cm^{-1}

Viscosity of water ~ 0.01 dyn $\cdot$ cm^{-2}

Heat of vaporization of water (at 373 K, 1 atm) ~ 2300 J/g

Velocity of sound in air (at 293 K) ~ 340 m/s

Si band gap ~ 1.1 eV

Ge band gap ~ 0.7 eV

Bibliography

Arfken, G., *Mathematical Methods for Physicists*, 3rd ed., Orlando: Academic Press, 1985

Ashcroft, N. W., and Mermin, N. D., *Solid State Physics*, Philadelphia: Saunders, 1976

Callen, H. B., *Thermodynamics*, New York: John Wiley and Sons, Inc., 1960

Chen, M., *University of California, Berkeley, Physics Problems, with Solutions*, Englewood Cliffs, NJ: Prentice-Hall, Inc., 1974

Cohen-Tannoudji, C., Diu, B., and Laloë, F., *Quantum Mechanics*, New York: John Wiley and Sons, Inc., 1977

Cronin, J., Greenberg, D., and Telegdi, V., *University of Chicago Graduate Problems in Physics*, Chicago, University of Chicago Press, 1979

Feynman, R., Leighton, R., and Sands, M., *The Feynman Lectures on Physics*, Reading, MA, Addison-Wesley, 1965

Goldstein, H., *Classical Mechanics*, 2nd ed., Reading, MA: Addison-Wesley, 1981

Halzen, F., and Martin, A., *Quarks and Leptons*, New York: John Wiley and Sons, Inc., 1984

Hill, T. L., *An Introduction to Statistical Thermodynamics*, Reading, MA: Addison-Wesley, 1960

Huang, K., *Statistical Mechanics*, 2nd ed., New York: John Wiley and Sons, Inc., 1987

Jeans, J., *An Introduction to the Kinetic Theory of Gases*, Cambridge: Cambridge University Press, 1940

Kittel, C., *Introduction to Solid State Physics*, 6th ed., New York: John Wiley and Sons, Inc., 1986

Kittel, C., and Kroemer, H., *Thermal Physics*, 2nd ed., New York: Freeman and Co., 1980

Kozel, S. M., Rashba, E. I., and Slavatinskii, S. A., *Problems of the Moscow Physico-Technical Institute*, Moscow: Mir, 1986

Kubo, R., *Thermodynamics*, Amsterdam: North Holland, 1968

Kubo, R., *Statistical Mechanics*, Amsterdam: North Holland, 1965

Landau, L. D., and Lifshitz, E. M., *Mechanics*, Volume 1 of *Course of Theoretical Physics*, 3rd ed., Elmsford, New York: Pergamon Press, 1976

Landau, L. D., and Lifshitz, E. M., *Quantum Mechanics, Nonrelativistic Theory*, Volume 3 of *Course of Theoretical Physics*, 3rd ed., Elmsford, New York: Pergamon Press, 1977

Landau, L. D., and Lifshitz, E. M., *Statistical Physics*, Volume 5, part 1 of *Course of Theoretical Physics*, 3rd ed., Elmsford, New York: Pergamon Press, 1980

Landau, L. D., and Lifshitz, E. M., *Fluid Mechanics*, Volume 6 of *Course of Theoretical Physics*, 2nd ed., Elmsford, New York: Pergamon Press, 1987

Liboff, R. L., *Introductory Quantum Mechanics*, 2nd ed., Reading, MA: Pergamon Press, 1977

Ma, S. K., *Modern Theory of Critical Phenomena*, Reading, MA: Benjamin, 1976

MacDonald, D. K. C., *Noise and Fluctuations*, New York: John Wiley and Sons, 1962

Messiah, A., *Quantum Mechanics*, Volume 1, Amsterdam: North Holland, 1967

Newbury, N., Newman, M., Ruhl, J., Staggs, S., and Thorsett, S., *Princeton Problems in Physics, with Solutions*, Princeton: Princeton University Press, 1991

Pathria, R. K., *Statistical Mechanics*, Oxford: Pergamon Press, 1972

Reif, R., *Fundamentals of Statistical and Thermal Physics*, New York: McGraw-Hill, 1965

Sakurai, J. J., *Modern Quantum Mechanics*, Menlo Park: Benjamin/Cummings, 1985

Sakurai, J. J., *Advanced Quantum Mechanics*, Menlo Park: Benjamin/Cummings, 1967

Schiff, L. I., *Quantum Mechanics*, 3rd ed., New York: McGraw-Hill, 1968

Sciama, D. W., *Modern Cosmology*, New York: Cambridge University Press, 1971

Sciama, D. W., *Modern Cosmology and the Dark Matter Problem*, New York: Cambridge University Press, 1994

Schiff, L. I., *Quantum Mechanics*, 3rd ed., New York: McGraw-Hill, 1968

Shankar, R., *Principles of Quantum Mechanics*, New York: Plenum Press, 1980

Sze, S. M., *Physics of Semiconductor Devices*, New York: John Wiley and Sons, Inc., 1969

Tinkham, M., *Introduction to Superconductivity*, New York, McGraw-Hill, 1975

Tolman, R. C., *The Principles of Statistical Mechanics*, Oxford: Oxford University Press, 1938

Ziman, J. M., *Principles of the Theory of Solids*, 2nd ed., Cambridge, Cambridge University Press, 1972